Medieval Scotland
Crown, Lordship and Community

Medieval Scotland
Crown, Lordship and Community

Essays presented to
G.W.S. Barrow

edited by
ALEXANDER GRANT *and* KEITH J. STRINGER

EDINBURGH UNIVERSITY PRESS

Edinburgh University Press Ltd
22 George Square, Edinburgh

Typeset in 10/12 CGTimes
at Lancaster University, and
printed in Great Britain by
The University Press, Cambridge

A CIP record of this book is
available from the British Library

ISBN 0 7486 0418 9

Contents

Acknowledgements

The editors wish to record their gratitude to all their fellow contributors, and also, for kind help in various ways with the production of this volume, to Dr Michael Lynch of Edinburgh University and to Miss Wendy Francis, Mrs Alison Grant, Mrs Margaret Lynch and Mr Stewart McFarlane of Lancaster University.

                                                                        A.G.
                                                                        K.J.S.

Contributors

JOHN BANNERMAN is a Senior Lecturer in Scottish History at the University of Edinburgh.

JULIA BARROW is a Lecturer in Medieval History at the University of Nottingham.

ARCHIBALD A.M. DUNCAN is Professor of Scottish History at the University of Glasgow.

ELIZABETH EWAN is an Associate Professor of History at the University of Guelph, Canada.

ALEXANDER GRANT is a Senior Lecturer in History at the University of Lancaster.

ALAN MACQUARRIE was formerly a Lecturer in Scottish History at the University of Glasgow.

HECTOR L. MacQUEEN is a Senior Lecturer in Scots Law at the University of Edinburgh.

NORMAN H. REID is University Archivist at Heriot-Watt University.

WILLIAM W. SCOTT is an independent scholar.

GRANT G. SIMPSON is a Senior Lecturer in Scottish History at the University of Aberdeen.

KEITH J. STRINGER is a Senior Lecturer in History at the University of Lancaster.

DONALD E.R. WATT is Emeritus Professor of Scottish Church History at the University of St Andrews.

BRUCE WEBSTER is a Senior Lecturer in History at the University of Kent at Canterbury.

ALAN YOUNG is a Principal Lecturer in History at the College of Ripon and York St John, York.

Abbreviations

Aberdeen-Banff Colls.	*Collections for a History of the Shires of Aberdeen and Banff* (Spalding Club, 1843).
Aberdeen-Banff Ills.	*Illustrations of the Topography and Antiquities of the Shires of Aberdeen and Banff* (Spalding Club, 1847-69).
Aberdeen Burgh Chrs.	*Charters and other Writs illustrating the History of the Royal Burgh of Aberdeen*, ed. P.J. Anderson (Aberdeen, 1890).
Aberdeen Burgh Recs.	*Early Records of the Burgh of Aberdeen, 1317, 1398-1407*, ed. W.C. Dickinson (SHS, 1957).
Aberdeen Council Reg.	*Extracts from the Council Register of the Burgh of Aberdeen* (Spalding Club, 1844-8).
Aberdeen Reg.	*Registrum Episcopatus Aberdonensis* (Spalding and Maitland Clubs, 1845).
Acts of the Lords of the Isles	*Acts of the Lords of the Isles, 1336-1493*, ed. J. and R.W. Munro (SHS, 1986).
Ancient Burgh Laws	*Ancient Laws and Customs of the Burghs of Scotland* (Scottish Burgh Records Society, 1868-1910).
Anderson, *Aberdeen Friars*	*Aberdeen Friars: Red, Black, White, Grey*, ed. P.J. Anderson (Aberdeen, 1909).
Anderson, *Early Sources*	*Early Sources of Scottish History 500 to 1286*, ed. A.O. Anderson (Edinburgh, 1922; reprinted Stamford, 1990).
Anderson, *Scottish Annals*	*Scottish Annals from English Chroniclers 500 to 1286*, ed. A.O. Anderson (London, 1908; reprinted Stamford, 1991).
Ann. Tigernach	'The Annals of Tigernach', *Revue Celtique*, xvi (1895), pp. 374-419; xvii (1896), pp. 6-33, 119-263, 337-420; xviii (1897), pp. 9-59, 150-97, 267-303.
Ann. Ulster	*The Annals of Ulster (to A.D. 1131)*, ed. S. Mac Airt and G. Mac Niocaill (Dublin, 1983); also *Annala Uladh: Annals of Ulster ... a Chronicle of Irish Affairs from A.D. 431 to A.D. 1541*, ed. W.M. Hennessy and B. MacCarthy (Dublin, 1887-1901).
APS	*The Acts of the Parliaments of Scotland*, ed. T. Thomson and C. Innes (Edinburgh, 1814-75).
Arbroath Lib.	*Liber S. Thome de Aberbrothoc* (Bannatyne Club, 1848-56).
Bamff Chrs.	*Bamff Charters 1232-1703*, ed. J.H. Ramsay (Oxford, 1915).
Barbour's Bruce	(John Barbour), *Barbour's Bruce*, ed. M.P. McDiarmid and J.A.C. Stevenson (Scottish Text Society, 1980-5).
Brechin Reg.	*Registrum Episcopatus Brechinensis* (Bannatyne Club, 1856).
Cambuskenneth Reg.	*Registrum Monasterii S. Marie de Cambuskenneth* (Grampian Club, 1872).
Carnwath Court Bk.	*The Court Book of the Barony of Carnwath 1523-42*, ed. W.C. Dickinson (SHS, 1937).

Cawdor Bk.	*The Book of the Thanes of Cawdor* (Spalding Club, 1859).
CDS	*Calendar of Documents relating to Scotland*, ed. J. Bain *et al.* (Edinburgh, 1881-1986).
Chron. Bower (Goodall)	*Joannis de Fordun Scotichronicon cum Supplementis et Continuatione Walteri Boweri*, ed. W. Goodall (Edinburgh, 1759).
Chron. Bower (Watt)	*Scotichronicon by Walter Bower*, general editor D.E.R. Watt (Aberdeen and Edinburgh, 1987-).
Chron. Fordun	*Johannis de Fordun, Chronica Gentis Scotorum*, ed. W.F. Skene (Edinburgh, 1871-2).
Chron. Lanercost	*Chronicon de Lanercost* (Maitland Club, 1839).
Chron. Melrose	*The Chronicle of Melrose*, ed. A.O. Anderson *et al.* (London, 1936).
Chron. Picts-Scots	*Chronicles of the Picts, Chronicles of the Scots, and other Early Memorials of Scottish History*, ed. W.F. Skene (Edinburgh, 1867).
Chron. Pluscarden	*Liber Pluscardensis*, ed. F.J.H. Skene (Edinburgh, 1877-80).
Chron. Wyntoun	*The Original Chronicle of Andrew of Wyntoun*, ed. F.J. Amours (Scottish Text Society, 1903-14).
Chron. Wyntoun (Laing)	Androw of Wyntoun, *The Orygynale Cronykil of Scotland*, ed. D. Laing (Edinburgh, 1872-9).
Coldstream Cart.	*Chartulary of the Cistercian Priory of Coldstream* (Grampian Club, 1879).
Coupar Angus Chrs.	*Charters of the Abbey of Coupar Angus*, ed. D.E. Easson (SHS, 1947).
Dryburgh Lib.	*Liber S. Marie de Dryburgh* (Bannatyne Club, 1847).
Dunfermline Reg.	*Registrum de Dunfermelyn* (Bannatyne Club, 1842).
ER	*The Exchequer Rolls of Scotland*, ed. J. Stuart *et al.* (Edinburgh, 1878-1908).
ESC	*Early Scottish Charters prior to 1153*, ed. A.C. Lawrie (Glasgow, 1905).
Family of Rose	*A Genealogical Deduction of the Family of Rose of Kilravock* (Spalding Club, 1848).
Fife Court Bk.	*The Sheriff Court Book of Fife 1515-22*, ed. W.C. Dickinson (SHS, 1928).
Foedera	*Foedera, Conventiones, Litterae, et Cuiuscunque Generis Acta Publica*, ed. T. Rymer (Record Commission, 1816-69).
Fraser, *Carlaverock*	W. Fraser, *The Book of Carlaverock* (Edinburgh, 1873).
Fraser, *Douglas*	W. Fraser, *The Douglas Book* (Edinburgh, 1885).
Fraser, *Facsimiles*	*Facsimiles of Scottish Charters and Letters prepared by Sir William Fraser* (Edinburgh, 1903).
Fraser, *Melville*	W. Fraser, *The Melvilles Earls of Melville and the Leslies Earls of Leven* (Edinburgh, 1890).
Fraser, *Menteith*	W. Fraser, *The Red Book of Menteith* (Edinburgh, 1880).
Fraser, *Southesk*	W. Fraser, *History of the Carnegies, Earls of Southesk, and of their Kindred* (Edinburgh, 1867).
Glasgow Reg.	*Registrum Episcopatus Glasguensis* (Bannatyne and Maitland Clubs, 1843).
HMC	Royal Commission on Historical Manuscripts. *1st* to *9th Reports* cited by number; others by name of owner or collection.
Holyrood Lib.	*Liber Cartarum Sancte Crucis* (Bannatyne Club, 1840).

Inchaffray Chrs.	*Charters, Bulls and other Documents relating to the Abbey of Inchaffray* (SHS, 1908).
Kelso Lib.	*Liber S. Marie de Calchou* (Bannatyne Club, 1846).
Laing Chrs.	*Calendar of the Laing Charters 854-1837*, ed. J. Anderson (Edinburgh, 1899).
Lawrie, *Annals*	*Annals of the Reigns of Malcolm and William, Kings of Scotland*, ed. A.C. Lawrie (Glasgow, 1910).
Lennox Cart.	*Cartularium Comitatus de Levenax* (Maitland Club, 1833).
Lindores Cart.	*Chartulary of the Abbey of Lindores*, ed. J. Dowden (SHS, 1903).
May Recs.	*Records of the Priory of the Isle of May* (Society of Antiquaries of Scotland, 1868).
Melrose Lib.	*Liber Sancte Marie de Melros* (Bannatyne Club, 1837).
Moray Reg.	*Registrum Episcopatus Moraviensis* (Bannatyne Club, 1837).
Morton Reg.	*Registrum Honoris de Morton* (Bannatyne Club, 1853).
Nat. MSS Scot.	*Facsimiles of the National Manuscripts of Scotland* (London, 1867-71).
Newbattle Reg.	*Registrum S. Marie de Neubotle* (Bannatyne Club, 1849).
NLS	National Library of Scotland, Edinburgh.
North Berwick Chrs.	*Carte Monialium de Northberwic* (Bannatyne Club, 1847).
Paisley Reg.	*Registrum Monasterii de Passelet* (Maitland Club, 1832; New Club, 1877).
Panmure Reg.	*Registrum de Panmure*, ed. J. Stuart (Edinburgh, 1874).
Patrick, *Statutes*	*Statutes of the Scottish Church*, ed. D. Patrick (SHS, 1907).
PRO	Public Record Office, London.
PSAS	*Proceedings of the Society of Antiquaries of Scotland.*
Raine, *North Durham*	Appendix to J. Raine, *The History and Antiquities of North Durham* (London, 1852).
RCAHMS	Royal Commission on the Ancient and Historical Monuments of Scotland.
Retours	*Inquisitionum ad Capellam Domini Regis retornatarum ... Abbrevatio*, ed. T. Thomson (London, 1811-16).
RMS	*Registrum Magni Sigilli Regum Scotorum*, ed. J.M. Thomson *et al.* (Edinburgh, 1882-1912).
Rot. Scot.	*Rotuli Scotiae in Turri Londinensi et in Domo Capitulari Westmonasteriensi asservati*, ed. D. Macpherson *et al.* (Record Commission, 1814-19).
RRS	*Regesta Regum Scottorum*, ed. G.W.S. Barrow *et al.* (Edinburgh, 1960-).
RSCHS	*Records of the Scottish Church History Society.*
St Andrews Copiale	*Copiale Prioratus Sanctiandree*, ed. J.H. Baxter (Oxford, 1930).
St Andrews Lib.	*Liber Cartarum Prioratus Sancti Andree in Scotia* (Bannatyne Club, 1841).
Scalacronica	*Scalacronica by Sir Thomas Gray of Heton Knight* (Maitland Club, 1836).
Scalacronica (Maxwell)	*Scalacronica: The Reigns of Edward I, Edward II and Edward III*, translated by H. Maxwell (Glasgow, 1907).
Scone Lib.	*Liber Ecclesie de Scon* (Bannatyne and Maitland Clubs, 1843).
Scots Peerage	*The Scots Peerage*, ed. J.B. Paul (Edinburgh, 1904-14).

SES	*Concilia Scotiae* (Bannatyne Club, 1866).
SHR	*Scottish Historical Review.*
SHS	Scottish History Society.
SHS Misc.	*Miscellany of the Scottish History Society* (SHS, 1893-).
Spalding Misc.	*Miscellany of the Spalding Club* (Spalding Club, 1841-52).
SRO	Scottish Record Office, Edinburgh.
Stevenson, *Documents*	*Documents Illustrative of the History of Scotland 1286-1306*, ed. J. Stevenson (Edinburgh, 1870).
Stones, *Anglo-Scottish Relations*	*Anglo-Scottish Relations: Some Selected Documents*, ed. E.L.G. Stones (Oxford, 1970 reprint).
TCWAAS	*Transactions of the Cumberland and Westmorland Antiquarian and Archaeological Society.*
TDGNHAS	*Transactions of the Dumfriesshire and Galloway Natural History and Antiquarian Society.*
Wigtown Charter Chest	*Charter Chest of the Earldom of Wigtown* (Scottish Record Society, 1910.
Wigtownshire Chrs.	*Wigtownshire Charters*, ed. R.C. Reid (SHS, 1960).

Foreword

Geoffrey Wallis Steuart Barrow was born in Headingley, Leeds, in 1924, and was initially educated at St Edward's School, Oxford. A family move brought him to the Highlands of Scotland when he was fourteen; his love-affair with Scotland's history developed at Inverness Royal Academy and continued at the University of St Andrews, and then, south of the Border, at Pembroke College, Oxford, where he wrote a B.Litt. thesis on 'Scottish royal ecclesiastical policy, 1107-1214'. In 1950 his academic career took him further south, to a lectureship at University College, London, where unlike several other medievalists he survived the regime of J.E. Neale, leaving only in 1961 on his appointment as Professor of Medieval History at King's College, Durham University, subsequently the University of Newcastle upon Tyne. That brought him north to the Anglo-Scottish Border;[1] in 1974 he crossed it once more to become the first Professor of Scottish History in the University of St Andrews. Finally, in 1979 he succeeded Gordon Donaldson as Sir William Fraser Professor of Scottish History and Palaeography in the University of Edinburgh. This volume is published to commemorate his retirement from that chair in 1992.

It is a long and distinguished career, which is far from over yet. High points include his award of a D.Litt. by the University of St Andrews in 1971, election to Fellowships of the British Academy in 1976 and of the Royal Society of Edinburgh in 1977, election as Ford's Lecturer in the University of Oxford in 1977, and award of an Honorary D.Litt. by the University of Glasgow in 1988. He has, in addition, served on numerous committees and learned bodies, most notably as Chairman and President of the Scottish History Society, President of the Saltire Society, Literary Director and Vice-President of the Royal Historical Society, Chairman of the British Academy/Royal Historical Society Committee for publishing Anglo-Saxon charters, member of the Royal Commission on Historical Manuscripts, member of the Bureau of the *Commission Internationale de Diplomatique*, and editor of the *Scottish Historical Review*.

But his most obvious, and enduring, academic contribution lies in his publications. A reputation made initially with his Royal Historical Society Alexander Prize Essay, 'Scottish rulers and the religious orders, 1070-1153'

---

1   G.W.S. Barrow, *The Border*, Inaugural Lecture, King's College, Newcastle upon Tyne (Durham, Durham University, 1962).

(1953), was firmly consolidated by that still admired textbook *Feudal Britain* (1956), which for arguably the first time since John Mair's *Historia Majoris Britanniae* (1521) provided a genuinely even-handed study of medieval Britain. From Mair's footsteps he jumped forwards into those of the great nineteenth-century editors of Scotland's medieval records; but his volumes initiating the *Regesta Regum Scottorum* series, *Malcolm IV* (1960) and *William I* (1971), set new standards for Scottish charter scholarship. Meanwhile, painstaking research was combined with delightful readability in his most famous work, *Robert Bruce and the Community of the Realm of Scotland* (1965). The sheer eye-opening thrill of encountering *Robert Bruce*, vividly remembered by one scholar from distant Sixth-Year days, is something that very few books are able to engender; now deservedly in a fourth edition, it has justly been described as 'the first modern book on medieval Scottish history'.[1] That same combination of highest-quality research and readability was exhibited later in his complementary studies of the centuries before Robert Bruce, *The Anglo-Norman Era in Scottish History* (1980: his Ford Lectures) and *Kingship and Unity: Scotland 1000-1306* (1981): the one a meticulous scholarly analysis of 'Norman' Scotland with important lessons for English and Continental historians as well; the other an elegant, economical textbook distilling the deepest of learning for the student and the man in the street. Not content with these major works, his pen has produced a stream of shorter studies — many of which are collected in his *The Kingdom of the Scots* (1973) and *Scotland and its Neighbours in the Middle Ages* (1992) — in which quality is never once sacrificed to quantity. At his retirement presentation, the assembled company was staggered to hear it announced that the Barrow publication list extended to over ninety items. The precise total, to date, is eight books and eighty-six learned articles, essays, part-books and pamphlets, plus numerous encyclopedia entries and countless reviews;[2] and more are yet to come.[3] Truly he bestrides his world like a Colossus.

His writings embrace almost the entire spectrum of Scottish medieval history — political, constitutional, administrative, ecclesiastical, aristocratic, social and economic — spanning the period from the Dark Ages to the fourteenth century, across the Highlands, Islands, Lowlands and Borders. All of them display an utter mastery of the broadest range of material, encompassing not only the traditional sources but also, for instance, place-name and linguistic evidence. The brilliant use of the latter particularly

---

1   A.A.M. Duncan, *The Nation of Scots and the Declaration of Arbroath* (Historical Association, 1970), p. 38.
2   See below, pp. 297-307. There are also many GWSB items scattered through various editions of the *Encyclopaedia Britannica*.
3   Including, at last, two of the first pieces he ever wrote, contributions on 'Faringdon' and 'Kencot with Radcot' in Bampton Hundred, Oxfordshire, for the Victoria County History; these are to appear in the next-but-one volume of *Victoria County History: Oxfordshire*. For his newest project, see 'The Charters of David I', in *Anglo-Norman Studies XIV: Proceedings of the Battle Conference, 1991* (Woodbridge, 1992).

illustrates his immense knowledge of, and passion for, the actual country and landscape of Scotland. It is a passion which is fiercely exhibited in his more general papers, especially his Edinburgh inaugural lecture 'The Extinction of Scotland' (1980);[1] and also, less formally but more exhaustingly, in actions such as his tramp up Dunsinnan Hill in 1991, accompanied by students, celebrities and members of the media puffing in his wake, to protest against the quarrying of an ancient landmark.

The Dunsinnan climb gives us the other side of Geoffrey Barrow. He is a rarity in academic life: a scholar with an international reputation, but one without any trace of pomp or circumstance.[2] As a head of department he spread an atmosphere of kindness and good-natured bonhomie, which encouraged both a personal loyalty and genuinely close working relationships. It was an atmosphere where neither harsh words nor raised voices ever belonged, not even in the difficult years that the university world has recently been experiencing. Characteristically, his own reflections on his contribution to Edinburgh's Scottish History Department, in his valedictory speech, focused on his early acquisition of a coffee machine, which turned that department's small staff room into a daily forum for lively debate and academic chat, always in a kind-hearted vein. There, on any day, might be found not only members of the department but scholars from all parts of the country, dropping in for a brief word with him or to seek his advice. Such advice is always given wholeheartedly; and even when criticising, he is invariably constructive and stimulating. Thus publications alone do not reflect his contribution to Scottish history; it lies as much in the encouragement he has always given to others, as teacher, research supervisor, colleague and above all as friend. He is always first, and warmest, in his congratulations; 'That was smashing' is typical of many fondly-remembered comments on academic works and lectures, while individual letters to his Finals students are still treasured to this day. Constantly advising, encouraging, and stimulating; that is the essence of Geoffrey Barrow.

Such magnanimity naturally engenders wide friendship in return. So great is the regard for him within the academic community — outwith as well as within Scotland — that had all his well-wishers contributed to a *Festschrift* in his honour, the wide-ranging result would have filled several volumes. Therefore, to keep it manageable, the present collection of essays was focused on medieval Scotland, the subject which has benefited most from Geoffrey Barrow's pen, and the contributors were recruited from scholars who have had some formal relationship with him, as

1   Published 1982 (Stirling).
2   Cf. '... and Galbraith ... whose incredulous exclamation, uttered when I first came to Oxford to work for a higher degree, still makes my ears burn for very shame: "What! You want to do research in medieval Scottish history and you've never heard of Lawrie's *Charters*! That's a *pretty bad show*!"' (G.W.S. Barrow, *The Anglo-Norman Era in Scottish History* [Oxford, 1980], p. v). With his new project on the charters of David I the wheel has come full circle.

undergraduates, graduate students, colleagues, and collaborators in the *Regesta* series. The resulting essays explore some of the central themes in the development of the kingdom of the Scots. In particular, they discuss the interplay between Celtic and feudal influences, the political definition of the kingdom, crown-magnate relations, and the relationship between the local and national communities. All these essays are thus very much on 'Barrow' themes — as their footnotes graphically demonstrate. They are offered here as a tribute and with our deepest affection.

1

The Kings of Strathclyde, c.400-1018

ALAN MACQUARRIE

Of the four peoples who made up early historical Scotland - Picts, Scots, Britons and Angles - it is the Britons about whom least is known, and about whom least has been written. The best recorded group is the Scots; their dynasty ultimately was the most successful, eliminating the others to rule over a disparate medieval kingdom to which they gave their name. They have left annals, chronicles, poems, king-lists and genealogies as proof of their lineage and lore. The Angles have their great histories, Bede's *Ecclesiastical History* and the *Anglo-Saxon Chronicle*, which deal, among other things, with northern events in what has since become Scotland. The Pictish record is more slender, consisting of a king-list (in several versions) as a meagre supplement to the enigma of the symbol-stones; but the latter, with their strange symbols and incomprehensible inscriptions, and certain other apparently anomalous features of Pictish society, have given rise to a considerable literature.[1] But, until recently, the Britons of the south-west attracted relatively little attention.

This situation, however, has now begun to change. It remains true that the historical legacy of the Britons of the North is slender; it consists chiefly of royal pedigrees, a few annalistic entries, and some poems attributed to northern writers.[2] Yet, although this material is difficult to use, it is attracting increasing attention, and scholars are now attempting to do justice to the importance of the northern British kingdoms, especially Strathclyde, for the history of early Scotland.[3]

---

1   Especially *The Problem of the Picts*, ed. F.T. Wainwright (Edinburgh, 1955), and I. Henderson, *The Picts* (London, 1967). Picts and Scots — but not Britons — are covered in *Chron. Picts-Scots*, and in M.O. Anderson, *Kings and Kingship in Early Scotland* (Edinburgh, 1973); while the Scots are the main focus of J.W.M. Bannerman, *Studies in the History of Dalriada* (Edinburgh, 1974).
2   E. Phillimore, 'The *Annales Cambriae* and Old Welsh Genealogies from Harleian MS 3859' *Y Cymmrodor*, ix (1888); R. Bromwich, *Trioedd Ynys Prydein* (Cardiff, 1961); K.H. Jackson, *The Gododdin: The Oldest Scottish Poem* (Edinburgh, 1969), is a translation of the most substantial of the poems.
3   Notably A.P. Smyth, *Warlords and Holy Men: Scotland, AD 80-1000* (London, 1984), and more recently in his contributions to A. Williams *et al.*, *A Biographical Dictionary of Dark Age Britain* (London, 1991). Among the more important articles are: D.P. Kirby, 'Strathclyde and Cumbria', *TCWAAS*, new ser., lxii (1962); P.A. Wilson, 'On the Use of the Terms "Strathclyde" and "Cumbria"', ibid., lxvi (1966); M. Miller, 'Historicity and the Pedigrees of the Northcountrymen', *Bulletin of the Board of Celtic Studies*, xxvi (1976); see also the bibliography in Smyth, *Warlords and Holy Men*, pp. 242-4.

The present essay is intended as a contribution to our changing perception of the place of the British kingdoms in early historical Scotland. Its concentration is, unavoidably, on the kingdom of Strathclyde, with its great citadel at Dumbarton Rock. Strathclyde was the only one of the northern kingdoms to survive long into the historical period, and indeed remained distinct, to some degree at least, until the eleventh century. This study attempts to provide some detail about its individual rulers, together with a narrative of the development and decline of the kingdom from the post-Roman period until its incorporation into the wider kingdom of the Scots.

Of the origins of the Dumbarton dynasty we have no knowledge at all. Ptolemy's *Geography* places a tribe called the *Damnonii* at the western end of the Antonine Wall, on the shores of the River *Clota* or Clyde. They were clearly a Celtic people, whose name is closely related to that of the *Dumnonii* of Devon and Cornwall. Their neighbours were the *Novantae* and *Selgovae* to the south, the *Votadini* (later known as the Gododdin) to the east, and the various Caledonian tribes to the north.[1] How far they fell under Roman influence is unclear; Roman roads and fortifications were built in their region, but seem to have been used intermittently after the mid-second century.[2] A fifth-century Votadinian king had a treasure which included portable silver objects of Roman origin, some even bearing Christian symbols; but it is not clear how these were acquired, so it is not easy to use them as evidence for cultural influence.[3] It has been suggested that some names in the Strathclyde pedigree in its earliest part represent garbled forms of Latin names, but this is very doubtful.[4] It is clear, however, that fifth- and sixth-century British kings within and on the fringes of the old imperial province of Britain retained a thin veneer of *Romanitas*, and it is possible that the kings of the *Damnonii* did so too.[5]

The first positive evidence about the early kings of Dumbarton comes in the writings of Saint Patrick, a fifth-century British bishop who conducted a mission in northern Ireland. The exact date of Patrick's mission is uncertain, but it occurred some time between the 430s and the 490s. One of Patrick's two surviving letters is addressed to the 'milites Corotici', the war band of Ceredig, whom he denounces as allies of Scots and Picts, not fellow-citizens

1   For discussions of Ptolemy's map, see Henderson, *Picts*, pp. 15-19; Smyth, *Warlords and Holy Men*, pp. 40-2, 46-50; K.H. Jackson, 'The Pictish Language', in Wainwright, *Problem of the Picts*, pp. 135-7.
2   A.A.M. Duncan, *Scotland: The Making of the Kingdom* (Edinburgh, 1975), pp. 19-26; M. Todd, *Roman Britain* (London, 1981), pp. 138-80; L. Keppie, *Scotland's Roman Remains* (Edinburgh, 1986), pp. 17-19.
3   E. Burley, 'A Catalogue and Survey of the Metalwork from Traprain Law', *PSAS*, lxxxix (1955-6); A.O. Curle, *The Treasure of Traprain* (Glasgow, 1923); L. Alcock, *Arthur's Britain* (London, 1971), p. 254.
4   J. Morris, *The Age of Arthur* (London, 1973), pp. 17-19.
5   J. Campbell *et al.*, *The Anglo-Saxons* (Oxford, 1982), pp. 22-44; Alcock, *Arthur's Britain*, pp. 99-113; Morris, *Age of Arthur*, pp. 14-19, 414-6.

of Romans and Christians.[1] The implication is that Ceredig and his soldiers would have liked to think of themselves as Roman citizens and Christians, for otherwise the rebuke would have had little point.

Saint Patrick does not identify Ceredig's kingdom or sphere of influence, but the reference to Pictish allies singles him out as a northerner. The *Book of Armagh* called him 'Coirthech rex Aloo', Ceredig king of *Ail*; this is probably short for *Ail Cluaide*, 'Rock of the Clyde', i.e., Dumbarton.[2] So by *c.*800 it was believed that Patrick's Coroticus or Ceredig had been a king of Dumbarton. A king of that name appears in the pedigree of the Dumbarton kings in a version probably belonging to the ninth century: the grandfather of Dyfnwal Hen is named as 'Ceretic Guletic map Cynloyp'.[3] Dyfnwal Hen appears in many pedigrees of northerners as an obscure ancestral figure, as does his rough contemporary, the equally obscure Coel Hen 'Guotepauc'; Coel's epithet, in later Welsh *godebog*, may be equivalent to the Latin title *protector*, used by at least one sixth-century Welsh tyrant.[4] Dyfnwal and his ancestors may also have considered themselves to be Roman *protectores*, or federate captains. The identity of Patrick's Coroticus with Ceredig king of Dumbarton has not been universally accepted,[5] but since he was undoubtedly a Briton and a northerner, allied with Picts and raiding Patrick's northern Irish churches, there seems little reason to doubt it.

The identification does not make either Ceredig or Saint Patrick himself any easier to date. In Ceredig's case, the only way to date him (other than by Patrick) is through the hazardous process of generation-counting from known and datable descendants. In the earliest known genealogy, Ceredig is placed eight generations before Beli (d. 722), seven generations before Elffin (d. 693), six generations before Ywain (*fl.* 642), five generations before Beli (the Beli who died in *c.*626?), and four generations before Nwython (the Pictish king Nechton who died in *c.*620?).[6] He is also placed five generations before Rhydderch Hael (d. *c.*614?), which would hardly be consistent. If Rhydderch's pedigree had added an extra generation by mistake, or if Nwython's had omitted one, that would resolve the inconsistency; but this does not help us to say anything more definite than that Ceredig can be assigned either to the middle or to the second half of the fifth century. It has been argued, on the basis of the *obits* of his known contemporaries, that Patrick belongs towards the end of the fifth century

1   *St Patrick: His Writings and Muirchú's Life*, ed. A.B.E. Hood (London, 1978): *Epistola Militibus Corotici*, cap. 2.
2   Ibid.: *Vita Patricii*, chapter headings; also in *The Patrician Texts in the Book of Armagh*, ed. L. Bieler (Dublin, 1979).
3   Phillimore, '*Annales Cambriae*', pp. 172-3 [hereafter the Harleian genealogies printed here are referred to as H].
4   Campbell, *Anglo-Saxons*, p. 21.
5   The argument against the identification is set out in E.A. Thomson, 'St Patrick and Coroticus', *Journal of Theological Studies*, new ser., xxxi (1980).
6   H, pp. 172-3; for the dates of these later kings, see below, pp. 8ff.

rather than in the middle;[1] but there is no universal agreement on this point.

Saint Patrick's letter does provide a few clues about Ceredig and the society over which he ruled. It is clear that he controlled a powerful war band, capable of serious piratical raids on the Irish coast. Clearly that indicates some measure of sea power. He was able to form alliances with Picts and with *Scoti*, Gaels either of northern Ireland or settled on the west coast of Scotland. It is implied that Ceredig and his men considered themselves to be Roman citizens and Christians - hence Patrick's especial bitterness about their preying on an Irish Christian community. Gildas's denunciations of sixth-century British kings makes an interesting comparison.

Patrick does not denounce Ceredig himself for having participated in raiding, but states that it was carried out 'by these freebooters at the command of the aggressive Ceredig'. His by-name, 'guledig', 'wealthy', suggests a long and successful career of plunder, and it seems that at times he was able to send his war band out on raids without taking part himself.

The suggestion that there was a veneer, albeit thin, of Christianity as well as *Romanitas* in south-west Scotland in the fifth century has a number of implications. It is consistent with the finding of fifth- or sixth-century Christian stones at Kirkmadrine and Whithorn in Galloway, and at other places in southern Scotland (though none has been found in Strathclyde itself). It lends credence to the hint in Jocelin of Furness's *Life of Kentigern* that there were vestiges of Christianity in the area on Kentigern's arrival there, some time in the second half of the sixth century. Finally, it raises intriguing possibilities that Patrick himself may have been a native of the south of Scotland, as was believed by Irish writers of the ninth century. Sadly, Muirchú's form of the name of Patrick's birthplace,'Ventre', appears to be corrupt, and has not been satisfactorily identified.[2]

1   St Patrick is the subject of a vast literature, ancient and modern. Among the most important studies are: J.B. Bury, *The Life of St Patrick* (London, 1905); E. MacNeill, *St Patrick* (Dublin, 1964 reprint); T.F. O'Rahilly, *The Two Patricks* (Dublin, 1942); J. Carney, *The Problem of St Patrick* (Dublin, 1961); C. Mohrmann, *The Latin of St Patrick* (Dublin, 1961); D.A. Binchy, 'St Patrick and his Biographers, ancient and modern', *Studia Hibernica*, ii (1962); R.P.C. Hanson, *St Patrick* (Oxford, 1968); A.C. Thomas, *Christianity in Roman Britain* (London, 1981); R. Sharpe, 'St Patrick and the See of Armagh', *Cambridge Medieval Celtic Studies*, iv (1982).
2   On the stones of Galloway, see C.A.R. Radford and G. Donaldson, *Whithorn and Kirkmadrine* (Edinburgh, 1953); J. Wall, 'Christian Evidences in the Roman Period', *Archaeologia Æliana*, 4th ser., xliii (1965), p. 211; Campbell, *Anglo-Saxons*, p. 21; A. Macquarrie, 'The Date of St Ninian's Mission: A Reappraisal', *RSCHS*, xxiii (1987). On St Kentigern, see A. Macquarrie, 'The Career of St Kentigern of Glasgow: *Vitae, Lectiones*, and Glimpses of Fact', *Innes Review*, xxxvii (1986). The location of St Patrick's birthplace at Old Kilpatrick is first made in the *Tripartite Life of St Patrick* of *c*.900 (ed. W. Stokes [Rolls Ser., 1887], pp. 8-17). See also A. Boyle, 'The Birthplace of St Patrick', *SHR*, lx (1981). Boyle's argument in favour of Fintry is attractive, but unacceptable for several reasons. The name Kilpatrick cannot have been in existence before the immigration of Gaelic-speaking settlers into Strathclyde, which is unlikely to have been extensive before the 9th century: see A. Macquarrie, 'Early Christian Govan: the Historical Context', *RSCHS*, xxiv (1990), esp. pp. 14-15. The name Fintry, from British *Venn tref*, 'white house', was still written and so presumably pronounced with its final *f* in the 12th century: W.J. Watson, *The History of the Celtic Place-Names of Scotland* (Edinburgh, 1926), p. 364. So

Between Ceredig and the point in the mid-sixth century when the persons named in the genealogies become truly historical figures, there are a number of obscure and shadowy figures. Among them are Cynwyd, his son Dyfnwal Hen ('Hen' = 'old'), and their contemporary Coel Hen. A 'triad' interpolated in *Bonedd Gwyr y Gogledd* names Cynwyd, Coel and Cynfarch as the leaders of three north British war bands.[1] The genealogy of Cynfarch is not included in the Harleian genealogies [hereafter referred to as H], but Coel is made the ancestor of Urien Rheged, Gwallawg and Morgan. This parallels the three sons of Dyfnwal Hen from whom kings were descended, though Coel and Dyfnwal are given distinct ancestries. Coel's other epithet, *godebog* ('protector'), has been mentioned above. In the *Gododdin* there is talk of enmity between the sons of 'Godebog' (Coel) and the men who fought at Catraeth.[2] His descendants included the kings of Rheged; and it has been suggested that the district of Kyle in Ayrshire is named after him.

Dyfnwal Hen is another obscure, but doubtless important, ancestor figure. The sons credited to him in H are Gwyddno, Clynog, and Cynfelyn. *Bonedd Gwyr y Gogledd* gives him different descendants; it is consistent in naming 'Ryderch Hael map Tutwal Tutclyt' as his great-grandson,[3] but differs in also naming Gwyddno as a great-grandson. The same source also connects Dyfnwal with the family of Aedán mac Gabráin, king of the Dál Riata at the end of the sixth century, a connection which may be doubted. One of Dyfnwal's grandsons is called 'Clydno Eidyn' ('Eidyn' = Edinburgh), and Clydno's son Cynon is said to have fought with the Gododdin at Catraeth in c.600.[4]

Another obscure but important figure of this period named in H is Morgan, a descendant of Coel and possibly grandfather of the Morgan who assassinated Urien Rheged. The latter Morgan is presumably the same as the Morgan who was traditionally an enemy of St Kentigern.

From about the middle of the sixth century, the kings of Strathclyde emerge as historical figures, and are worthy of individual consideration. An attempt to reconstruct their genealogical tree is given on the following page.

TUDWAL (*fl*. mid-sixth cent.):

He is called 'Titagual son of Clinoch and father of Riderch Hen' in H; Adomnán confirms that Roderc (Rhydderch) was son of Tothail (that is, Tudwal).[5] A tyrannical king, who was contemporary with Saint Ninian, is

---

Muirchú's 'Ventre' (if that be the correct reading) cannot be an earlier form of the same name. It has also been suggested that the first element in Fintry could be Gaelic *fionn*, 'white': W.F.H. Nicolaisen, *Scottish Place-Names* (London, 1976), pp. 169-70.

1   *Trioedd Ynys Prydein*, p. 238.
2   *Gododdin*, pp. 121-2.
3   H, pp. 172-3; *Trioedd Ynys Prydein*, pp. 238-9.
4   *Gododdin*, pp. 131, 139, 146.
5   H, p. 173; *Adomnan's Life of Columba*, ed. A.O. and M.O. Anderson (Edinburgh, 1961), lib. i, cap. 8.

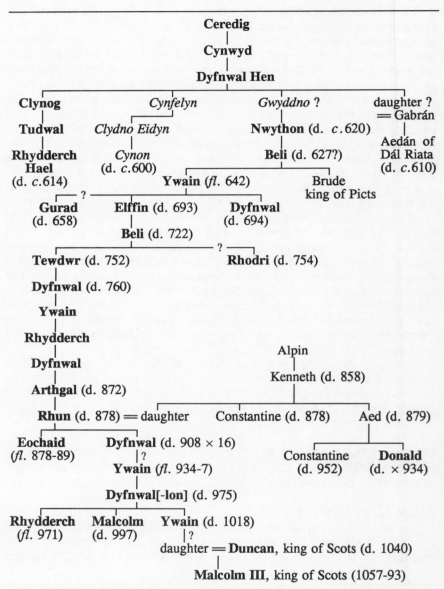

Ceredig
|
Cynwyd
|
Dyfnwal Hen

Clynog   *Cynfelyn*   *Gwyddno* ?   daughter ?
| | | = Gabrán
Tudwal   *Clydno Eidyn*   Nwython (d. *c*.620) |
| | | Aedán of
Rhydderch   *Cynon*   Beli (d. 627?)   Dál Riata
Hael   (d. *c*.600) | (d. *c*.610)
(d. *c*.614)   Ywain (*fl*. 642)   Brude
       ┌── ? ──   king of Picts
Gurad   Elffin (d. 693)   Dyfnwal
(d. 658) | (d. 694)
       Beli (d. 722)
       ┌───── ? ─────
Tewdwr (d. 752)   Rhodri (d. 754)
|
Dyfnwal (d. 760)
|
Ywain
|
Rhydderch
|
Dyfnwal   Alpin
| |
Arthgal (d. 872)   Kenneth (d. 858)
|
Rhun (d. 878) = daughter   Constantine (d. 878)   Aed (d. 879)
|
Eochaid   Dyfnwal (d. 908 × 16)
(*fl*. 878-89) | ?   Constantine   Donald
       Ywain (*fl*. 934-7)   (d. 952)   (d. × 934)
       |
       Dyfnwal[-lon] (d. 975)
Rhydderch   Malcolm   Ywain (d. 1018)
(*fl*. 971)   (d. 997) | ?
       daughter = Duncan, king of Scots (d. 1040)
       |
       Malcolm III, king of Scots (1057-93)

(This table seems to me the most likely reconstruction, but it should be compared with other tables given in the works cited on p. 1, n. 3, above. The positioning of Gurad (d. 658), Rhodri (d. 754), Ywain (*fl*. 934-7), and Ywain's successors is conjectural but likely. Certain or probable kings of Strathclyde are in bold; other persons named in the Harleian genealogies are in italics.)

Tentative Genealogy of the Kings of Strathclyde

named as 'Thuuahel' or 'Tuduael' in the eighth-century poem *Miracula Nyniae Episcopi*; and as 'Tuduvallus' or 'Tudwaldus' in Ailred of Rievaulx's twelfth-century *Vita Niniani*.[1] This is one of the very small number of details found in these works which is not also found in Bede's brief account in the *Historia Ecclesiastica*. No previous writer has suggested that Ninian's Tudwal could be identified with the father of Rhydderch Hael, thus making Ninian a figure of the mid-sixth century rather than of the fifth. But the only chronological pointer for Ninian is Bede's statement that he lived a long time before ('multo ante tempore') the conversion of the northern Picts by Saint Columba (*c*.521-97); and Bede on occasion uses 'multo tempore' to separate events within the lifetime of one man.[2] Saint Patrick's reference to Picts as 'apostatae' cannot be taken as evidence of Christianity among fifth-century Picts, since *apostata* means simply 'renegade, defector' (in the Vulgate, it can mean 'criminal, evildoer'), without any implication of religious backsliding such as it later came to have. Patrick also calls the Picts 'a foreign people who do not know God' ('gens extera ignorans Deum'), which clearly suggests that their conversion came later.[3] For what it is worth, an odd episode in Jocelin's *Life of Kentigern* makes Kentigern (d. *c*.614) perform the first burial in a cemetery formerly ('quondam') consecrated by Ninian, in which nobody had previously been buried ('in quo nunquam quisdam positus fuit').[4] This would be credible only if Ninian had lived in the mid-sixth century, and impossible if he were a fifth-century figure. The present writer has argued at length elsewhere that the possibility of a radical revision of Ninian's chronology is worth considering.[5]

RHYDDERCH HAEL (*fl. c.*580; d. *c.*614?):

H calls him Riderch son of Titagual. According to Adomnán, he was a contemporary of Saint Columba; he calls him 'Roderc son of Tothail, king of Petra Cloithe' (i.e., *Ail Cluaide*). Adomnán adds that he was a friend of Columba, to whom he sent a secret message *via* one of Columba's monks,

---

1   The poem is edited by K. Strecker in *Poetae Latini Aevi Carolini*, iv (*Monumenta Germaniae Historica* [Berlin, 1923]), pp. 943-62, and by W. MacQueen in *TDGNHAS*, 3rd ser., xxxvii (1960); translation in J. MacQueen, *St Nynia* (2nd edn, Edinburgh, 1990), pp. 88-101. Ailred's *Vita Niniani* is edited by A.P. Forbes in *The Lives of SS Ninian and Kentigern* (Edinburgh, 1874), pp. 137-57.
2   *Bede's Ecclesiastical History of the English People*, ed. B. Colgrave and R.A.B. Mynors (Oxford, 1969), lib. iii, cap. 4. For the arguments about the date of Ninian's activity, see Macquarrie, 'Date of St Ninian's Mission', esp. pp. 7-8.
3   *Epistola Militibus Corotici*, cap. 2, 14, 15; Macquarrie, 'Date of St Ninian's Mission', pp. 10-11.
4   Jocelin's *Vita Kentigerni* in *Lives of SS Ninian and Kentigern*, cap. 9; see also Macquarrie, 'Career of St Kentigern', pp. 10-12.
5   Macquarrie, 'Date of St Ninian's Mission'. Cf. the criticisms in MacQueen, *St Nynia*, pp. 22-31, which disagree with my conclusions without, it seems to me, fully considering the grounds on which they are based. Surprisingly, Professor MacQueen is happy to accept my argument that the dedication of Candida Casa to St Martin belongs to the 6th rather than the 5th century (ibid., p. 31). Since the argument for an early, 5th-century, Ninian rests entirely on a connection between Ninian and St Martin, this admission would appear to be fatal to that argument.

asking whether he would be slaughtered by his enemies or not. Columba
replied that he would die in his own house and bed, and not by the hand of
his enemies.[1] This incident may imply that Rhydderch was seeking
assurances that Aedán mac Gabráin would not attack him, and that Aedán's
counsellor Columba reassured him. A Welsh 'triad', however, records that
'Aedán the traitor went into *Ail Cluit*, to the court of Rhydderch Hael; after
[his plundering] there remained neither food nor drink, nor any living
thing'. In later times the plundering of Strathclyde became the subject of an
Irish folk tale (now lost).[2] Does this imply that Aedán broke a promise made
through Columba?

The *Historia Brittonum* names Rhydderch as an ally of Urien Rheged
against a king of Bernicia at Lindisfarne (possibly Hussa, 585 × 593).[3]
Jocelin's *Life of Kentigern* makes him a contemporary and benefactor of
Kentigern, whose *obit* in the *Annales Cambriae* is *sub anno* 612 (perhaps
representing *c*.614). Not all of Jocelin's information about Rhydderch is
likely to be genuine, but some of it may come from an early source written
within a century of Kentigern's death. In particular, Jocelin's account of
Kentigern's old age and death has been shown to be early,[4] and he goes on
to state that Rhydderch died within a year, at his royal estate of 'Pertnech'.
It is known that Partick was royal demesne in Strathclyde during the twelfth
century, but Jocelin is unlikely to have known this or to have invented the
detail. So it may cautiously be suggested that Jocelin's information about
Rhydderch's death is early and factual.

NWYTHON (*fl. c*.600; d. *c*.620?):

H names Neithon as son of Guipno (for which read 'Guiþno', i.e.
Gwyddno) and grandson of Dumnagual Hen; and as father of Beli and
grandfather of Eugein (Ywain). It has been suggested that he is identifiable
with Nechtan son of Cano, king of the Picts.[5] It is difficult to reconcile the
single generation between Dyfnwal Hen and Nwython and the two
generations between Dyfnwal and Nwython's earlier contemporary
Rhydderch Hael. But it is chronologically possible that Nwython king of
Dumbarton is identical with Nechton *nepos* Uerb, king of Fortriú, and
Nechtan son of Cano, whose death is placed in the Irish annals at *c*.621.
After suitable emendation, an interpolated stanza in *The Gododdin* may

1   *Adomnan's Life of Columba*, lib. i, cap. 8.
2   *Trioedd Ynys Prydein*, p. 147; P. Mac Cana, *The Learned Tales of Medieval Ireland*
(Dublin, 1980), pp. 47, 100-1. Although it is there suggested that the incident referred to in
the tale is identifiable with the great siege of 870 (see below, p. 12), it is perhaps more
probable that the tale concerned the same incident as that referred to in the 'triad'.
3   A readily accessible edition of the Harleian MS of the *Historia Brittonum* is by J.
Morris, *Nennius: British History and the Welsh Annals* (London, 1980); see cap. 63.
4   *Vita Kentigerni*, cap. 30-8, 42-5; D. MacRoberts, 'The Death of St Kentigern of
Glasgow', *Innes Review*, xxiv (1973), pp. 43-50; Macquarrie, 'Career of St Kentigern', pp.
17-18.
5   Smyth, *Warlords and Holy Men*, pp. 64-5; Williams, *Biographical Dictionary*,
pp. 183-4.

possibly name 'the grandson of Nwython' as victor in a battle in which Dyfnwal Frych was slain. Domnall Brecc, king of the Dál Riata, was defeated and killed at the battle of Strathcarron in 642 by Ywain, grandson of Nwython (for whom see below).[1] If Nechtan mac Canonn and Nechton *nepos* Uerb are the same, it may be that Uerb was the name of Nwython's maternal grandfather; this might be taken as evidence that Nwython's claim to the kingship of Fortriú was derived through his mother. His claim to the kingship of Strathclyde presumably came from his father, prompting the suggestion either that the genealogies are unreliable in naming his father as Gwyddno, or that Cano is an Irish mistake, or misformation of this name; both are equally possible. It has also been pointed out that the Irish cognate of the Pictish name Uerb, 'Ferb' (gen. 'Feirbe'), is a woman's name, and that Uerb might be Nwython's grandmother or possibly his aunt.[2] Despite recent questioning of the theory of Pictish matrilineal succession, it must be remembered that Bede, writing in *c*.730, believed that to be the practice in his own day, 'when the matter came into doubt'.[3]

BELI (d. *c*.627?, certainly × 642):

H calls Beli son of Neithon and father of Eugein. Brude son of Beli, king of Fortriú (d. 693), is called 'son of the king of Dumbarton' ('mac rígh Ala Cluaithi') in a verse attributed to Adomnán.[4] Brude and Ywain (Eugein) must, therefore, have been brothers, or at least half-brothers. Beli may be identifiable with a Belin whose *obit* is given as *c*.627 in the Welsh annals; he must have been dead by 642, the year of the battle of Strathcarron. Brude is called 'fratruelus' of Ecgfrith king of Northumbria, whom he defeated and slew at the battle of Dunnichen in Angus in 685; so Beli may have been married to a lady of Northumbrian royal descent — though Brude presumably had some qualification to rule in Fortriú.

YWAIN (*fl*. 642; d. × 658):

H makes Ywain son of Beli. An interpolated stanza in *The Gododdin* may, after emendation, mean that 'the grandson of Nwython' was leader of an army which defeated and slew Dyfnwal Frych. Irish versions of these events state that 'Oan' or 'Hoan', king of the Britons, killed Domnall Brecc, king of the Dál Riata, in the battle of Strathcarron in 642.[5] The date of Ywain's death is not recorded, though it must have been before 658 (see Gurad below).

GURAD (d. 658):

The Irish annals record the death of 'Guret king of Dumbarton' in 658.[6]

---

1   *Gododdin*, pp. 98-9; Anderson, *Early Sources*, i, p. 145.
2   Ibid., pp. 122, 145.
3   *Bede's Eccles. Hist.*, lib. i, cap. 1.
4   *Chron. Picts-Scots*, pp. 408-9; Anderson, *Early Sources*, i, p. 201.
5   *Ann. Ulster*, *s.a.* 642.
6   Ibid., *s.a.* 658.

He does not appear in H. He was possibly a son, or perhaps more likely a brother, of Ywain.

ELFFIN (d. 693):
    Elffin is son of Eugein in H; but the Irish annals record the death of 'Alpin son of Nechtan' in 693.[1] If Elffin was the son of Neithon son of Guipno, he must have outlived both Beli and Ywain to die in the same year as his probable nephew, Brude son of Beli. H is perhaps more likely to state the relationship correctly, and the annals probably erred in calling him son of Nechtan (if this was Nechton king of Fortriú, he had been dead since c.621). Some writers suggest two Elffins, one son of Nwython, the other son of Ywain.[2] Both must have died around the same time; but it is inconceivable that one great-grandson of Dyfnwal Hen (Elffin) can have died in 693, while another (Rhydderch Hael) had a *floruit* a century earlier. Perhaps descent from Nwython was regarded as important; the Irish annalists may have known that Elffin was a descendant of Nwython, but may have mistakenly called him son rather than great-grandson, the relationship stated (probably correctly) in H.

DYFNWAL (d. 694):
    He was the son of Ywain. The Irish annals record the death of 'Domnall son of Auin, king of Dumbarton', in 694.[3] He is not mentioned in H.

BELI (d. 722):
    He is called son of Elfin and father of Teudebur in H, and 'son of Elfin' in the Welsh annals. The Irish annals call him 'Bile son of Elphine, king of Dumbarton', and place his death in 722.[4]

TEWDWR (d. c.752):
    He is called Teudebur son of Beli in H. The Welsh and Irish annals also call him 'son of Beli' or 'Bile'; the latter name him as 'rex a lochlandaid', 'king of Lochlann', for which should be read 'rex alo cluaide', 'king of Dumbarton'. His death is placed in c.752. In 744 an unnamed British king fought against the Picts; in c.750 a British king overthrew Talorgen king of the Picts and slew him and many of his men at the battle of 'Mygedawc' or 'Moce-tauc'. This is presumably the present Mugdock, between Milngavie and Strathblane. The Irish annals record 'the waning of the power of Angus' in 750; this was Onuist son of Uurguist, king of the Picts, and Talorgen was his brother. Presumably the waning of his power was the result of the British victory.[5] Also in 750, however, English sources record the seizure

---

1   Ibid., *s.a.* 693; also in *Ann. Tigernach*, p. 213.
2   Smyth, *Warlords and Holy Men*, pp. 64-5 (obscuring the issue slightly by giving one of them the Irish form of the name, Alpín).
3   *Ann. Ulster*, *s.a.* 694; also in *Ann. Tigernach*, p. 213.
4   H, pp. 160, 172; *Ann. Ulster*, *s.a.* 722; *Ann. Tigernach*, p. 228.
5   H, p. 161; *Ann. Tigernach*, p. 253; *Ann. Ulster*, *s.a.* 750.

of the plain of Kyle by Eadberht king of Northumbria, presumably at Tewdwr's expense. The Picts and Angles may have been acting in concert, as they did a few years later.[1]

RHODRI (d. *c*.754):

The death of 'Rotri king of the Britons' (presumably Welsh 'Rhodri', and not identifiable with Tewdwr) is entered in the Welsh annals at *c*.754.[2] He does not appear in H. He may have seized the kingship on Tewdwr's death in *c*.752, and have been ousted and killed soon after by Tewdwr's son Dyfnwal (see below).

DYFNWAL (d. *c*.760):

He is called Dumnagual son of Teudebur in H, and likewise in the Welsh annals, where his death is placed in *c*.760.[3] Symeon of Durham records that in 756 Eadberht of Northumbria and Onuist king of the Picts united to lead an army to the city of 'Alcluth'; there they received the submission of the Britons on 1 August. But on 10 August almost the whole of Eadberht's army was destroyed while he was leading it from 'Ouania' to 'Niwanbirig'.[4] This may have been as a result of treachery by Onuist, who, according to Bede's continuator, 'from the beginning of his reign right to the end perpetrated bloody crimes, like a tyrannical slaughterer'.[5] Presumably it was Dyfnwal who submitted in 756. 'Niwanbirig' (Newburgh) is an Anglian name, and so was presumably somewhere in Northumbria. 'Ouania' must have been located between Dumbarton and Northumbria. A possible identification is the River Avon in Lanarkshire, which falls into the Clyde near Cadzow. Cadzow was royal demesne of the Scottish crown in the twelfth century, and presumably had earlier been demesne of the kings of Strathclyde.[6] A very late medieval life of Saint Kentigern associates Rhydderch Hael's queen (called 'Languoreth' by Jocelin of Furness) with Cadzow.[7] An alternative, and perhaps more likely suggestion, would be the River Avon in West Lothian, which would have been on the Anglian army's homeward march if it followed the Roman road along the Antonine Wall. It would also have been closer to Onuist's territories, and he is more likely to have been the perpetrator of the attack on his erstwhile ally than Dyfnwal and the recently cowed Britons. After Dyfnwal's death, there are very few mentions of the kings of Dumbarton for more than a century, leaving only the pedigree in H. Despite the death of

1   *Continuatio Baedae* in *Bede's Eccles. Hist.*, p. 574
2   H, p. 161.
3   H, pp. 161, 172.
4   *Historia Regum Angliae* in *Symeonis Dunelmensis Opera Omnia*, ed. T. Arnold (Rolls Ser., 1882), ii, pp. 40-1.
5   In *Bede's Eccles. Hist.*, p. 576.
6   King David granted the church of Cadzow to Glasgow Cathedral in *c*.1150; the place seems also to have been a royal residence (*ESC*, pp. 96, 178-9).
7   In *Breviarium Aberdonense* (Spalding and Maitland Clubs, 1854), *Pars Hiemalis*, 13 January; see Macquarrie, 'Career of St Kentigern', p. 9.

Onuist in 761,[1] Strathclyde seems to have remained subject to foreign control.

YWAIN (*fl.* late 8th cent.):
Ywain is son of Dumnagual and father of Riderch in H; he is not otherwise known. He may have been king when Dumbarton was again sacked in 780.[2]

RHYDDERCH (*fl.* early 9th cent.):
Rhydderch is son of Eugein and father of Dumnagual in H; he is not otherwise known.

DYFNWAL (*fl.* mid-9th cent.):
Dyfnwal is son of Riderch and father of Arthgal in H; he is not otherwise known. The Britons burned Dunblane in *c.*849, perhaps under his leadership, or that of his son Arthgal.[3]

ARTHGAL (d. 872):
In H, Arthgal is son of Dumnagual and father of Run. The Irish annals, Welsh annals, and Mac Firbis record that in 870 Olaf and Ivar, kings of the Norse of Dublin, besieged 'Ailech Cluathe' and reduced it after a siege of four months when the well on the rock dried up. The Welsh annals state that 'Arx Alt-Clut', the summit of Dumbarton Rock, was destroyed. This is probably a reference to the White Tower Craig, the higher and more precipitous of the two craigs. The well lies today between the craigs, and this has probably always been the case. Perhaps what happened was that, after a lengthy siege, the Norse stormed and occupied the saddle and the lower craig (the Beak), driving the defenders onto the White Tower Craig and cutting off their water supply.[4] Resistance thereafter would not have been prolonged. The same sources record that in 871 Olaf and Ivar returned to Dublin with many prisoners; the Irish annals state that in 872 'Artgal king of the Britons of Strathclyde was slain by counsel of Constantine son of Kenneth' (Constantine I).[5] The plundering of Strathclyde entered into Irish folklore, but it is perhaps more likely that this was an earlier raid carried out by Aedán mac Gabráin.[6] Arthgal was presumably one of Olaf's and Ivar's prisoners in 871; Constantine's sister was, or subsequently became, married

---

1   *Bede's Eccles. Hist.*, p. 576; *Ann. Ulster, s.a.* 761; *Ann. Tigernach*, p. 259.
2   H, p. 172; *Ann. Ulster, s.a.* 780.
3   This is the first recorded aggressive exploit by the British for a long period. Its date is uncertain, but the incident probably belongs to the later part of the reign of Kenneth mac Alpin (d. *c.*858): *Chron. Picts-Scots*, p. 8; Anderson, *Kings and Kingship*, p. 250; Anderson, *Early Sources*, i, p. 288.
4   *Ann. Ulster, s.a.* 870; H, p. 166; *Fragmentary Annals of Ireland*, ed. J.N. Radner (Dublin, 1978), p. 142; Anderson, *Early Sources*, i, pp. 301-2.
5   *Ann. Ulster, s.a.* 872.
6   Mac Cana, *Learned Tales of Medieval Ireland*, pp. 47, 100-1 and n. 90; see above, p. 8 and n. 2.

to Arthgal's son Rhun.[1] If the marriage had already taken place, or was being arranged, the death of Arthgal would have allowed Constantine to assume control of Strathclyde in the name of his (possibly young) brother-in-law.

RHUN (d. 878):
He is the last king named in H, where he is called Run son of Arthgal; Rhun married a daughter of Kenneth mac Alpin, and so was brother-in-law of Constantine mac Kenneth, who had consented to Arthgal's death (probably at the hands of his Norse captors). He succeeded his father in 872, and died in 878, the same year in which his over-king Constantine died in battle. The foundation of the important church of St Constantine at Govan probably belongs to this period, or shortly after.[2]

EOCHAID (*fl.* 878-89):
He was son of Run king of the Britons, and grandson of Kenneth mac Alpin by his daughter, according to the Chronicle of the Kings of Scotland. Accounts of his reign as king of Alba are confused. He may have ruled jointly with Giric (878-89) and have been expelled by him. According to 'Berchan's Prophecy', he was the first Briton to rule over the Gael; he is called 'an Britt a Cluaide, mac mna o Dhún Guaire' ('the Briton from the Clyde, son of the woman from *Dún Guaire*'). Berchan gives him a reign of thirteen years and states that he was dispossessed by 'Mac Rath' (i.e., 'son of fortune' — probably Giric), who abased Britain and exalted Alba. Mac Rath is said to have had English, Norse and British as slaves in his house (i.e., as subject rulers?). The early annals in the Chronicle of Melrose describe Giric as very powerful, dominating (the north of) England.[3]

In 890, according to *Brut y Tywyssogion*, 'the men of Strathclyde, those who refused to unite with the English, had to depart from their country, and go into Gwynedd'. They were settled there by Anarawd king of Gwynedd whom they aided in defeating the Saxons. It is likely that they had refused to unite with the Scots, rather than with the English.[4]

What may have happened is that Eochaid had attempted to assert his own kingship of Alba in *c*.878, through his descent from Kenneth. He was unsuccessful, and accepted joint status with, or client status under, Giric until *c*.889. In that year Giric was expelled, and probably Eochaid too. The

1   Eochaid son of Rhun was Kenneth's grandson: *Chron. Picts-Scots*, p. 9; Anderson, *Kings and Kingship*, p. 250.
2   The precise date of Rhun's death is not recorded, but Eochaid probably succeeded him in 878; Anderson, *Early Sources*, i, pp. 363-4. On the foundation of Govan, see Macquarrie, 'Early Christian Govan', esp. pp. 6-9.
3   The sources are listed in Anderson, *Early Sources*, i, pp. 363-8. The statement that Giric was 'alumpnus ordinatorque Eochodio' (Anderson, *Kings and Kingship*, pp. 250-1) indicates that Giric was the over-king and Eochaid the sub-king, even though the classical meaning of *alumnus* should be 'foster-son'.
4   Anderson, *Early Sources*, i, p. 368; discussed in Smyth, *Warlords and Holy Men*, pp. 217-18.

date of Eochaid's death is not recorded, nor is it known whether he led his war band into Gwynedd, or whether the migration happened after his death.

DYFNWAL/DONALD (d. 908 × 916):

According to the Chronicle of the Kings of Scotland, Donald king of the Britons died in the time of Constantine mac Aeda, king of Scots (900-43). The Chronicle describes the events of Constantine's reign in chronological sequence, placing Donald's death between events dated 908 and 916.[1] He may have been a descendant of the old British dynasty of Dumbarton; Ywain (Owen), king of the Cumbrians by 934, may have been his son, and traditional names reappear in the royal house of Strathclyde: Ywain, Dyfnwal, Rhydderch. His immediate successor was Donald son of Aed.[2]

DONALD (d. × 934):

Donald son of Aed (brother of Constantine mac Aeda, king of Scots [900-43]) was chosen as king of the Britons in succession to Donald (Dyfnwal).[3] A.O. Anderson has suggested that he was still king of Strathclyde at the time of Saint Catroe's visit in c.941, and a few years later, when 'Strathclyde was wasted by the Saxons' in 945, and King Edmund 'commended it to Malcolm king of Scots'. But in 934 Symeon of Durham names the king of the Cumbrians as 'Owen' (Ywain).[4] It may be that the Donald mentioned in the Life of Catroe was the Dyfnwal (Dyfnwallon) who appears below.

YWAIN (fl. 934):

Owen king of the Cumbrians was put to flight along with his ally King Constantine by Athelstan in 934. The king of the Cumbrians, probably this same Owen or Ywain, was with Constantine and Olaf of Dublin at the battle of Brunanburh in 937.[5] It is possible that he was son of Donald son of Aed, and therefore Constantine's nephew. On the other hand, the revival of traditional British names in the royal dynasty of Strathclyde suggests that Ywain may have been descended from Rhun, probably through Dyfnwal/ Donald (d. 908 × 916). A.P. Smyth identifies Ywain with Owen king of Gwent, one of the kings who submitted to Athelstan in 926 or 927;

---

1   Anderson, *Kings and Kingship*, p. 251; Anderson, *Early Sources*, i, pp. 444-6; *Chron. Picts-Scots*, p. 9.
2   Anderson, *Kings and Kingship*, p. 251; Anderson, *Early Sources*, i, p. 446; *Chron. Picts-Scots*, p. 9.
3   Anderson, *Kings and Kingship*, p. 251; Anderson, *Early Sources*, i, p. 446; *Chron. Picts-Scots*, p. 9.
4   The *Life of Catroe* is translated in Anderson, *Early Sources*, i, pp. 431-43; text in *Chron. Picts-Scots*, pp. 109-16. The *Life* states that Catroe was a relative of Donald king of the Cumbrians, but does not make the same claim for Constantine king of Scots. The presumption must be that Donald and Constantine were unrelated, and so the Donald in question cannot have been Donald son of Aed, who was Constantine's brother. See under Ywain and Dyfnwal/Dyfnwallon below.
5   *Historia Dunelmensis Ecclesiae* in *Symeonis Dunelmensis Opera Omnia*, i, p. 76. The sources are listed in Anderson, *Scottish Annals*, pp. 67-73.

for this there seems to be no certainty.[1]

Equally doubtful is Fordun's assertion that from the late ninth century onwards each Scottish king placed his successor-designate or *tanaise* as king of Strathclyde/Cumbria. Donald son of Aed is the only tenth-century king of Strathclyde who was demonstrably a member of the dynasty of Kenneth mac Alpin; the assertion that Indulf, Dub and other Scottish kings of this period had previously been sub-kings of Strathclyde cannot be traced further back than Fordun. It is true, however, that Duncan king of Scots (1034-40) was king of Strathclyde from 1018 under his grandfather Malcolm II, and that David I had held a great lordship in southern Scotland before his accession to the kingship of the Scots in 1124. The temporary grant of the earldom of Lennox (formerly part of the Strathclyde kingdom) to Earl David in 1174 is suggestive of a similar arrangement. But, despite the arguments of Smyth and A.A.M. Duncan, there is insufficient evidence that such an arrangement worked systematically in the tenth century.[2] For this period, Fordun's narrative bristles with improbable stories and unhistorical personages, and cannot be treated as reliable evidence; it is possible that he was aware of an occasional arrangement practised at a later period, and has projected it further back and interpreted it in an over-systematic way. Even Smyth, who for the most part accepts Fordun's version of the arrangement, concedes that in the second half of the tenth century the 'system' broke down and that Dyfnwal (d. 975) 'established a new dynastic segment in a tribal area reserved hitherto for heirs to the Scottish throne'.[3] But the presence of traditional British names (Ywain, Dyfnwal, Rhydderch) in Strathclyde in the tenth and eleventh centuries suggests not so much a 'new dynastic segment' as the continuation of the old Cumbrian dynasty.

DYFNWAL/DYFNWALLON *(fl. c.*941-73; d. 975):

He is called 'Donald son of Eogan, king of the Britons', in the Irish annals, where his death 'on pilgrimage' is recorded in 975. He is called 'Dunguallon king of Strathclyde' in *Brut y Tywyssogion*, where he is said to have gone to Rome in 975 and received the tonsure.[4] He was presumably the father of Malcolm and Ywain the Bald. He may have been king of Strathclyde as early as *c.*941. The tenth-century continental *Life of Catroe* says that, on leaving King Constantine at the church of St Brigid (Abernethy?), Catroe visited the land of the Cumbrians, which was ruled over by King Donald, and 'because he was a relative of Catroe, he came to

---

1   *The Anglo-Saxon Chronicle*, ed. B. Thorpe (Rolls Ser., 1861), *s.a.* 926; see Anderson, *Scottish Annals*, p. 66; Smyth, *Warlords and Holy Men*, pp. 201-4, 222. The identification is based on the assumption that at this time all Strathclyde kings were clients of the kings of Scots. There is no clear evidence to support the contention that this Ywain was killed at *Brunanburh*.
2   Ibid., pp. 222-8; Williams, *Biographical Dictionary*, pp. 103, 105, 199, and *passim*; A.A.M. Duncan, 'The Kingdom of the Scots', in *The Making of Britain: The Dark Ages*, ed. L.M. Smith (London, 1984), p. 137.
3   Smyth, *Warlords and Holy Men*, p. 224
4   Anderson, *Early Sources*, i, pp. 478-80.

meet him with all joy'.[1] The *Life* does not name Constantine king of Scots (*c.*900-43) as a relative of Catroe, so there is a strong presumption that Constantine and Dyfnwal were unrelated. One of the kings who rowed Edgar on the Dee at Chester in 973 was called 'Dufnal' (Dyfnwal); but on this occasion Malcolm, probably Dyfnwal's son, was recognised by Edgar as king of the Cumbrians.[2] If Rhydderch, another probable son, was king of Strathclyde, Dyfnwal may have resigned as early as 971.

RHYDDERCH (?) (*fl.* 971):

The slayer of Culen king of Scots (d. 971) is called 'Radharc' or 'Amdarch' (or slight variants of these) in Scottish sources. He is called 'son of Donald' in the Chronicle of the Kings of Scotland. 'Berchan's Prophecy' states that Culen was slain by the Britons. Culen is said to have carried off and raped his slayer's daughter, and to have been killed in Lothian at a place called 'Ybandonia'. A.O. Anderson's tentative suggestion that this was Abington in upper Strathclyde is doubtful.[3] Culen's successor Kenneth II 'immediately plundered Britain', but 'his footsoldiers were slain with very great slaughter in Moin Vacornar'. This place, presumably in Strathclyde, remains unidentified.[4] It would seem that this Rhydderch, if that was his name, was a son of Dyfnwal son of Ywain; it is not certain that he was himself king, but if he was it must have been before 973 (see Malcolm below). Perhaps Dyfnwal's abdication was in some way connected with these events.

MALCOLM (*fl.* 973; d. 997):

Malcolm king of the Cumbrians is named as one of the kings who rowed Edgar on the Dee in 973; the Dufnal who is named as participating in the same exercise was perhaps his father. In the Irish annals he is named as 'Malcolm son of Donald, king of the Britons of the North'; his death is placed in 997. Presumably he was the son of Dyfnwal son of Ywain; he may have succeeded Dyfnwal directly on his abdication, or after a brief reign by his brother Rhydderch (?), while Dyfnwal was still living.[5]

YWAIN (d. 1018):

The Welsh annals call him 'son of Dumnagual' (Dyfnwal), and place his death in *c.*1015. Symeon of Durham calls him 'Owen the Bald, king of the

1   Ibid., pp. 440-1.
2   Ibid., p. 478; cf. Anderson, *Scottish Annals*, pp. 76-7.
3   *Chron. Picts-Scots*, p. 151; Anderson, *Early Sources*, i, p. 476. Perhaps for 'in Ybandonia' we should rather read 'in Laudonia': other sources locate Culen's death 'in Lownes, Loinas, Lennas, etc.' This name is likely to represent Lothian, or perhaps (less likely) Lennox, the Gaelic *Lemnach*.
4   *Chron. Picts-Scots*, p. 10. Anderson, *Early Sources*, i, p. 512, is right to criticise the identification proposed by Skene in *Celtic Scotland* (2nd edn, Edinburgh, 1886-90), i, p. 368, with a moss near Abercorn. Mrs Anderson reads 'Moin Uacoruar': *Kings and Kingship*, p. 252.
5   *Ann. Tigernach, s.a.* [997], p. 351; Anderson, *Early Sources*, i, pp. 476, 480, 517.

men of Strathclyde', present with King Malcolm II at the battle of Carham in 1018.[1] It is a possible inference that he was killed at Carham. His father was presumably the Dyfnwal who died on pilgrimage to Rome in 975, in which case Ywain must have succeeded his brother Malcolm in 997. He may have been an old man by the time of his death. Ywain was the last of the old line of the kings of Strathclyde. It is possible that he had a daughter who was married to Duncan, grandson of Malcolm II. In spite of the doubtful testimony of Fordun, Duncan is the only king of Strathclyde who can be demonstrated to have succeeded to the kingship of the Scots on his predecessor's death. As pointed out already, apart from Duncan the only king of Strathclyde who was certainly not a member of the old native dynasty was Donald son of Aed (d. × 934), whose aunt had been married to Rhun son of Arthgal; he probably would have had a good chance of succeeding to the kingship of the Scots had his brother Constantine not been so long-lived. But he seems to have been an isolated example before the time of Duncan. It may be suggested that Duncan likewise had some female connection which entitled him to the kingship of Strathclyde.

A number of concluding points emerge from the above narrative. A successful British dynasty with an active war band had been established on Dumbarton Rock by the mid- or late fifth century, at which time it was carrying out plundering raids against north-east Ireland. Little is known of the kings between Ceredig and Rhydderch Hael, the latter of whom encouraged the episcopate of Saint Kentigern and died about the same time as the saint (c.614). He was also active in an alliance of north British kings against Anglian Bernicia in c.590, and was subjected to raids from Aedán mac Gabráin, king of the Dál Riata, in spite of assurances given to him by Aedán's adviser Saint Columba. Rhydderch and his successors maintained the integrity of their kingdom despite pressure from Bernicia and the Dál Riata, and despite the destruction of the neighbouring British kingdom of the Gododdin, based on Edinburgh, in c.640. In the seventh century, Strathclyde seems to have reached the height of its power; Nwython (d. c.620?) may also have been king of the Picts; Ywain defeated the Dál Riata and killed their king at the battle of Strathcarron in 642, and drove them out of central Scotland; his brother or half-brother Brude son of Beli (d. 693), 'son of the king of Dumbarton', was a very powerful king of the Picts who defeated the Angles of Northumbria at the battle of Dunnichen (685), as a result of which Anglian power waned in the north and the Britons of Strathclyde regained territory previously lost to them.

During the seventh century other north British kingdoms, such as

1 *Symeonis Dunelmensis Opera Omnia*, ii, pp. 155-6; Anderson, *Scottish Annals*, pp. 81-2. On the battle of Carham, see two articles in *SHR*, lv (1976): B. Meehan, 'The Siege of Durham, the Battle of Carham and the Cession of Lothian'; and A.A.M. Duncan, 'The Battle of Carham, 1018'.

Rheged, Gododdin and Elfed (Elmet), disappeared, swallowed up in the
Northumbrian advance. Strathclyde escaped this fate, but by 730
Northumbria was again advancing, and the Angles occupied the north
Solway shore and appointed a series of bishops at Whithorn. In 750 they
overran the plain of Kyle in Ayrshire. In the early 750s Strathclyde may
have been further weakened by dynastic strife between Rhodri map Beli and
Dyfnwal map Tewdwr, in which the latter was victorious. In 756
Dumbarton was besieged by a combined army of Angles and Picts, and
Dyfnwal was forced into a humiliating surrender and exile. The destruction
of the Northumbrian army immediately thereafter does not seem to have
benefited Strathclyde; for more than a century, nothing is known of the
kings of Strathclyde except some of their names, and the kingdom seems to
have been eclipsed.

Dumbarton may have been revived under Arthgal (d. 872), but only
sufficiently to attract the attention of the Norse kings of Dublin. They
captured Dumbarton after a great siege in 870, carrying Arthgal and many
others off to slavery and death in Ireland. It may be that for a time
Dumbarton became, like Dublin, 'a Viking headquarters';[1] this would
account for the hogback tombstones and other signs of Norse influence in
the stone-carving of Govan and its neighbourhood in the period following
870. But from the mid-ninth century, the family of Kenneth mac Alpin was
the dominant power in Scotland, and the kings of Dumbarton were usually
its clients. Nevertheless, the old British dynasty seems to have continued in
the persons of Rhun map Arthgal, his son Eochaid who co-reigned with the
powerful King Giric son of Dungal, and their successors; many of these had
traditional Welsh names such as Ywain, Dyfnwal and (if the reading is
correct) Rhydderch. Only Donald son of Aed and grandson of Kenneth mac
Alpin was definitely not a member of this family. Fordun's claim that each
Strathclyde king succeeded the preceding king of Scots on a regular basis
does not receive support from the earliest and most reliable evidence.
Although the genealogy cannot be reconstructed with total certainty, it
seems that the old line continued down to the time of Ywain the Bald (d.
1018), who appears in the military service of King Malcolm II; after his
death, Malcolm's grandson Duncan was king of the Cumbrians, and from
1034 king of Scots. He was the first person to unite the two kingships in
himself. His grandson David was lord of Cumbria while *tanaise* or heir-
apparent to his brother Alexander I, and there are later examples of the
earldom of Northumbria, and in one case of the earldom of Lennox, being
held by the heir to the Scottish throne. Fordun may have projected such
arrangements into the more remote past and portrayed them in a more
systematic way than would ever have been possible in practice.

The precise boundaries of the kingdom at any time are difficult to
determine. *Clach nam Bretann* above the head of Loch Lomond is supposed

1   Duncan, *Making of the Kingdom*, p. 90.

to have been the northern boundary-marker of the kingdom,[1] but its southern extent certainly varied. If the kingdom of Rheged was centred on Carlisle, Strathclyde may not have extended beyond Beattock while that kingdom lasted; but after the Anglian defeat at Dunnichen, the kingdom may have extended to the Solway for a few decades, only to contract again in the 720s. During the very obscure century following the first great siege (756), the kings may have been in control of only a small area around Dumbarton itself; but after the second siege (870), Rhun and his son Eochaid may have enlarged Strathclyde into something like its earlier size, though as clients of the kings of Scots. From the mid-tenth century, Strathclyde seems to have extended further south, as far as the Yorkshire-Westmorland border at Stainmore. This was due to the collapse of Viking power in the north of England, and to an acceptance by the West Saxon kings that they could not fill the vacuum. Perhaps the headquarters and focus of the Strathclyde kingdom moved further south; Dumbarton itself receives barely a mention between 870 and the thirteenth century.[2] The presence at Dumbarton, however, of 'Govan School' stone-carvings of tenth or eleventh century date is evidence of continuing occupation of the Rock during this period.[3]

In earlier centuries, the Strathclyde dynasty had been most closely associated with Dumbarton, but it must have had other important residences and royal centres as well. Jocelin of Furness names 'Pertnech', Partick, as one of these. A manor frequented by Rhydderch Hael and the place of his death, Partick was still royal demesne when David I gave it to the bishop of Glasgow in 1134. The kings must have had other centres elsewhere in the Clyde valley. Cadzow was royal demesne in the twelfth century, and the late medieval *Life of Kentigern* associates it with the wife of Rhydderch Hael. Rutherglen was probably a royal centre from very early times. A close analysis of the feudal colonisation of Clydesdale in the mid-twelfth century would probably reveal others. Further south, Carlisle was a favourite residence of David I from 1136, and may have been important to the kings of Strathclyde from the mid-tenth century.

Perhaps the most striking feature about the dynasty of Strathclyde is its longevity. Only for relatively brief periods did its kings exercise great influence in central and southern Scotland; for long periods, especially from the mid-eighth century onwards, Strathclyde was in subjection to Angles, Scots or Picts. But, until its final amalgamation into the kingdom of Scotland in the eleventh century, Strathclyde never suffered the total eclipse which was the fate of all the other northern British kingdoms.

1   Watson, *Celtic Place-Names*, p. 15.
2   Duncan, *Making of the Kingdom*, p. 90.
3   I. MacIvor, *Dumbarton Castle* (HMSO, 1981). More recently the carved fragments have been removed from storage in the Guardhouse, cleaned, and put on display in the Governor's House at Dumbarton Castle. For a general discussion of the historical significance of the 'Govan School' stone-carvings, see Macquarrie, 'Early Christian Govan', pp. 1-14.

2

MacDuff of Fife

JOHN BANNERMAN

The name 'MacDuff' and its usage in historical record have presented something of a problem to historians in the past.[1] If, however, they are approached from the point of view of the kin-based society and of Gaelic, the original language of that society, the difficulties, perceived or not, that surround the name in its various non-Gaelic contexts readily dissolve and disappear. In addition, the later examples of usage are instructive in that they give a glimpse of the accommodation made by the incoming feudal system with existing society; an accommodation which common sense tells us must have occurred more frequently and in greater depth than the feudally orientated written record of the twelfth and thirteenth centuries tends to reveal.[2]

'MacDuff' is the Scotticised form of two Gaelic words, *mac* and *Dub*, which together mean literally 'son of *Dub*', *Dub* being the adjective *dub*, 'black', used substantively as a forename meaning 'the black one'. There are examples of 'MacDuff' as a patronymic, a style, a surname and as part of a kindred name. The one thing it is not and cannot be is a forename, although this is perhaps what it was most often thought to be in later times.[3] It is, of course, true that forenames in *mac*, pre-Christian in origin, were still quite common in Scotland in the twelfth century, especially *Mac-bethad* (*betha* = 'life') and *Mac-raith* (*rath* = 'fortune'); both are still with us, although in a reduced form as part of the surnames MacBeth and MacRae.[4] The word *mac* in such compounds means 'pupil, devotee, disciple', or some such, and there could be no linguistic objection to *Mac-Duib* as a forename.[5] But the connotations of *Dub* are made quite explicit in another pagan forename which survived in Scotland into the sixteenth century at least. *Dub-shíde*

1   *ESC*, pp. 244-5; G.W.S. Barrow, *Robert Bruce and the Community of the Realm of Scotland* (3rd edn, Edinburgh, 1988), p. 58.
2   A remarkably late and well-documented instance of this is the continuing importance of the bloodfeud in the society of lowland Scotland into the 17th century: J. Wormald, 'Bloodfeud, Kindred and Government in Early Modern Scotland', *Past and Present*, 87 (1980); K.M. Brown, *Bloodfeud in Scotland 1573-1625* (Edinburgh, 1986).
3   Certain Scottish surnames do duty as forenames today but rarely surnames in *mac*. In any case, *MacDuff MacDuff*, as would result in the latter instances (below, pp. 32-8), is surely an impossible combination.
4   J. Bannerman, *The Beatons: A medical kindred in the classical Gaelic tradition* (Edinburgh, 1986), p. 1.
5   *Dictionary of the Irish Language* (Royal Irish Academy, 1913-76), 'M', pp. 7-8.

means 'the black one of the fairy mound'. *MacDuffie*, the early Scotticised form of its appearance in a *mac* surname, is sometimes and quite erroneously associated with 'MacDuff'. Later, *MacDhuibh-shídhe* was reduced in Gaelic to *Mac a Phì*, which is reflected in the present-day Scotticised forms MacPhee or MacFie.[1] Although *Dub* and *Dub-shíde* were acceptable as forenames, *Mac-Duib* is not on record as such in either Scotland or Ireland. The reason is surely that to give a child such a forename was tantamount to dedicating him to the supernatural, indeed, in Christian terms, to the Devil himself.

In Scotland the only person of note to bear the forename *Dub* was the king of Scots who began to reign in 962 and four years later, in 966, was killed in feud probably in favour of his successor Cuilean.[2] This uniqueness makes it likely that it was his name that appears in 'MacDuff' in all its relevant manifestations, and there is, as we shall see, other evidence to support this view.[3] There can be no doubt about the patronymic, for when the Annals of Ulster record the death in 1005 of 'Cinaed mac Duib', this was Kenneth III, king of Scots and son of the aforementioned Dub.[4]

In the later Gàidhealtachd the relationship between kindred name, style and surname is usually clear. For instance, the head of the *Clann Domhnuill* or Clan Donald is identified as such by his style *MacDhomhnuill* (MacDonald). The style in turn becomes a surname not only for the use of the chief and his immediate family but eventually also for other members of the kindred. Donald, the eponymous ancestor, was a son of Raghnall, *Rí Innse Gall*, who died in *c.*1207.[5]

In 1384 the earl of Fife was described as 'capitalis legis de Clenmcduffe', 'chief of the law of Clan MacDuff'.[6] This is the earliest contemporary record of the kindred name and at first glance it does not conform to the pattern presented above. But the difference is more apparent than real, for it represents an early example in a non-Gaelic context of the surname replacing the forename in the name of the kindred. So today 'Clan MacDonald' is to be heard in English as often as or more often than 'Clan Donald'. This probably developed from the pre-eminent position afforded surnames in the Lowlands of Scotland in the medieval period.[7] In time it influenced Gaelic usage also, for although *Clann MheicDhomhnuill* rarely does duty for *Clann Domhnuill*, *Clann MheicLeòid* (Clan MacLeod), for

1   K.A. Steer and J. Bannerman, *Late Medieval Monumental Sculpture in the West Highlands* (RCAHMS, 1977), pp. 119, 122; A. MacBain, *An Etymological Dictionary of the Gaelic Language* (Stirling, 1911), p. 409; G.F. Black, *The Surnames of Scotland* (New York, 1962 reprint), pp. 488, 493.
2   Anderson, *Early Sources*, i, pp. 472-4.
3   See below, p. 24.
4   Anderson, *Early Sources*, i, pp. 521-4.
5   Steer and Bannerman, *Monumental Sculpture*, pp. 88, 183, 202. See also M.D.W. MacGregor, 'A Political History of the MacGregors before 1571' (Edinburgh University Ph.D. thesis, 1989), pp. 23-8.
6   *APS*, i, p. 187.
7   J. Wormald, *Court, Kirk and Community* (London, 1981), p. 30.

instance, is more frequently spoken and written thus in Gaelic today than is *Clann Leòid*. In other words what lies behind Clan MacDuff of 1384 is *Clan Duff* in Scots and ultimately *Clann Duib* in Gaelic.

The association of Clan MacDuff with the earl of Fife in 1384 is a reflection of the fact that all other instances of 'MacDuff' to be discussed in this paper, whatever their context, are tied firmly to the ruling family of the province of Fife.[1] Andrew of Wyntoun in his *Original Chronicle*, completed 1420 × 1424, claims that MacDuff thane of Fife requested of Malcolm III the privilege for himself and his successors of enthroning the king of Scots at his inauguration.[2] This was certainly the stated situation in an indenture of agreement made between Duncan IV earl of Fife and Robert I king of Scots in *c*.1313; and it can be shown to have pertained at the inauguration of Alexander III at Scone on 13 July 1249, when the earls of Fife and Strathearn enthroned Alexander at the beginning of the ceremony and the earl of Fife also performed the final act, which was to present Alexander with the sword of his ancestors.[3] John of Fordun (d. *c*.1387), Wyntoun's older contemporary, claimed further that at the inauguration of Alexander II in 1214, seven earls headed by Fife and Strathearn (in that order) brought him to Scone where 'they raised him to the throne'.[4] Indeed, it is probable that one of David I's reasons for sending his grandson Malcolm on a circuit of the country in the care of the earl of Fife in 1152 to 'gert pronownsse [hym] ... kynge of lauche to be', as Wyntoun put it, was because Fife was already regarded as the senior inaugural official.[5]

Without referring specifically to the inauguration, Fordun makes Malcolm III promise MacDuff that 'he will be the first in the kingdom, after the king'.[6] There can be no doubt about the premier position held by the earls of Fife in twelfth-century Scotland. Duncan I and II, successive earls of Fife from *c*.1133 to 1204, witnessed royal documents far more frequently than any others of the native Scottish nobility.[7] Often they were the first of the lay witnesses, whether Scottish or Anglo-Norman, to be named. It is surely no coincidence that Duncan I is the first native Scot on record to have

---

1   The suggestion by W.F. Skene (*Celtic Scotland* [2nd edn, Edinburgh, 1886-90], iii, p. 306) that the Clan MacDuff arrived in Fife from the north as followers of MacBeth was based on his belief that the genealogy of the ruling family of the men of Moray in Trinity College, Dublin, MS H.2.18, was entitled *Genealach Clann Dubh*, 'Pedigree of Clann Duib'. But that manuscript is in fact the Book of Leinster which, as Skene himself pointed out (*Celtic Scotland*, iii, p. 476), entitles the genealogy 'Genelach Clainde Lulaig', 'Pedigree of Clan Lulaigh'. By the middle of the 12th century, when the Book of Leinster began to be compiled, Lulach was a likely eponym for this kindred, while there is no Dub visible in their ancestry.
2   *Chron. Wyntoun*, iv, p. 303.
3   *RRS*, v, p. 355; J. Bannerman, 'The King's Poet and the Inauguration of Alexander III', *SHR*, lxviii (1989), pp. 124-6, 132-3.
4   *Chron. Fordun*, i, p. 280.
5   *Chron. Wyntoun*, iv, p. 407; *RRS*, i, pp. 5-6; W.D.H. Sellar, 'Celtic Law and Scots Law: Survival and Integration', *Scottish Studies*, xxix (1989), pp. 13-14.
6   *Chron. Fordun*, i, p. 203.
7   G.W.S. Barrow, *David I of Scotland (1124-1153): The Balance of New and Old* (Reading, 1985), p. 16; *RRS*, i, p. 6; ii, *passim*.

the territory that he controlled within the existing kin-based society turned into a feudal fief. When David I granted the earldom of Fife to Duncan on that basis in *c.*1136, he was doubtless hoping that other native Scots would be persuaded to follow the earl of Fife's lead in the matter.[1]

Another indication of the standing of the earls of Fife in twelfth-century society and of their importance to the crown in promoting acceptance of the new ideas that flowed from Norman England was the appointment of Duncan II (1154-1204) certainly, and probably also of his predecessor Duncan I (*c.*1133-54), as justiciars of Scotia, that is, of Scotland north of the Forth-Clyde line.[2] Little is known of Gille-Mícheil, earl for some three years only from *c.*1130 to *c.*1133, but his predecessor Constantine, who first appears on record in 1095, was entitled in *c.*1128 'comes de Fyf' and 'magnus judex in Scotia'.[3] Geoffrey Barrow has already suggested that *justitia*, 'justiciar', may have been genuinely confused with *judex*.[4] And in fact an intermediate Gaelic stage makes this the most likely explanation. The account of the boundary dispute in Fife between the Culdees of Loch Leven and Robert of Burgundy, which contains this reference to Constantine, is the final item in a series of *notitiae* in the Register of the Priory of St Andrews which were at some point translated from Gaelic into Latin.[5] Besides Constantine, two other *judices* were involved, Mael-Domhnaich son of Macbethad, whom we know from other evidence to have been a *judex regis* at about this time,[6] and Dubgall son of Mocche, to whom the other two deferred 'on account of his age and skill in law', but also perhaps on account of his local knowledge, for he is likely to have been *judex* of the province of Fife.[7] If the Gaelic scribe recording the proceedings was faced with the Latin designation *justitia in Scotia* applied to Constantine and descriptive of an office which could only have been introduced into Scotland very recently, he would be most likely to translate *justitia* by *brithem*, the normal Gaelic term for a lawman. But in order to distinguish Constantine from the other two lawmen involved, especially from Mael-Domhnaich, who as *brithem ríg* or *judex regis* occupied the highest rank possible in the native legal profession,[8] he may have added the adjective *mór*, so *brithem mór i nAlbain* is likely to have been the Gaelic text. The later Latin translator, failing to identify Constantine as justiciar in Scotia, then gave a literal translation of what was before him.

It has been suggested that Constantine must have attended a native law school to allow him to participate in the proceedings.[9] But while this is still

1   G.W.S. Barrow, *The Kingdom of the Scots* (London, 1973), p. 283; *The Anglo-Norman Era in Scottish History* (Oxford, 1980), pp. 84-90.
2   Barrow, *Kingdom of the Scots*, p. 105.
3   *ESC*, p. 67.
4   Barrow, *Kingdom of the Scots*, p. 105.
5   *St Andrews Lib.*, p. 113.
6   J. Bannerman, *Studies in the History of Scotia* (forthcoming).
7   Barrow, *Kingdom of the Scots*, p. 78.
8   F. Kelly, *A Guide to Early Irish Law* (Dublin, 1988), pp. 51-7.
9   Bannerman, 'The King's Poet', p. 139.

possible, it is much more likely that as earl of Fife his participation stemmed from his office of justiciar in Scotia. Certainly at a later period, justiciars were frequently involved in boundary matters, and the proceedings could have been those of an early court of justiciary, at which members of the existing legal profession also pleaded. On the other hand, it is perhaps more probable, especially as all those identifiable of the 'multitude of men' said to be present seem to have been from Fife, that it was the court of the *brithem* or *judex* of Fife, perhaps presided over by the justiciar.[1] Whatever the case, it is an interesting and early conjoining of the old and the new. That the decision of the court went in favour of the existing church and against the incoming knight Robert of Burgundy may be itself testimony to the careful, not to say diplomatic, nature of David's promotion of the new ideas from England and the Continent.[2] In the person of Constantine we can extend backwards into the eleventh century the premier position of the earldom of Fife and the close association between the earls and the royal dynasty, the ultimate expression of which was the earl's function of enthroning the king of Scots at his inauguration ceremony.

Such inaugural nobles were often heads of dynastic segments rewarded in this way for demitting their own legitimate claims to kingship within the kin-based system of succession.[3] That the ruling family of the province of Fife was closely related to the reigning royal house is indicated by their shared forenames *Donnchadh* (Duncan), *Mael-Coluim* (Malcolm) and perhaps particularly the otherwise rare *Causantín* (Constantine).[4] Dub king of Scots (d. 966) belonged to this royal dynasty and is the only one of that name on record, which makes it all the more certain that he was the eponymous ancestor of the *Clann Duib* of Fife.

The earliest MacDuff of whom we have any knowledge is placed in the mid-eleventh century by the story or stories surrounding the succession to the kingship of Malcolm III made famous in William Shakespeare's *MacBeth*. The chronicles of John of Fordun and Andrew of Wyntoun, compiled independently of one another, contain the earliest record of these traditions[5] — Wyntoun, who hailed from Fife, giving the fuller if less wordy account. In both sources 'MacDuff' is used in the sense of a style and the bearer is also entitled 'thane of Fife'. But despite the apparently lowly title, MacDuff is given the same role that the later earls of Fife are seen to possess. Wyntoun's claim that MacDuff and his successors were created inaugural officials by Malcolm III, as a direct result of his support against Macbeth, has already been noted. It could be argued, of course, that this was an anachronism, but the known *floruit* of Constantine (*c*.1095-*c*.1130), apparently already functioning as premier earl and therefore presumably

1   Barrow, *Kingdom of the Scots*, pp. 114-19.
2   Barrow, *David I of Scotland*, pp. 13-16.
3   F.J. Byrne, *Irish Kings and High-Kings* (London, 1973), p. 21; D. Ó Corráin, 'Irish Regnal Succession', *Studia Hibernica*, xi (1971), p. 32.
4   G.W.S. Barrow, 'The Earls of Fife in the Twelfth Century', *PSAS*, lxxxvii (1952-3), p. 54.
5   *Chron. Fordun*, i, pp. 189-90, 197-204; *Chron. Wyntoun*, iv, pp. 280-305.

senior inaugural official, is only some forty years removed from MacDuff's participation in the events of 1057-8.

Was there time for the style *MacDuib* to have come into being between the death of Dub in 966 and MacBeth's reign (1040-57)? The emergence of style, surname and kindred name seems not only to have become a rapid process in itself but to have happened close to the lifetime of the eponymous ancestor.[1] For example, in 'MS 1467' the pedigree of the *Clann Somhairle* of Monydrain, a branch of the larger Clan Lamont, is headed 'Genelach Cloinne Somhairle' only four generations removed from the eponym.[2] Somhairle's great-grandson Gilleasbuig is referred to as 'Celestinus Angusii dictus Maksowirle', which is his style in a charter of 1410, and in 1414 he is identified by his forename plus surname as 'Celestinus McSowerle'.[3] Other evidence indicates that the process actually began, and may sometimes have been completed, in the second generation from the eponym, that is, with the eponym's grandson. It has recently been pointed out that the kindred name *Clann Griogair* was probably in use by *c*.1400, within two generations of Griogair himself.[4] The grandson of the eponym of *Clann Ruairi*, Ruairi son of Alan, son of Ruairi, was styled 'MacRuairi' at his death in 1318 by the contemporary Annals of Ulster,[5] while in the Annals of Loch Cé his second cousin Alexander son of Angus Mor, son of Donald, eponym of *Clann Domhnuill*, was surnamed 'MacDomnaill' in 1299, which, of course, implies the existence of the style.

We will look at the mechanics of the evolution of the style later.[6] Suffice it to say at this point that the first bearer thereof seems to be the grandson of the eponym. This means that, chronologically speaking, a MacDuff so styled from Dub, who died in 966, had to have been in existence by the end of MacBeth's reign; nor was he likely to have been the first MacDuff, that is, a grandson of Dub. As we have seen, Dub had a son Kenneth III, king of Scots from 997 until his death in 1005, and therefore the first to bear the style MacDuff ought to have been Kenneth's son and successor. This would seem to be Giric, whom some versions of the Chronicle of the Kings of Scotland actually put into the kingship of the Scots in place of Kenneth.[7] The explanation for that may be either that he ruled jointly with his father or that he ruled some part of Scotia, including perhaps the province of Fife,

---

1 MacGregor, 'Political History', pp. 23-8; Steer and Bannerman, *Monumental Sculpture*, pp. 88, 104.
2 NLS, MS Advocates 72.1.1, fo. lv bll.
3 *Highland Papers*, ed. J.R.N. Macphail (SHS, 1914-34), iv, p. 236; Inveraray Castle, Argyll Transcripts (made by 10th duke of Argyll), 4 June 1414.
4 MacGregor, 'Political History', pp. 25-7.
5 Barrow, *Robert Bruce*, p. 347 (n. 104).
6 See below, pp. 28-30.
7 If, however, Boite was a son of Kenneth III rather than of Kenneth II, then he might be the first MacDuff (A.A.M. Duncan, *Scotland: The Making of the Kingdom* [Edinburgh, 1975], pp. 113, 628). Malcolm son of Dub, king of Cumbria (d. 997), seems to be a mistake for Malcolm son of Donald (Anderson, *Early Sources*, i, pp. 478, 517; see also Alan Macquarrie's essay, 'The Kings of Strathclyde, *c*.400-1018', above, p. 16).

under his father.[1]

In fact there is every reason to accept not only the existence of MacDuff so styled by the time of MacBeth but also the authenticity in large measure of the traditions associating him with MacBeth and Malcolm III as reported by Fordun and Wyntoun. The only seriously discordant note is their insistence on entitling him 'thenus de Fif' and 'thayne of Fyfe' respectively. *Thenus* in Latin and *thayne* in Scots both derive from Anglo-Saxon *þegn*. In its Latinised form it was used by scribes of the twelfth century and later to equate with Gaelic *toísech túaithe*, ruler of a *túath*, the lowest rank in the threefold ruling hierarchy of the kin-based society.[2] In eastern Scotland the other two ranks were the *mormaer*, ruler of a province, and *rí Alban* or king of Scots. In the mid-eleventh century the ruler of Fife was surely a *mormaer*. However, leaving aside the known Latin precursors of the Fordun/Wyntoun account, none of which mentions MacDuff, and assuming that its ultimate and perhaps oral source was in Gaelic,[3] it may be that MacDuff was not given a title therein. The style would be a sufficient means of identification. So, for instance, Donald, Lord of the Isles, signed the Gaelic charter of 1408 simply as 'McDomhnaill'.[4] On the other hand *mormaer*, continuing in Gaelic as a pre-Anglo-Norman title, but early on equated with the Norman *comes*, which in turn came to be translated 'earl' in Scots,[5] may have been lost sight of by non-Gaelic speakers and replaced by *thenus* or *thayne*, the one title of that period still familiar to them.

Wyntoun makes the three 'weird sisters' of MacBeth's dream hail him as 'thane of Cromarty', 'thane of Moray' and 'king of Scots' in that order, which, given that 'thane' in the title 'thane of Moray' is once again doing duty for what ought to have been represented in the fourteenth century by *comes* or 'earl', surely indicates progression from *toísech* up through *mormaer* to *rí*.[6] This is much more believable and logical in the context of the kin-based society than the titles 'thane of Glamis' and 'thane of Cawdor' with which 'Cromarty' and 'Moray' were replaced by Hector Boece in his *Chronicles of Scotland*.[7] One of the Gaelic *notitiae* in the Book of Deer tells

1   Anderson, *Early Sources*, i, p. 522; Duncan, *Making of the Kingdom*, pp. 97, 113, 628.
2   J. Bannerman, 'The Scots Language and the Kin-based Society', in *Gaelic and Scots in Harmony: Proceedings of the Second International Conference on the Languages of Scotland*, ed. D.S. Thomson (Glasgow, 1990), pp. 6-7; Kelly, *Early Irish Law*, pp. 17-18.
3   N.K. Chadwick, 'The Story of MacBeth', *Scottish Gaelic Studies*, vi (1949), pp. 189-211.
4   W.J. Watson, *Rosg Gàidhlig* (Glasgow, 1915), pp. 182-3. See also J. Bannerman, 'Literacy in the Highlands', in *The Renaissance and Reformation in Scotland*, ed. I.B. Cowan and D. Shaw (Edinburgh, 1983), pp. 231-2.
5   Bannerman, 'Scots Language and Kin-based Society', p. 7.
6   *Chron. Wyntoun*, iv, p. 275. Skene (*Celtic Scotland*, iii, p. 249) listed a thanage of Moray, but his sources were retours of 1608 and 1696, which mentioned the lands of the earldom of Moray, 'viz. thanagii de Murray', and 'comitatum seu thanagium de Murray' (*Retours*, Elgin, nos. 25, 178); this is hardly convincing evidence for the existence of a medieval thanage of Moray, to which there seem to be no medieval references apart from the story by Wyntoun. I would repeat, moreover, that the progression from *toísech* to *toísech* makes little sense and should not even be possible in the context of the kin-based society.
7   *The Chronicles of Scotland compiled by Hector Boece*, translated into Scots by John Bellenden, 1531 (Scottish Text Soc., 1938-41), ii, p. 150.

us that a *mormaer* was also a *toísech*, presumably, that is, of his own kindred.[1] Once he was king of Scots, MacBeth, already mormaer of Moray, had become the embodiment of all three ruling grades of the kin-based society. True, Cromarty is not otherwise recorded as a thanage and, more importantly perhaps, was in Ross rather than in Moray. It is possible therefore that it was an error for Cromdale, which was a thanage in the province of Moray.[2] Cromarty, however, was a remarkably small sheriffdom, and in that respect should be compared to other small sheriffdoms which originated in thanages, namely Kinross, Auchterarder, and (perhaps) Clackmannan.[3] Furthermore, Ross and Moray in Pictish times probably formed a single kingdom or province and politically they remained closely associated with one another well into the medieval period.[4]

Folk motifs have crept into the Wyntoun/Fordun account, notably the three weird sisters of MacBeth's dream and the three tests laid on MacDuff by Malcolm III to prove his good intentions, and perhaps also the three requests granted to MacDuff by Malcolm on the successful conclusion of his bid to succeed to the kingship.[5] But so much else either receives direct confirmation from other sources or else conforms to what is known of contemporary society that what is presented by Wyntoun as the first of the three requests made to Malcolm III by MacDuff may indeed be an historically authentic record of the elevation of MacDuff and his descendants to the position of inaugural officials to the king of Scots, and therefore of premier mormaers in the kingdom of Scotland. This would be partly in reward for his efforts on Malcolm III's behalf, as Wyntoun claims, but partly also no doubt in return for demitting on behalf of himself and his successors his recent, and therefore strong and legal, claim to the kingship itself within the kin-based system of succession.

Professor Duncan has made a good case for regarding the charter containing the earliest possible contemporary occurrence of 'MacDuff' as a largely authentic copy of an original dating to 1095.[6] It records a grant of lands to Durham by Edgar, already claiming to be king of Scots, and is witnessed by among others 'Constantinus filius Magdufe'. This could represent Gaelic *Causantín mac meic Duib*, 'Constantine son of the son of Dub', in which case it is an example of the simple patronymic; we shall see that the *mac meic* formula became a common alternative in this position to *ua*, 'grandson', in the Irish annals of the second half of the eleventh century.

---

1   K. Jackson, *The Gaelic Notes in the Book of Deer* (Cambridge, 1972), p. 30.
2   *Moray Reg.*, no. 286; see also item no. 8 in the Appendix to Alexander Grant's essay on 'Thanes and Thanages', below, p. 72.
3   G.W.S. Barrow, *Kingship and Unity: Scotland 1000-1306* (London, 1981), pp. 16-17.
4   G.W.S. Barrow, 'MacBeth and other mormaers of Moray', in *The Hub of the Highlands*, ed. L. Maclean (Inverness, 1975).
5   *Chron. Wyntoun*, iv, pp. 275, 291-5, 303-5; *Chron. Fordun*, i, pp. 197-203.
6   A.A.M. Duncan, 'The Earliest Scottish Charters', *SHR*, xxxvii (1958). But a note of caution is introduced by J. Donnelly, 'The Earliest Scottish Charters?', *SHR*, lxxiii (1989).

We have, however, already demonstrated that Constantine was almost certainly not the grandson of Dub and not, therefore, the first MacDuff. In that case, the Gaelic lying behind the Latin must be *Causantín mac MeicDuib*, 'Constantine son of MacDuff', that is, the style rather than the patronymic. Its usage here represents an intermediate stage on the way to its becoming a surname.

We have to go to Ireland to discover the mechanics of the evolution of the style in *mac* and its further development as a surname. The Annals of Ulster and the Annals of Innisfallen provide us with a long series of more or less contemporary entries covering much of the relevant period,[1] and it emerges, as might be expected, that the style and surname in *mac* seem to have their origins in the earlier style and surname in *ua* (later *ó*). In the Annals of Ulster there is a marked increase from the beginning of the tenth century in the use of *ua* as a means of identification in the formula 'X grandson of Z' and this is matched by a corresponding increase, although less obviously so, in the parallel 'X son/daughter of Y son of Z'. In the second half of the tenth century we meet the first styles and surnames in *ua*,[2] and there is a corresponding and immediate decrease in its literal meaning of 'grandson', although people continue as before to be identified as 'X son/daughter of Y son of Z'. The year 1028 in the Annals of Ulster sees the first example from this period of the formula *mac meic*, that is, 'X son of the son of Z',[3] which increasingly thereafter becomes an alternative to *ua* in its literal meaning of 'grandson'. The first appearance of the style in *mac* in the Annals of Ulster is the application of 'MacCarthaig' to Tadg son of Muiredach, son of Carthach, king of Desmond, in 1118, in which year in the Annals of Innisfallen he is identified as 'Tadg mac meic Carthaig', 'Tadg son of the son of Carthach'. Staying with the Annals of Innisfallen, much fuller in their treatment of the MacCarthys of Munster than are the Annals of Ulster, it is not until 1177 that the style in *mac* is used again in a reference to 'Diarmait mac MeicCarthaig', 'Diarmait son of MacCarthaig', exactly corresponding to the formula used by the scribe of the 1095 charter to identify Constantine. By 1200 it had become a surname when Domhnall MacCarthaig king of Desmond was so described. Just as the style and surname in *ua* derive from its literal use as 'grandson', so the style and surname in *mac* derive from the *mac meic* formula originally intended to serve as an alternative to *ua* in the meaning of 'grandson'. One reason for the long gestation of the progression from style to surname in *mac* was doubtless the competition provided by the older style and surname in *ua*, which continued to be productive, and in the 1160s and '70s *Ua Carthaig* actually took over for a time. The MacLochlainn kings of the Cenél nEógain

---

1   *Ann. Ulster*; *The Annals of Innisfallen*, ed. S. Mac Airt (Dublin, 1951).
2   See also B. Ó Cuív, 'Aspects of Irish Personal Names', *Celtica*, xviii (1986), pp. 180-3.
3   *Ann. Ulster, s.a.* 1080: 'ingen meic', 'daughter of the son of'; 1130: 'mac ingine', 'son of the daughter of'. There are occasional early examples (Ó Cuív, 'Aspects of Irish Personal Names', p. 180).

provide another, although not precisely similar, example of the process in action. Although 'Conchobar mac MeicLochlainn', 'Conchobar son of MacLochlainn', who was great-grandson of Lochlann, was identified thus in 1128 — providing the earliest Irish example from the annals of the intermediate stage as exemplified by the identification of Constantine in 1095 — Conchobar's father Donald, grandson of the eponym, was styled not *MacLochlainn* but 'Ua Lochlainn' in the Annals of Ulster in 1102 and 1109.[1] It was not until 1167 that 'MacLochlainn' was first recorded in the annals as a surname identifying Niall MacLochlainn thus, while in the interval 'Ua Lochlainn' was frequently in use. However, the pace of the progress from style to surname in Ireland soon quickened to match the later Scottish examples already quoted. Thus, although in 1122 Enna son of Donnchad, son of Murchad and king of Leinster, was styled 'MacMurchada' more or less at the same point as the styles 'MacCarthaig' and 'MacLochlainn' came into being, his brother and successor Diarmait, styled 'MacMurchada' in 1156 and 1161, was surnamed Diarmait 'MacMurchada' in 1162.[2] The style and surname in *ua* are nowhere in evidence.

It has been pointed out that the style and surname in *ua* first appeared at the highest levels of society in Ireland.[3] It is clear that the earliest Irish examples of the style and surname in *mac* were also first adopted at this level. This, too, is where the MacDuffs of Fife belong in a Scottish context. Indeed, since the first of the family to bear the style was probably Giric grandson of Dub, king of Scots, and himself possibly joint king of Scots with his father Kenneth III, there could be none higher. Thus, as with the MacCarthaig kings of Desmond and the MacLochlainn kings of Cenél nEógain in Ireland, so with the MacDuff *mormaers* of Fife we are at the beginning of the process that produced the style and surname in *mac* in Scotland. It is in keeping with these circumstances that the evolutionary progress from style to surname should be slower than it later appears to be, and therefore that the intermediate stage represented by *Causantín mac MeicDuib* should be visible. We shall see that by *c.*1128 the process was complete, for *MacDuib* had become a surname. Doubtless the ruling families of other provinces followed suit, and indeed there is evidence to suggest that *MacAeda* was the style and surname of the mormaers of Ross by the early twelfth century.[4] There is something of a conundrum in the fact that, according to the evidence so far presented, the process was begun in Scotland about a hundred years before it began in Ireland, and yet it is clear from the Irish evidence that the surname in *mac* evolved out of the surname in *ua*, which, judging by the apparent lack of indigenous surnames in *ua*,

1   'Ua Lochlainn', of course, also means literally 'grandson of Lochlann' which Donald was; but in both entries 'Ua Lochlainn' is set against 'Ua Briain' and that was Muircheartach, great-grandson of the eponym Brian Bóroma.
2   Indeed, there is evidence from elsewhere that he may have been surnamed *MacMurchada* as early as 1138: B. Ó Cuív, 'Two Notes', *Éigse*, xvi (1975-6), p. 136.
3   D. Ó Corráin, 'Nationality and Kingship in pre-Norman Ireland', *Historical Studies*, xi (1978), p. 33.
4   *RRS*, i, p. 209; ii, pp. 11-13.

was never part of the Scottish experience.[1] The explanation may be that the fashion for identifying people by their grandfather's name did not catch on in Scotland until *ua*, in its literal meaning of grandson, was being replaced by the *mac meic* formula in Ireland.[2] Uninhibited by an existing style and surname in *ua*, the evolution of a style and surname in *mac* may therefore have got under way more rapidly in Scotland. It is possible to point to supporting evidence for this view in the Gaelic *notitiae* in the Book of Deer, which range in date from early in the eleventh century through to the middle of the twelfth. The word *ua* is never used, but along with the name of Donnchad son of Mac-bethad, son of Ided, are those of 'Domnall mac meic Dubbacín' and 'Cainnech mac meic Dobarcon'.[3] Whatever the case, the fact that Constantine was identified by the intermediate evolutionary form *mac MeicDuib* in 1095 adds to the authenticity of the witness-list, if not the content, of Edgar's charter.[4]

It is in the witness-list of the important confirmation charter of *c.*1128 by David I recording grants of lands and privileges to the church of

1   The Clan Campbell, so-called from the second half of the 13th century at least, had what was presumably an earlier style, surname and kindred name, the eponym of which was *Duibhne*. The style and surname normally appear in the form *Ua Duibhne* and most frequently in a literary and/or antiquarian context which probably also encouraged the replacement of an original *mac* by *ua*: W.D.H. Sellar, 'The Earliest Campbells — Norman, Briton or Gael', *Scottish Studies*, xvii (1973), pp. 111-12; J. Bannerman, 'Two early post-Reformation inscriptions in Argyll', *PSAS*, cv (1972-4), p. 308. Significantly, the surname is written 'M'Duine' for *MacDhuibhne* in a charter of 1369: HMC, *4th Report*, p. 477. But see H.L. MacQueen, 'The laws of Galloway: a preliminary survey', in *Galloway: Land and Lordship*, ed. R.D. Oram and G.P. Stell (Edinburgh, 1991), p. 140 (n. 13), for possible Galloway surnames in *ua*.
2   I am indebted to Dr David Brown for this suggestion.
3   Jackson, *Gaelic Notes*, pp. 31-2. Previously *Dubbacín*, a form of *Dubacán*, and *Dobarchú*, commonly on record as such (M.A. O'Brien, *Corpus Genealogiarum Hiberniae* [Dublin, 1976], pp. 590, 603), were considered to be forenames compounded with *mac* (Jackson, *Gaelic Notes*, pp. 53, 56), neither of which is otherwise recorded. In the once Gaelic *notitia* for *c.* 1128, in the Register of the Priory of St Andrews, 'Douinalde nepos Leod' surely translates *Domnull mac meic Leóid* (*ESC*, no. 80).
4   Although Constantine seems to have been *mormaer* of Fife by 1095, Aethelred, third son of Malcolm III and still alive in the reign of his brother Edgar (1097-1107), judging by the order of listing the royal grants to the church of Dunfermline (*ESC*, no. 74), was entitled 'Abbas de Dunkeldense et insuper comes de Fyf', 'Abbot of Dunkeld and also earl of Fife', in a royal confirmation (1114 × 24) of a grant of lands in Fife made by him to the Culdee community of Loch Leven (*ESC*, no. 14). It should be noted that this is the only record of Aethelred so entitled; furthermore, the confirmation is one of the *notitiae* from the Register of the Priory of St Andrews, translated from Gaelic into Latin, and we have already seen that the translator was capable of misinterpreting his Gaelic original (above, p. 23). This is likely to be a grant of lands made jointly by Aethelred as abbot of Dunkeld and by the *mormaer* of Fife, in whose province the lands were. Joint grants to the Church were characteristic of the kin-based society (Bannerman, *Studies in the History of Scotia* [forthcoming]), and taking the contemporary Gaelic *notitiae* of such grants in the Book of Deer as our model (Jackson, *Gaelic Notes*, pp. 30-2), the Gaelic verb underlying the Latin 'contulit' of the confirmation would probably be *do-rat*, third person singular preterite independent of *do-beir*, 'to give', which, even when the subject is plural, is used in such constructions as often as, or more often than, the third plural *do-ratsat*. The translator, unfamiliar with joint grants of this kind and taking the singular form of the verb at face value, may have thought that Aethelred was at once abbot of Dunkeld and *mormaer* of Fife. In some surprise that this could be so — considering that the document further describes Aethelred as a 'vir venerandae memoriae', implying that he was a bona-fide clergyman — the translator added the word 'insuper', 'also', before 'comes de Fyf', perhaps even thereby replacing the latter's forename in the original.

Dunfermline that 'MacDuff' next occurs.[1] Immediately following the names of five *comites* or earls, including that of Constantine earl of Fife, is the name 'Gillemichel Mac duf'. It has generally been accepted that he was the Gille-Mícheil who succeeded Constantine as earl of Fife in *c*.1130, and it has been assumed further that he was Constantine's son.[2] But W.F. Skene pointed out long ago the unlikelihood, if he was Constantine's son, of his being identified in this way in a Latin witness-list which also named Constantine.[3] In any case, the province of Fife did not become a feudal fief until it was granted as such by David I to Duncan, Gille-Mícheil's successor, in *c*.1136, when, of course, it became heritable in terms of the system of primogeniture. Within the kin-based system of succession, operative up to that point, Gille-Mícheil, surely tanist or heir to Constantine (hence his presence in the witness-list) was much more likely to have been a brother, nephew or cousin.

In this instance *Mac duf* cannot be a style, but was it a patronymic or a surname? We have already noted that the forename *Dub* was uncommon in Scotland, but, more importantly, it was never used again by the Clan MacDuff in all its recorded ramifications. It is true that some kindreds continued to use their eponym, no matter how unusual or rare it otherwise was, as a characteristic forename in later generations; the Clan MacDuffie of Colonsay is a case in point. But others like, for instance, *Clann Leòid* or Clan MacLeod seem deliberately to have refrained from ever using their eponym as a forename, perhaps ascribing some special status to it thereby.[4] *Dub* is therefore unlikely to have been the name of Gille-Mícheil's father; what we have is an example of 'MacDuff' used as a surname. In any case this is more likely in the context of identifying Gille-Mícheil as heir to Constantine in the witness-list, especially as there cannot yet have been many people eligible to use 'MacDuff' as a surname.[5]

Frater and *nepos*, for 'brother' and 'nephew' of Constantine respectively, are as easy to use as *filius*, 'son', in a Latin witness-list, so Gille-Mícheil was almost certainly neither of these but a cousin of Constantine. Again, although there are, of course, other possibilities, the kin-based system of succession makes it more likely that Duncan I, Gille-Mícheil's successor as earl of Fife, was a son of Constantine rather than of Gille-Mícheil himself, as has often been assumed.[6] Gille-Mícheil did have a son and grandson both called 'Hugo' in the Latin documents of the period.[7] On two occasions,

1   *ESC*, no. 74.
2   *Scots Peerage*, iv, p. 4. Gille-Mícheil, who was *comes* in *c*.1126 (*ESC*, no. 68), is perhaps the same as he who appears in the *notitia* in the Book of Deer dated 1131 × 2 (Jackson, *Gaelic Notes*, p. 31).
3   Skene, *Celtic Scotland*, iii, p. 63.
4   Steer and Bannerman, *Monumental Sculpture*, pp. 119, 122; I.F. Grant, *The MacLeods* (London, 1959), *passim*.
5   A similar example may be Malcolm 'Maceth' (*MacAeda*), identified thus in a witness-list of 1157 × 60, but as 'Malcolmus comes de Ros' in 1160 × 2, his earldom having been restored to him in the interval (*RRS*, i, pp. 209, 222, 287).
6   *Scots Peerage*, iv, p. 5.
7   *RRS*, i, p. 223; ii, pp. 138, 182-3, 424.

however, the grandson's name was given as 'Egu' or 'Eggu', which is a rendering of *Aeda*, genitive of the Gaelic forename *Aed* for which *Hugo* was clearly being used as the Latin equivalent.[1] With the transformation of the province of Fife into a feudal fief to be held of the crown by Duncan I, the system of primogeniture took over from tanistry in terms of governing the succession to both territory and title, and so Aed son of Gille-Mícheil, the likely heir as far as the kin-based society was concerned, was passed over at Duncan I's death in 1154 in favour of the latter's son Duncan II, even although, as we shall see, he was probably a minor at the time.[2] Thereafter inheritance by primogeniture leaves little doubt as to the genealogical antecedents of succeeding earls of Fife. A pedigree of the relevant members of the ruling family of Fife descended from Dub king of Scots — albeit necessarily incomplete in its upper reaches — is provided in tabular form on the following page.

Although the rulers of Fife from Duncan I onwards held their lands and title as tenants-in-chief of the crown, and appeared as 'comites de Fyf' in the Latin documents of the twelfth and thirteenth centuries, they would continue to be entitled *mormaer* and styled *MacDuib* in the vernacular. For example, as late as the year 1424 a contemporary Gaelic source could describe the earl of Lennox as 'mormaer'[3] while Geoffrey Barrow has pointed out that *Beinn MacDuibh*, 'MacDuff's mountain', was probably named thus in Gaelic some time after 1187×1204, when Duncan II earl of Fife was granted an extensive territory in the north-east whose western boundary was marked by the summit of the mountain.[4] However, there are two periods covered by the feudal part of the *Clann Duib* pedigree when the records reveal that the person styled *MacDuib* was not the same as the holder of the title *comes de Fyf*. The first instance appears in the copy of a charter which has only just come to light granting lands in Fife to a certain MacDuff some time between 1165 and 1172.[5] Duncan II earl of Fife was the first of two witnesses named. Nothing further is known about this MacDuff.[6] The second instance concerns the MacDuff who made his now famous appeal for justice to

1   *RRS*. ii, pp. 138, 424. The spelling 'g' ('gh') for original 'd' ('dh') appears elsewhere in the genitive of this name (Steer and Bannerman, *Monumental Sculpture*, pp. 100-1, 156). The more or less contemporary spelling 'th' as seen in *Maceth* (above) is how 'dh' was sometimes represented before it changed from a dental to a guttural spirant (T.F. O'Rahilly, *Irish Dialects Past and Present* [Dublin, 1976], p. 65). Judging by the evidence, this sound shift was beginning in the Gaelic of eastern Scotland by the mid-12th century. Later *Odo* became the normal Latin equivalent of *Aed*, while *Hugo* seems to have been reserved for Gaelic *Úisdean* (Steer and Bannerman, *Monumental Sculpture*, pp. 101, 118, 125, 156).
2   See below, pp. 34-5.
3   *Ann. Ulster, s.a.* 1424. 'Earl' or *comes* came to be represented by *iarla* in Gaelic, originally a borrowing of Old Norse *jarl*. *Mormaer*, now *morair*, took on the generic meaning of 'lord' (*Dictionary of the Irish Language*, 'I', p. 22; Jackson, *Gaelic Notes*, p. 103).
4   *Moray Reg.*, nos. 16, 62; Barrow, *Anglo-Norman Era*, p. 86.
5   SRO, J. Maitland Thomson's Notebooks, GD.212/15/42. I am indebted to Geoffrey Barrow for drawing my attention to this document.
6   By an oversight, kindly brought to my attention by Geoffrey Barrow, I omitted to record the fact that MacDuff was still alive in 1198×9 when, along with Duncan II earl of Fife, he witnessed an agreement between the prior and convent of St Andrews and the Culdees thereof (*St Andrews Lib.*, pp. 318-19; Barrow, *Kingdom of the Scots*, p. 224).

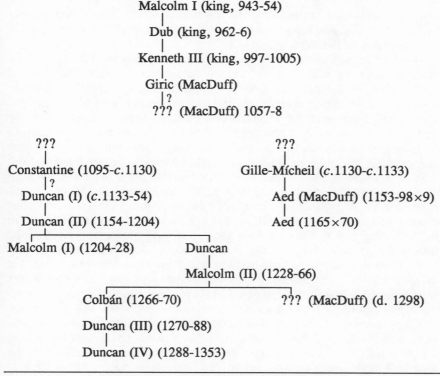

Malcolm I (king, 943-54)
|
Dub (king, 962-6)
|
Kenneth III (king, 997-1005)
|
Giric (MacDuff)
| ?
??? (MacDuff) 1057-8

???
|
Constantine (1095-*c*.1130)
| ?
Duncan (I) (*c*.1133-54)
|
Duncan (II) (1154-1204)

Malcolm (I) (1204-28)   Duncan

???
|
Gille-Mícheil (*c*.1130-*c*.1133)
|
Aed (MacDuff) (1153-98×9)
|
Aed (1165×70)

Malcolm (II) (1228-66)

Colbán (1266-70)   ??? (MacDuff) (d. 1298)
|
Duncan (III) (1270-88)
|
Duncan (IV) (1288-1353)

Pedigree of the Ruling Family of Fife

Edward I of England in 1293. He was a brother of Colbán earl of Fife (1266-70), and he claimed that their father Malcolm II earl of Fife (1228-66) had granted him lands in Fife of which he had been unjustly dispossessed, first by William Wishart bishop of St Andrews, keeper of the earldom of Fife during the minority of Duncan earl of Fife (1270-88), and secondly by John Balliol's parliament held at Scone on 10 February 1293 during the minority of Duncan IV earl of Fife (1288-1353).[1] Nowhere are we ever told his forename.

An unambiguous example of the separation of the kin-based and feudal systems of society in the thirteenth century comes from the province of Carrick. Using the feudal instruments to hand, Niall (Neil) earl of Carrick, presumably some time shortly before his death in 1256, ensured that the headship of the ruling kindred of Carrick would go to his nephew Lachlann (Roland) of Carrick. This grant was confirmed by Alexander III in January

1   *APS*, i, p. 445; *Rot. Scot.*, i, pp. 18, 20; *Rotuli Parliamentorum* (London, 1783-1832), i, pp. 110-11; *Foedera*, i, p. 788; Barrow, *Robert Bruce*, pp. 54-60.

1276 and again in 1372 by Robert II. The title, earl of Carrick, and the lands associated with it passed to Marjorie, the eldest of Niall's four daughters, and from her to her eldest son Robert Bruce, who became earl of Carrick in c.1292 and king of Scots in 1306. Thereafter the title remained in the possession of the royal dynasty.[1]

In the sixteenth century what had been until recently the Lordship of the Isles provides us with another and similar example. Niall MacNeill of Gigha, chief of his kindred, died in c.1530 leaving a daughter as his sole legitimate offspring. Annabelle, by right of the feudal rules of inheritance, succeeded to the lands of Gigha, and the designation de Gigha, but the leadership of the kindred devolved on Annabelle's second cousin Torquil MacNeill, who was described in 1531 as 'cheif and principale of the clan and surname of Maknelis'. Even although a crown grant of the MacNeill lands was made in 1542 to Niall, a natural son of Niall MacNeill, with all claim to them having been resigned by his half-sister Annabelle, Torquil continued to be regarded as chief, for he was 'principalis seu primarius tribus sive familie de MacNeille' in 1553. Presumably not until his death in the following year did leadership of the kindred and feudal ownership of the clan lands formally coincide again in the person of Niall MacNeill of Gigha. Niall, however, may have been de facto leader for some time before this, for in 1553 Torquil, 'by reason of his bad health and great age and of the weakness of his body', is represented as 'hoping soon to die'.[2]

It was certainly in part because the previous chief's legitimate successor was a female in c.1530 that Torquil became chief of the kindred, but it is also likely that Niall, the natural son of the previous chief, was a minor at the time of his father's death. Indeed Annabelle's resignation of the MacNeill lands to her half-brother in 1542 may have marked his coming of age. We have already noted that there were two minorities in the earldom of Fife during the adult lifetime of the second MacDuff, that of his nephew Duncan III, who succeeded in 1270, and then that of his grand-nephew Duncan IV, who was only three years old on his succession in 1288. In the eyes of the kin-based society there was clearly a need for a MacDuff at this time. Moreover, it can probably be assumed that the first MacDuff was so styled in 1165 × 1172 because Duncan earl of Fife had been a minor when he succeeded his father in 1154, although, like the ongoing minority of Malcolm IV in the kingship of the Scots, it was perhaps not an official one.[3] Duncan II, for fifty years earl of Fife, did not die until 1204, and the first certain occasion on which he witnessed a royal charter was in the year 1159, five years after he had succeeded to the earldom. Again, although he had married and had a child by 1160 × 1162, it is a further indication of his

---

1   RMS, i, nos. 508-9; see also Hector MacQueen's essay, 'The Kin of Kennedy, "Kenkynnol" and the Common Law', below.
2   Registrum Secreti Sigilli Regum Scottorum (Edinburgh, 1908-), ii, no. 790; Inveraray Castle, Argyll Transcripts, 23 May 1553; Steer and Bannerman, Monumental Sculpture, pp. 147-8.
3   RRS, i, no. 6.

relative youth that his name did not take the earl of Fife's accustomed position at the head of the comital witnesses to royal documents until 1163.[1]

The person most likely to be *MacDuib* during this period is surely the one most likely to have succeeded as ruler of the province of Fife under the kin-based system of succession; as far as the records take us, that was Aed son of Gille-Mícheil. He was an adult by the time Duncan I died in 1154, and he was one of five men of Fife, including Duncan I, addressed in a royal brieve of 1153 × 1154 concerning matters in Fife.[2] It is implicit in their retention of the style that the two MacDuffs continued to be leaders of their kindred beyond the point at which Duncan II and Duncan III respectively attained adulthood. We have seen that Torquil MacNeill retained leadership of his kindred, in name at least, until his death, and the second MacDuff was killed at Falkirk in 1298 as leader of his kindred in both name and fact. Duncan IV was at once a minor and in England at the time. The memory, if imperfectly recalled, of the possible separation of title and style, especially perhaps when earls of Fife named Duncan were involved, is surely present in the spurious record of a grant of lands in Fife by Malcolm III king of Scots to the church of Dunfermline.[3] It was allegedly witnessed by both 'Mackduffe comes' and 'Duncanus comes'. Interestingly they head the list of lay witnesses, in that order. In the eyes of the kin-based society the person bearing the style *MacDuib* would be seen as the more prestigious of the two.

Yet another minority in the earldom of Fife was that of Colbán, who succeeded his father Malcolm II as earl of Fife in 1266. There is no mention of a MacDuff during Colbán's tenure of the earldom, but it is only by chance that the other two appear in the records. Leaving aside the chronicle references to the second MacDuff siding with Wallace, it was their possession of lands in Fife on a feudal basis that brought them into view. But even that is surprising. As late as the fourteenth and fifteenth centuries both the style and surname *MacDhomhnuill* are to be seen only in contemporary Gaelic sources relating to the Lordship of the Isles. They were never used in parallel feudal Latin documents associated with the Lordship, which almost always identify members of the ruling family by forename and title and/or by the Latin designations 'de Ile' or 'de Insulis'.[4] Because there is no Gaelic documentation for the province of Carrick the style and surname in *mac* of its ruling family seem forever hidden behind the Latin designation 'de Carryck', at least until the Kennedys took over that role in the second half of the fourteenth century.[5] Indeed, we can count ourselves fortunate that there are, relatively speaking, so many examples of 'MacDuff' in all its possible usages. One reason was doubtless the premier position of

1   *RRS*, i, nos. 131, 190, 243.
2   *RRS*, i, no. 181.
3   *ESC*, no. 10.
4   *Acts of the Lords of the Isles*, p. lxxxi; Steer and Bannerman, *Monumental Sculpture*, p. 154.
5   MacQueen, 'Kin of Kennedy', below. 'Kennedy' is a Scotticised form of the Gaelic surname *MacCennédigh*.

the earls of Fife in the country. It is possible that the given forename of the second MacDuff, uncle and great-uncle of Duncan III and IV respectively, will come to light as we become more familiar with the Latin documentation of the second half of the thirteenth century.

In both the Lordship of the Isles and the Carrick cases of the separation of the headship of the kindred and the feudal ownership of land and title, the crown acquiesced in the prevailing situation. So, too, it was clearly acceptable to the crown that the style *MacDuib* and the title *comes de Fyf* should be held by two different people in the second half of the twelfth century. It was William I who granted lands to MacDuff in 1165 × 1172. This simply mirrored the acceptance of this situation within the province of Fife which was itself implicit in the lands in question being in Fife and also in that the grant was witnessed by Duncan II earl of Fife, by then an adult.

This was probably the first occasion on which such an accommodation between the kin-based and feudal systems was required below the level of the kingship of the Scots, for the minority in the earldom began only one year after Malcolm IV's accession as a minor to the kingship in 1153 which is in turn the first certain evidence of the system of primogeniture in operation at any level. Had a precedent been set on that occasion? Were David I's native subjects persuaded that a leader was appointed to conduct the kin-based affairs of the country at least until Malcolm came of age? If so, it is tempting to see Duncan I earl of Fife as that person. He was, as we have seen, the premier *mormaer* or earl in the country. He was also closely related to the existing royal dynasty and yet represented no threat because he is likely to have belonged to a segment that had formally renounced all claim to the kingship of the Scots.[1] David I's motive in sending Malcolm on circuit round the country in the care of Duncan in the months before his own death was perhaps to show the populace at large not only that Malcolm's eventual accession had the support of the premier earl and senior inaugural official, important as that doubtless was, but also that in the person of Duncan they were to have an experienced adult leader drawn from the kin-based society in the not unlikely event of Malcolm's succeeding to the kingship as a minor. When Duncan I earl of Fife died in the year following Malcolm's accession, it might have been thought that Aed son of Gille-Mícheil, as the new MacDuff, would take on the dual responsibility of looking after the kin-based affairs of both Fife and the kingdom. But such a close association with the crown would surely have required his frequent presence in contemporary record, and nothing of the surviving royal documentation of the period is witnessed by him. Perhaps the most likely person to have succeeded Earl Duncan of Fife was Ferteth, mormaer or earl

---

1   Donald son of William, son of Duncan II, the most likely successor to David as king of Scots in terms of the kin-based system of succession, would doubtless be considered unsuitable for that reason. True, there is no evidence that he put in any bid for the succession at this point, but by 1181 he had made a number of attempts to win the kingship of the Scots for himself, and it was only his death in 1187 that put an end to his ambitions in that direction (*RRS*, ii, pp. 11-12).

of Strathearn (*c.*1141-71).[1] As already noted, the earl of Strathearn was the other, possibly the original, enthroning official,[2] and of the ten royal charters witnessed by Ferteth no fewer than nine belong in or about the period of Malcolm's minority.[3] This, too, would explain his leadership of the six *comites* who besieged Malcolm in Perth in 1160, 'being enraged against the king because he had gone to Toulouse' in the army of Henry II of England, thereby presumably compromising the independent status of Scotland.[4] One thing is certain: the acceptance, indeed active promotion by the ruling kindred of Fife, of the accession of Malcolm IV as a minor would make it impossible for them to do other than accept the feudal rules of inheritance and recognise in the following year Duncan II, almost certainly also a minor, as earl of Fife.

By the second half of the thirteenth century, it looks at first glance as if the crown and the kin-based society of Fife were at odds with one another. First of all, a *custos* or 'keeper' of the earldom of Fife, William Wishart bishop of St Andrews, who was not a member of the *Clann Duib*, was appointed during the minority of Duncan III in the 1270s. Secondly, Wishart, acting as keeper, dispossessed MacDuff of lands in Fife granted to him by his father Malcolm II earl of Fife (1228-66).[5] Third, that action was upheld by John Balliol's first parliament at Scone in February 1293.[6] These lands had recently been restored to MacDuff by the Guardians of Scotland at the prompting of Edward I of England, so it was certainly not surprising, and it was less reprehensible than has been generally represented, that he should make a second appeal to Edward I in 1293. The point at issue was land in Fife, and both *custos* and *custodia* seem originally to relate to the keepership of land.[7] Wishart should be seen as the keeper of the feudal lands of Fife during the minority of Duncan III, which would include, of course, those lands in Fife held feudally by MacDuff himself. And 'keepers of Fife' appointed by Edward I in the years 1293 to 1296 during the minority of Earl Duncan IV (following the second appeal of MacDuff, who was presumably then on good terms with the English king) seem to have been solely concerned with the feudal management of the earl's lands.[8]

In other words, the MacDuff episode as it concerned Fife, far from being an example of a clash between the kin-based and feudal systems, is a further illustration of how it was possible to accommodate both in the special circumstances of a minority. The precise duties of MacDuff,

1   I am indebted to Dr David Brown for this suggestion.
2   See above, p. 22; Bannerman, *Studies in the History of Scotia* (forthcoming).
3   *RRS*, i, nos. 118, 131, 138, 157, 159, 173, 176, 226, 227. In the tenth charter (no. 254), dated 1163 × 5, Ferteth was second in the list of lay witnesses behind Duncan II earl of Fife. It is true that in the first four Ferteth's name was preceded by that of Cospatric III earl of Dunbar, but Cospatric witnessed royal documents evenly over the whole period of his tenure of the earldom from 1138 until his death in 1166.
4   *Chron. Melrose*, p. 36; *RRS*, i, p. 12.
5   *Rot. Scot.*, i, p. 18.
6   *APS*, i, p. 445.
7   R.E. Latham, *Revised Medieval Latin Word-list* (London, 1965), p. 128.
8   *CDS*, ii, nos. 700-1, 708.

Wishart's opposite number as leader of the kin-based society of Fife, are probably best delineated in the terms of Lachlann's succession to the headship of his kindred after the death of his uncle, Niall earl of Carrick, in 1256.[1] He was to be 'head of all his kindred both in *calumpnie* and in other matters pertaining to *kenkynoll*', that is, the office of *cenn ceneóil* or *hede of kyn*. As Hector MacQueen has shown elsewhere in this volume, *calumpnie* were the accusations or charges of sergeants who were under the authority of the *cenn ceneóil* and whose basic role was to police the province by finding and accusing criminals.[2] But Lachlann was also to be bailie of Carrick which surely represents the function implicit in the title *mormaer*, that is, originally the collection on behalf of the king of Scots of the *cáin* or tribute due from the freemen of his province. Parliamentary legislation which seems to relate to the thirteenth century can still refer to the earl of Fife as king's 'maer' in Fife.[3] As it happens, this legislation determined fines to be paid for absence from the king's host, and *duccio* or military leadership of the men of Carrick was the final duty to be undertaken by Lachlann under the conditions of his uncle's grant. In short, his role would seem to be that of the *mormaer*, the ruler of a province in the kin-based society.[4] It is in respect of military leadership that MacDuff is seen fulfilling the same role for Fife, for we are specifically told that he fought and died at the head of the men of Fife in the battle of Falkirk in 1298.[5]

This MacDuff should surely be seen as a principled patriot because in 1297, despite the wide disparity in treatment meted out to him by John Balliol on the one hand and Edward I on the other, he and his two sons joined William Wallace, whose militant opposition to Edward's objectives in Scotland was undertaken in the name of John Balliol as rightful king of Scots.[6] It has often been claimed that by the time of Falkirk Wallace had been abandoned by the magnates of Scotland, but in the eyes of the kin-based society MacDuff was the greatest noble in the land. It would be interesting to know if the role of 'MacDuff' was reactivated on the death in 1353 of Duncan IV, the last of the native earls of Fife whose sole heir and successor was his daughter Isabella. It is certainly the case that aspects of the law of Clan MacDuff were still operating in Fife in the sixteenth century[7] — a telling indication of the tenacity of the kin-based element in the society of a province which had experienced the feudalisation of its land almost from the first introduction of the Anglo-Norman feudal system into Scotland.[8]

1   *RMS*, i, nos. 508-9.
2   See MacQueen, 'Kin of Kennedy', p. 281.
3   *APS*, i, p. 398; Duncan, *Making of the Kingdom*, p. 338.
4   Bannerman, *Studies in the History of Scotia* (forthcoming).
5   *Chron. Fordun*, i. p. 330; *Chron. Wyntoun*, v, p. 317.
6   Stevenson, *Documents*, ii, no. 462.
7   *The Practicks of Sir James Balfour of Pittendreich*, ed. P.G.B. McNeill (Stair Soc., 1962-3), p. 511; J. Skene, *De Verborum Significatione* (Edinburgh, 1597), under 'Clan-Makduf'.
8   I wish to thank Drs David Brown, Martin MacGregor, Donald Meek, and Mr David Sellar, who kindly read this paper in a first draft, for help and advice.

3

Thanes and Thanages,
from the Eleventh to the Fourteenth Centuries

ALEXANDER GRANT

From ancient times indeed kings had been in the habit of giving to their knights greater or smaller tracts from their own lands in feu-ferme, a portion of some province or a thanage. For at that time almost the whole kingdom was divided up into thanages. He [Malcolm II] apportioned these lands to each man as he thought fit either for one year at a time for ferme, as in the case of tenant farmers, or for a term of ten or twenty years or life with at least one or two heirs permitted, as in the case of certain freemen and gentlemen, and to some likewise (but these only a few) in perpetuity, as in the case of knights, thanes and magnates, with the restriction however that each of them should make a fixed annual payment to the lord king.[1]

That is how, in the 1360s, John of Fordun commented on the origins of the Scottish thanages. His interest is not surprising: as (probably) an Aberdeen chaplain, he would have witnessed continuous episcopal agitation over 'second teinds' from thanages;[2] while Fordoun parish in the Mearns, where he or his forebears doubtless originated, contained the thanage of Kincardine [31][3] and was within a few miles of at least ten others.[4] But the passage is chiefly a reflection on unwise royal patronage in Fordun's own time;[5] how much he and his contemporaries actually knew about the thanages' history

---

1   *Chron. Fordun*, i, p. 186; translation from *Chron. Bower* (Watt), ii, p. 417. Thanages might link Fordun with the author of the 'Laws of Malcolm MacKenneth', discussed below by A.A.M. Duncan (see especially pp. 262-70). If the author was connected with the Aberdeen sheriffs or justiciar ayres, he would inevitably have encountered thanages through the second teind issue; while Walter Moigne (below, pp. 265-7), was relieved of arrears from Aboyne thanage by David II: *RMS*, i, app. II, no. 1362. The author might have consulted Fordun about thanages, which could explain why he knew the chapter on Malcolm II.
2   *RRS*, i, no. 116, note; *Aberdeen Reg.*, i, pp. 5-6, 50, 54-8, 69-70, 72-4, 86-7, 138-41, 155-70; *Chron. Fordun*, i, pp. xiv, xvi, for Fordun as a chaplain.
3   Numbers in square brackets refer to the list of thanages given in the Appendix to this essay; they also locate the thanages on the accompanying maps. Sources for statements about thanages not specifically referenced in the footnotes are in the Appendix entries.
4   Aberluthnott [34], Arbuthnott [29], Cowie [27], Durris [26], Fettercairn [32], Kinnaber [36], Laurencekirk (or Conveth) [34], Morphie [35], Newdosk [33], and Uras [28], all lay within a 10-mile radius of the centre of Fordoun parish.
5   Cf. Duncan, 'Laws of Malcolm MacKenneth', below, pp. 242-4.

may be questioned. Nowadays, especially since Geoffrey Barrow illuminated their origins in his seminal article on 'Pre-Feudal Scotland: Shires and Thanes',[1] we are better informed. As its title states, however, this article focuses essentially on the period before c.1150. And although elsewhere both Barrow and A.A.M. Duncan have provided many important comments on thanes and thanages in the later twelfth and the thirteenth centuries,[2] no connected discussion of their history throughout the Middle Ages yet exists.[3] That is what is offered here, as a postscript, so to speak, to 'Shires and Thanes'.

Geoffrey Barrow has shown that Scotland's thanages derived from the 'multiple-estate' system which extended throughout the British mainland from very early times. It was a means of exploiting royal and princely demesnes through a network of local bases, to which tribute was brought from outlying farmsteads. The territorial units thus formed had various names; in northern England the most common was 'shire', and this came to be applied to the multiple estates of eastern Scotland, both lowland and highland, as well. With shires went thanes: twelfth-century records show many Scottish shires being run by thanes, and (as Barrow has argued persuasively) that might originally have been true of all of them. Be that as it may, the function of the Scottish thane in the immediate pre-feudal era was to look after a specific unit of demesne on behalf of his overlord: he administered it, led its inhabitants in war, supervised justice within it, and (having deducted his own share) paid its renders and dues to the king or to an earl.[4] The units of demesne run by thanes came eventually to be called thanages — though perhaps not until the end of the twelfth century, when the term thanage first appears in surviving records.[5]

Whatever the terminology, the thane's role was clearly managerial; he was not the proprietor of the territory. The distinction is brought out in twelfth-century legislation on teind: if peasants refused to pay, 'the thane

1   G.W.S. Barrow, 'Pre-feudal Scotland: Shires and Thanes', in his *Kingdom of the Scots* (London, 1973), pp. 7-68. See also his *Kingship and Unity: Scotland 1000-1306* (London, 1981), pp. 5-8, 15-17; and his 'Popular courts in early medieval Scotland: some suggested place-name evidence', *Scottish Studies*, xxv (1981).
2   E.g. *RRS*, i, pp. 45-9; *RRS*, ii, pp. 17-18, 48-51, 56; *RRS*, v, pp. 302-3; A.A.M. Duncan, *Scotland: The Making of the Kingdom* (Edinburgh, 1975), pp. 157-8, 161, 165, 169, 177-8, 208, 329-30, 392-4, 597; G.W.S. Barrow, 'Badenoch and Strathspey, 1130-1312, 1: Secular and Political', and '2: Ecclesiastical', *Northern Scotland*, viii (1988), ix (1989).
3   W.F. Skene, *Celtic Scotland* (2nd edn, Edinburgh, 1886-90), iii, pp. 246-83, is one exception; but this, while illuminating, is unconnected and not entirely reliable.
4   Barrow, 'Shires and Thanes', especially pp. 41-50. Cf. *APS*, i, p. 398, for the thanes' military function; and *RRS*, ii, pp. 50, 346-7, plus *Holyrood Lib.*, no. 65, for their judicial powers.
5   The earliest recorded use of thanage (*thanagium*) is in a charter of John bishop of Dunkeld, referring to the 'whole thanages' of Dalmarnock [60] and Findowie [59], in Atholl; it is probably to be dated 1189 × 92: *Scone Lib.*, no. 55, in conjunction with *RRS*, ii, no. 336. In royal documents, thanage is not used until 1236: *Lindores Cart.*, no. 22. Strictly speaking, the use of the term 'thanage' is anachronistic before the end of the 12th century.

under whom the peasant is, or his lord if he has a lord', must enforce it.[1] And, judging from the 1206 lawsuit involving Arbuthnott [29],[2] thanes did not have their territories for very long. One witness remembered seeing 'eight thanes or more' in Arbuthnott before the 1180s, another remembered thirteen. That suggests, on average, one thane every five or six years: interestingly, when the first feudal lord of Arbuthnott went on crusade, he fermed it for six years to Isaac of Benvie, who was implicitly called a thane. But in that case, the thanes of Arbuthnott must have been only short-term managers — and hence, no doubt, had other lands elsewhere. It would seem, therefore, that thanes were drawn from the local lairds, and acted as royal or comital estate managers on a temporary basis.

In the Arbuthnott evidence, the thanes (except Isaac) are consistently referred to in the abstract, while the feudal lords are always named. This is echoed in William I's charter freeing the men of Inverkeilor kirkton [42] of all dues 'that they used to render to the thanes of Inverkeilor and afterwards to Walter of Berkeley'. In Laurencekirk (or Conveth) [34], Humphrey son of Theobald had 'all the rights which the thane was accustomed to have' over the kirkton. Similarly, St Andrews Priory had Haddington kirk [71] and its lands 'free of all dues owed to me and to the thane', and the kirkton of Ecclesgreig (= Morphie) [35] with the same common grazing as 'my thane and my men'; while Urquhart Priory's grant of Fochabers [11] included 'the dues of fish which belong to the thane'. Such phraseology highlights the thane's ministerial role.

That corresponds with its etymology: the Anglo-Saxon *þegn* derived from *þegnian*, to serve, and was commonly translated into Latin as *minister*.[3] In the late Anglo-Saxon period, however, *þegn* could describe virtually any landholder; most of the thousand or so thegns in Domesday Book were unimportant tenants. But there were also king's thegns. Many of these were to be found on the royal demesnes; but there were several to each unit of territory or shire, and they were subordinate to royal managers or reeves. Other king's thegns, in contrast, had individual estates, with administrative and judicial powers over them; but they held them in 'bookland', that is as hereditary private landlords, without owing rents or renders.[4] Neither type of late Anglo-Saxon king's thegn, therefore, quite equates to the Scottish

1   *RRS*, ii, no. 281.
2   *Spalding Misc.*, v, pp. 209-13; cf. *RRS*, ii, no. 569.
3   H.R. Loyn, 'Gesiths and Thegns in England from the Seventh to the Tenth Century', *English Historical Review*, lxx (1955), pp. 529-40.
4   For the difficult subject of the late Anglo-Saxon thegn (which is oversimplified here) see, most recently, D. Roffe, 'From Thanage to Barony: Sake and Soke, Title, and Tenants-in-Chief', *Anglo-Norman Studies*, xii: *Proceedings of the Battle Conference, 1989*; and D. Roffe, 'Domesday Book and Northern Society: a reassessment', *Eng. Hist. Rev.*, cv (1990), pp. 329-3; also P. Stafford, 'The Farm of One Night and the Organisation of King Edward's Estates in Domesday Book', *Economic History Review*, xxxiii (1980); H.R. Loyn, *The Governance of Anglo-Saxon England* (London, 1984), ch. 6; J.E.A. Joliffe, 'Northumbrian Institutions', *Eng. Hist. Rev.*, xli (1920); R.R. Reid, 'Barony and Thanage', ibid., xxxv (1920). My thanks to Dr Pauline Stafford for her advice on this subject.

thane — who, as the manager of a unit of demesne with judicial powers over it, appears to have combined attributes of both. The implication is that the term 'thegn' did not simply spread northwards into Scotland, as shire may have done, but that it was introduced deliberately, presumably by one or more of the pre-feudal kings of Scots.

Whatever the thane/thegn relationship, the recorded names of early Scottish thanes, where they are known, are generally Gaelic: Hywan Macmallothen of Dairsie [68], and so on.[1] At the upper levels of pre-feudal Scotland's Gaelic society were the *mormaer* and *toísech*; since *mormaer* was anglicised into earl, did *toísech* similarly become thane? That is implied by 'The Laws of the Brets and the Scots', which gives the grades of king, earl, thane, *ócthigern*, and peasant; *ócthigern* means young, or lesser, lord (*thigern*), the lowest rank above the peasantry.[2] Moreover, thane certainly did translate into Gaelic as *toísech*, as is illustrated by the name 'Finlayus Toschoch thanus de Glentilt' [55] recorded in 1502; an estate acquired by the thane of Cawdor in 1476 became known as 'Ferintosh', 'the land of the toísech'.[3] But toísech was probably a rather looser concept than thane. It literally meant 'first or foremost person', or 'leader', and could apply both to the head of a kindred and to the subordinate officer of a king or mormaer; it is the latter sense which is probably meant in Gaelic notes of grants to Deer Abbey 'free of [obligations to] mormaer and toiseach (which parallel other grants 'free of all dues owed to [the king and] the thane').[4] And Professor K. Jackson has pointed out that we probably 'must distinguish quite clearly between' the two concepts of toísech; the latter may simply be 'the Anglo-Saxon thane borrowed and accommodated with a vaguely appropriate Gaelic title'.[5] Given that many thanes were Gaelic Scots, it is easy to see how that could have happened. But, while thane became *toísech* in Gaelic, not every toísech was necessarily a thane; it is probably safer to see thanes and local Gaelic lords existing side-by-side, with both perhaps having toísech's status. That might explain why, when thanes' names appear in records, they are generally accompanied by other Gaelic names with no functional designation — which brings us back to the distinction between ministerial thane and local landlord.

The records surviving from medieval Scotland provide the names of at least

1   Cf. Macbeath of Falkland [66], Gilys of Idvies [44], Dugall of 'Molen' [12], Ewen of 'Rathenech' [10], and Lorne of Uras [27].
2   *APS*, i, pp. 663-5; Duncan, *Making of the Kingdom*, pp. 107-11; J. Bannerman, 'The Scots Language and the Kin-based Society', in *Gaelic and Scots in Harmony: Proceedings of the Second International Conference on the Languages of Scotland*, ed. D.S. Thomson (Glasgow, 1990), pp. 6-8.
3   *RMS*, ii, no. 2655; cf. John 'Toschoch' thane of Glentilt, mentioned in 1434 (SRO, Register House Transcripts, RH.1/2/204, and also 1/2/312-5); *Cawdor Bk.*, index, at 'Ferrintosh', 'Mulquaich' and 'Drumurny'.
4   Bannerman, 'Scots Language and Kin-based Society', p. 6; K. Jackson, *The Gaelic Notes in the Book of Deer* (Cambridge, 1972), pp. 110-14; Duncan, *Making of the Kingdom*, pp. 109-10; *ESC*, no. 122.
5   Jackson, *Gaelic Notes*, pp. 113-14.

48 thanages, together with 19 places which had thanes and another four which almost certainly did so: a total of 71 (see the accompanying maps and Appendix). Although the evidence for them is often thirteenth- and fourteenth-century, they are unlikely to have been created after David I came to the throne; therefore, later references to thanages or thanes may be taken as demonstrating their existence in and before David's reign. The actual number of thanes at that time may have been considerably higher, but this minimum list provides a good basis for the present analysis.[1] For convenience, albeit anachronistically, all 71 territories are here simply called thanages.

They are located, together with the major provinces and earldoms of twelfth-century Scotland,[2] in Map (A), 'Thanages and Provinces' (overleaf). As this shows, thanages belonged almost entirely to eastern Scotia between the Forth and the Moray Firth — the heartland of the kingdom of the Scots, where members of the MacAlpin kindred wielded consolidated power during the tenth and eleventh centuries. In contrast, they can hardly be found at all in the kingdom's periphery. To the west and north, where Scottish authority was disputed with the Norse and was at best occasional, there was only Dingwall [1], from its name a Norse administrative centre which may not have been incorporated permanently into Scotland until the 1060s.[3] Similarly, thanages are conspicuously absent south of the Forth. In Strathclyde, there were multiple estates, at least two shires, and territories owing renders much like those of northern thanages, but no trace of thanes.[4] Much the same can be said of Lothian, where there were several shires,[5] but where thanes are found only at Callendar [70] and Haddington [71]. Here, Scottish control was probably achieved during the tenth century in the north

---

1   My list is not exhaustive, and no doubt references to other thanes or thanages will be discovered in the future. It is limited to the period before 1450; some references to thanes and thanages after that date have not been included, because it is not clear whether these reflect genuine continuity with the 12th century or not. The four places counted as almost certainly having had thanes are Coupar Angus [51] and Longforgan [52], Forfar [45], and Kincardine O'Neil [23].

2   The boundaries of the earldoms are based on the map prepared by Dr Keith Stringer (to whom my thanks) for the forthcoming revised edition of *An Historical Atlas of Scotland*.

3   B.E. Crawford, *Scandinavian Scotland* (Leicester, 1987), pp. 96, 206, and 61-74. But E.J. Cowan, 'The Historical MacBeth', in *Moray in the Middle Ages*, ed. W.D.H. Sellar (Scottish Society for Northern Studies, forthcoming), at n. 57, claims that in the 11th century 'the Orcadian empire never extended south of Strathoykel'; in which case Dingwall could have been Scottish at an earlier date. My thanks to Professor Cowan for permission to consult his article in advance of publication.

4   Barrow, 'Shires and Thanes', p. 36. For Cadzow-shire and Machan-shire, *ER*, i, pp. 40, 46; *Origines Parochiales Scotiae*, ed. C. Innes (Bannatyne Club, 1851-5), i, pp. 106-8. Cadzow probably rendered at least £80 plus 22 chalders of wheat and 6 chalders of oats (*RRS*, vi, no. 418); Cambusnethan, also in Lanarkshire, probably rendered at least 20 chalders of grain (*RMS*, i, no. 79). The point perhaps tallies with Alan Macquarrie's argument (in his essay, 'The Kings of Strathclyde, *c*.400-1018', above, pp. 16-19) that the Scottish takeover of Strathclyde did not begin until after 1018. Also, the feudal settlement of Clydesdale under Malcolm IV has the appearance of having taken place in something of a *tabula rasa*, at least with regard to native landowners (G.W.S. Barrow, 'The beginnings of military feudalism', in *Kingdom of the Scots*, pp. 288-91; *RRS*, i, no. 265).

5   Barrow, 'Shires and Thanes', pp. 28-36.

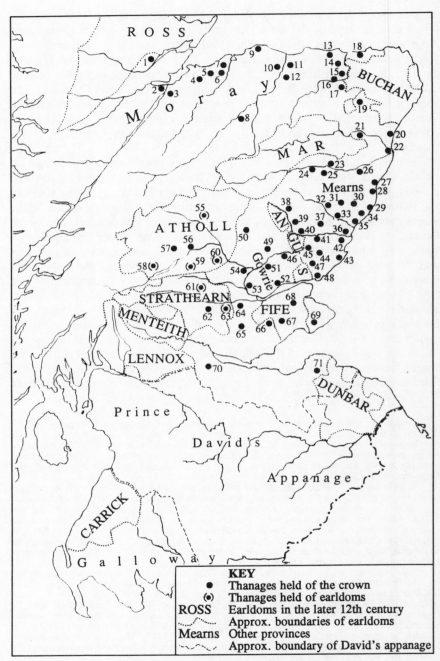

Map (A): Thanages and Provinces

of the region (roughly modern Lothian), which includes Callendar and Haddington; were they frontier outposts? Beyond Lammermuir, however, consolidated Scottish lordship does not seem to have reached the Tweed until much later; even in Alexander I's reign, the huge appanage of Strathclyde, Tweeddale, Teviotdale and southern Lothian held by his brother David suggests that this region was still not fully integrated into Scotland. That probably accounts for the absence of thanes from it.[1]

Admittedly, Earl David did address a charter to 'omnibus suis fidelibus tegnis et dregnis de Lodeneio et de Teuegetedale' [Lothian and Teviotdale]; and two charters of Cospatric earl of Dunbar have similar addresses.[2] But they may not have meant thanes like those of Scotia. 'Thegns and drengs' is Northumbrian terminology; they were common there, holding small portions of land (several to a shire) in return for various renders.[3] That was no doubt the case in south-eastern Scotland — Scottish Northumbria — as well. Thus, in these charters, Earls David and Cospatric were probably employing the late Anglo-Saxon concepts of the thegn, rather than that of the Scottish thane; this brings us back to the contrast between thane and thegn discussed above. The paradox that, despite its Anglo-Saxon derivation, the Scottish type of thane seems to be almost completely absent from Anglian Scotland highlights the point that thanes and thanages belong to Scotland's ancient heartland, not to its periphery.

In that heartland, the most significant feature of Map (A) is the relationship between thanages and earldoms. Although earls could have thanes — Earl Gilbert of Strathearn referred to 'Anechol, my thane' (of Dunning)[4] — there are few instances of this. With at most six exceptions (Dunning [63] and Strowan [61] in Strathearn, Crannach [58], Dalmarnock

---

1   The appanage of Earl David, who was sometimes styled 'prince of Cumbria' (*Glasgow Reg.*, i, no. 1), looks like an extension of the Strathclyde client kingdom held by his grandfather, Duncan, under Malcolm II; see Macquarrie, 'Kings of Strathclyde', above, pp. 16-19. As for Lothian, Edgar's famous charter to Durham of 1095 stated that he possessed 'the kingdom of Scotland and the land of Lothian', which indicates a distinction: *ESC*, no. 15. Under Edgar, part at least of Ednam, on the fringe of the Merse, was waste; so was territory higher up the Tweed which David I gave to the Melrose Cistercians (*ESC*, nos. 24, 141). In the 12th century, most of the Merse was held by the earls of Dunbar. Elsewhere, part of the region consisted of the inhospitable Ettrick and Selkirk forests; part was probably best suited to the monastic settlements of Kelso, Jedburgh and, especially, Melrose; parts were granted as large 'provincial' lordships; parts were turned into knights' feus; and very little, except for the hinterland of Peebles, Roxburgh, and Jedburgh, was retained as royal demesne (*Origines Parochiales*, i, pp. 177-496, *passim*). David I's burghs and monasteries were obviously very important to him, but it is hard to see the rest of the region as being at the heart of David's kingdom — and it is impossible to do so for his predecessors.
2   *ESC*, no. 30; *Coldstream Cart.*, nos. 8, 11.
3   Joliffe, 'Northumbrian Institutions'; cf. Barrow, 'Shires and Thanes', pp. 19-28. Roffe, 'Domesday Book and Northern Society', pp. 329-30, argues that there was no difference between drengs and most of the thegns. Unlike thegn/thane, dreng is rare in Scotland, though we should note a 'Drengysland' in Fowlis Easter, Angus, in 1261 (*Panmure Reg.*, ii, pp. 84-5). Given its sense of youth, it is worth asking whether it should be equated with the *óchtigern* of 'The Laws of the Brets and the Scots'. If so, that might indicate that thane had been assimilated into Celtic society, but dreng had not.
4   *Inchaffray Chrs.*, nos. 4, 14; cf. no. 19.

[60], Findowie [59] and Glentilt [55] in Atholl), the thanages recorded in medieval Scottish records were either certainly or presumably in crown hands in the early twelfth century.[1] Even Falkland [66] was then a royal property, with its thane, Macbeath, independent of Earl Constantine of Fife in the 1120s; it did not go to the earls of Fife until the later 1150s. In the first half of the twelfth century the known Scottish thanages were, in fact, overwhelmingly royal.

They also, as the map shows strikingly, penetrated all the earldoms from Fife to Moray. Within Fife there was not only Falkland but Kingskettle [67], while Dairsie [68] and Kellie [69] lay on its fringes. In Strathearn, the thanage of Auchterarder [62] — later a small sheriffdom — was inside the earldom's boundaries, and Forteviot [64] was on its edge. The extensive thanages of Dull [56] and Fortingall [57] were in the middle of the earldom of Atholl. The earldom of Gowrie contained the king's *maneria* of Scone [53], Strathardle [50], Coupar Angus [51] and Longforgan [52]; Gowrie was in royal hands, but the royal possessions in Gowrie antedated its acquisition by the crown.[2] Further north, Angus and Mearns had the highest concentration of thanages in the country. Thirteen are to be found throughout the province of Angus, completely overshadowing the meagre holdings of the later twelfth-century earls; similarly, eleven were scattered across the Mearns, no doubt intermingling with what before the end of the eleventh century had been mormaer lands.[3] Beyond the Mearns, Mar and Buchan did not contain royal thanages in quite the same way. But Kincardine O'Neil [23] (fully) and Aboyne [24] (partly) lay north of the Dee, and might be expected to have belonged to Mar; with Kintore [21] and Aberdeen [22] they ringed its eastern half. Similarly, Buchan was limited to the south by Belhelvie [20] and the large thanage of Formartine [19], which cut across the earldom's south-west corner;[4] to the north, Glendowachy [18] impinged on its coastline; and the other thanages in Banff [13-17] bounded it on the west. These Banff thanages also possibly look like a bulwark against the great earldom of Moray — which, itself, was full of thanages.

Moray's thanages, however, pose a problem. Before the earldom's forfeiture in 1130, did they belong to the crown or to the mormaers and earls of Moray? Geoffrey Barrow has suggested they were royal: that after 1130 the comital demesnes in Moray were used to endow new settlers, while the crown kept what were already royal thanages.[5] This is certainly plausible. On the map, most of the Moray thanages form two groups in the lower Spey and Findhorn valleys [10-12, 4-7]: they seem to have hemmed in

---

1   I.e. when first mentioned they are obviously crown possessions; see Appendix. Also, since the references to thanages held of earls all date from the later 12th century at the earliest, it is by no means certain that they were not previously alienated by the crown, as happened with Falkland; cf., for Glentilt, Barrow, 'Badenoch and Strathspey, 1', p. 9.
2   *RRS*, i, no. 245.
3   Duncan, *Making of the Kingdom*, p. 165; cf. p. 125 for the mormaer of Mearns.
4   For Buchan, see Map (A) in Alan Young's essay, 'The Earls and Earldom of Buchan in the Thirteenth Century', below, p. 201.
5   Barrow, 'Badenoch and Strathspey, 1', pp. 2-3.

the area between the rivers, the hinterland of Elgin, doubtless an important portion of the comital demesne. Similarly, Inverness (a major centre of the earldom) and the rich lands to the east on the south shore of the Beauly Firth, which were also probably significant comital lands, were ringed by thanages [2-7]. The pattern could well reflect the establishment of a royal presence in Moray long before 1130. As for Dingwall [1], this was presumably a royal base on the fringe of Scandinavian Scotland, but (as suggested above) perhaps not established before the late 1060s.

If so, Dingwall was a late thanage; the others were doubtless earlier. But how much earlier — in other words, exactly when units of royal demesne first came under the thanes — is unclear. We may presumably connect the development with an assertion of royal control, perhaps to save the management of the royal demesnes from being hereditary within the kindreds of Gaelic society. We might ponder a link with *mormaer*, for this term, meaning 'great steward', has connotations of overall royal authority. But *mormaer* is Gaelic, and first appears under the year 918; thane derives from Anglo-Saxon, and, judging by English usages, is unlikely to have been current in Scotland in the early tenth century.[1] Was the introduction of thanes, therefore, a later step in the growth of Scottish royal power? Since the concept came from England, can its introduction be attributed to Kenneth II (971-95), who witnessed Anglo-Saxon kingship at the height of its prestige at Chester in 973? Or should we look to his son, Malcolm II (1005-34)? He was certainly an innovator, going against the Gaelic principles of royal succession to make his daughter's son follow him as king; he diverted the main rival segment of the royal kindred, the descendants of King Dub, into the mormaership of Fife;[2] he greatly extended Scottish power in Lothian by his victory over the Northumbrians at Carham in 1018; through his grandson Duncan he in effect took over the kingdom of Strathclyde;[3] and, as Professor E.J. Cowan has recently stressed, he was called 'King of the Mounth', that is, unlike his predecessors he ruled north of the Grampians, in Moray, as well as in the south.[4] Malcolm II, one of the longest-surviving and toughest of the pre-feudal kings of Scots, looks the most likely ruler to have introduced thanes as royal agents in Scotia and in Moray. In that case, is Fordun's association of him with thanages merely a coincidence? Alas, we can only speculate; on this issue it is hard to improve on Fordun's remark that thanages had existed

1   Jackson, *Gaelic Notes*, p. 104 (and cf. pp. 108-9); Loyn, 'Gesiths and Thegns', pp. 540-2; Barrow, 'Shires and Thanes', p. 64.
2   See John Bannerman's essay, 'MacDuff of Fife', above, pp. 24-6, 33: the diversion of 'the MacDuffs' into Fife must have happened after the death of Dub's son Kenneth III in 1005, presumably under Kenneth's successor Malcolm II.
3   Cf. Macquarrie, 'Kings of Strathclyde', above, pp. 16-19.
4   Cowan, 'The Historical MacBeth', at n. 29 (citing K. Jackson, 'The Duan Albanach', *SHR*, xxxvi [1957], p. 133). Cowan argues that the men of Moray accepted Malcolm as over-king, perhaps as a result of 'a marriage between Malcolm's sister and Findlaech [mormaer of Moray] — the parents of MacBeth' (at n. 30; citing Anderson, *Early Sources*, i, p. 580).

'from ancient times indeed'.[1]

Whatever the case, there can be no doubt about the importance of the thanes and thanages in the early twelfth century. The country north of the Forth has been called the land of earls;[2] it was also, perhaps more significantly, the land of thanes. Their immediate value was fiscal. Originally, Scottish crown revenue came chiefly from 'cain' and 'conveth'. Cain can best be thought of as a regular tribute on all territory acknowledging the king's superiority; conveth was the obligation of hospitality owed to the immediate lord of an estate, or to the crown from the royal demesnes, and was either actual food and accommodation for so many nights, or food renders *in lieu*.[3] Since the royal thanes ran units of crown demesnes, they would have been responsible for delivering both cain and conveth. David I's endowment of Scone Abbey included a tenth of his cain of cheeses and hides from Scone, Coupar, Longforgan and Strathardle;[4] and there were actually two thanages named 'Conveth' [17, 34]. An impression of what conveth might involve can be gained from the surviving fragments of royal accounts for 1263-6 and 1290-1 (in which it is called 'waiting', an English term which had spread from Lothian, where it denoted similar hospitality dues[5]). The thanages of Forfar and Glamis [45, 46] are recorded as owing two and one-and-a-half nights' waitings respectively, made up as follows: 37½ cattle (24 from Forfar, 13½ from Glamis); 75 pigs; 225 cogals of cheese (144 from Forfar, 81 from Glamis, each cogal weighing six stones, which probably makes over eight tons!); 291 chickens; 32 chalders (each weighing roughly a ton) of malt; ten chalders of barley meal; about 80 chalders of fodder; and 960 eels.[6] Elsewhere, four nights' waitings were due from Kinross thanage [65], one night's from Fettercairn [31], and two nights' — or £40 — from Aberchirder [16].[7] In these cases the amounts of produce paid are unknown, except for 40 cattle from Kinross, which is roughly *pro rata* with the 37½ from Forfar and Glamis. The sum of £40 at which the Aberchirder waitings were reckoned also tallies reasonably with the value of the produce from Forfar and Glamis, which, judging by contemporary commodity prices, amounted to £75-85 (half from cheese) for the three-and-a-half nights' waitings, or around £23 per night. Possibly, therefore, in the later thirteenth century one night's waiting was the equivalent of about £20 — in which case, might a year of waitings have been worth some £7,300? The calculation is too neat, but the total is not far from the likely amount of crown revenue from its lands in late thirteenth-

1   *Chron. Fordun*, i, p. 186.
2   Duncan, *Making of the Kingdom*, p. 164.
3   Ibid., pp. 152-7; *RRS*, ii, pp. 52-3.
4   *RRS*, i, no. 57.
5   But for the opposite, waiting being called *conveis*, i.e. conveth, in Northumberland, see G.W.S. Barrow, 'The lost Gàidhealtachd of medieval Scotland', in *Gaelic and Scotland: Alba agus a' Ghàidhlig*, ed. W. Gillies (Edinburgh, 1989), p. 74.
6   *ER*, i, pp. 6-7, 49-50. The obvious mistakes in the 1264 account, as printed, can be corrected from the 1290 one.
7   *ER*, i, pp. 12, 16, 49; *Aberdeen Reg.*, i, p. 55.

century Scotland.[1] This shows, perhaps, how conveth or waiting might once have sustained the early kings and their followings. It is argued below that the thanages made an important contribution to royal finances even as late as Alexander III's reign; in the pre-feudal era, their fiscal contribution must have been vital.

In that era, of course, the renders would all have been in kind, and would have been consumed locally — which gives a picture of peripatetic kings moving constantly round the thanages. That may seem a primitive system, but the benefits would have been great. Throughout the Middle Ages, the function of royal demesnes was to sustain kings politically as well as physically, by providing bases for royal lordship.[2] In pre-feudal Scotland the thanages provided such bases from the Forth to beyond Moray. When not on campaigns, Scottish kings must have spent much of their time travelling around them, maintaining a royal presence in the localities. They would thus have been able to supervise their local agents, make new appointments, deal with problems of justice, and above all simply assert royal authority. The witnesses in the Arbuthnott lawsuit deduced episcopal possession of the kirkton from the fact that the bishops of St Andrews had repeatedly received hospitality in person from it;[3] on a much larger scale, that is the main point about the kings and the thanages. It is the essence of personal kingship, and it must have been a major factor in the unification of pre-feudal Scotland. In that context, the significance of the thanes is much more than simply fiscal. Indeed, it may be suggested, it was largely by means of the thanes and the thanages that royal power was consolidated, especially in the kingdom's heartland between the Forth and the Mounth. Thanes were derived, ultimately, from the Anglo-Saxon system of government: we are increasingly aware of the huge Gaelic contribution to the medieval kingdom of the Scots, but with the thanages we have a major Anglo-Saxon contribution as well.

On the eve of Scotland's feudalisation, therefore, the thane was obviously a vital link between the crown and the local communities north of the Forth. During David I's reign this did not alter. David's famous grants to the

---

1   The 'Valuation of the Sheriffdoms', probably made in *c.*1292, gives a total crown income of just under £8,100: *SHS Misc.*, ii, pp. 24-6 (and pp. 61-2, below).
2   The importance of this point has become particularly clear for late medieval England, where successive kings dissipated their demesnes and hence lost their own power bases throughout the country — a major contributory factor in the Wars of the Roses. See, e.g., B.P. Wolffe, *The Royal Demesne in English History* (London, 1971); C. Given-Wilson, *The Royal Household and the King's Affinity, 1360-1413* (New Haven, 1986); R.L. Storey, *The End of the House of Lancaster* (London, 1966); B.P. Wolffe, *Henry VI* (1971); R. Horrox, *Richard III: A Study of Service* (Cambridge, 1989); A.J. Pollard, *North-Eastern England during the Wars of the Roses* (Oxford, 1990); and, for the re-establishment of royal authority under Henry VII, a king supremely aware of the significance of royal lands and agents in the localities, A. Grant, *Henry VII: The Importance of his Reign in English History* (London, 1985). *Mutatis mutandis*, the thanes of pre-feudal Scotland look not unlike the 'receivers' of Yorkist and early Tudor England: vital royal agents who collected revenues from royal demesnes and, more broadly, helped to run the localities on the crown's behalf.
3   *Spalding Misc.*, v, pp. 211-13.

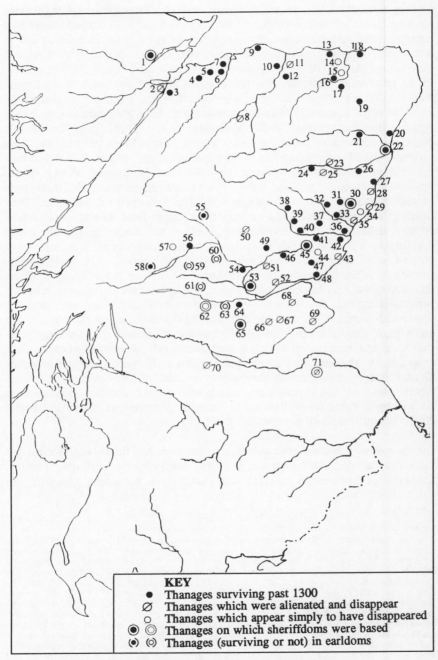

KEY

- ● Thanages surviving past 1300
- ∅ Thanages which were alienated and disappear
- ○ Thanages which appear simply to have disappeared
- ◉ ◎ Thanages on which sheriffdoms were based
- (●) (◦) Thanages (surviving or not) in earldoms

Map (B): The Survival of the Thanages

Church and to 'Normans' were hardly at all at the expense of known royal thanages. He only gave one to the Church: Fochabers [11], to Urquhart Priory in Moray.[1] When he added to the endowments of Scone Abbey, it was with the teinds of Scone parish and a tenth of his revenue from his four *maneria* of Gowrie, but no lands apart from Cambusmichael to the north were transferred; the *manerium*, or thanage, of Scone [53] stayed in his own possession.[2] Similarly, his 'Norman' followers did not receive any thanages; only Haddington [71] was alienated, temporarily, as part of the dower of his son's wife Ada de Warenne. In David I's Scotland — where 'the balance of new and old' was so strikingly maintained[3] — the thanes were a most significant aspect of the old.

Thereafter, things changed: the later twelfth and the thirteenth centuries witnessed the feudalisation of Scotland north of the Forth, and as a result the thanages were no longer so important. Nevertheless — as Map (B), 'The Survival of the Thanages', demonstrates — they did still have a place. Some became the bases for sheriffdoms. In the south, there was a sheriff of Haddington [71] in 1184 and probably much earlier. To the north, Scone [53] had a sheriff under David I; later in the twelfth century they are found at Forfar [45], Kincardine [30] and, probably, Aberdeen [22]; and in the thirteenth at Kinross [65], Auchterarder [62] and Dingwall [1]. Indeed in northern Scotland outside Moray (where former comital centres were perhaps used), over half the twelfth- and thirteenth-century sheriffdoms were centred on thanages. In their function as the crown's main local agents, were the sheriffs very different from the thanes? Their position, of course, was superior, but should they not be regarded as 'super-thanes' rather than as entirely new agents of government? Although Aberdeen [22] had both thane and sheriff, elsewhere that was not so, and the two offices may have merged. Was John of Kinross, sheriff of Kinross in 1264, a descendant of earlier thanes? He was, at any rate, a kinsman of the former thanes of Callendar.[4] Further north, some of the earliest recorded sheriffs have Gaelic names, and at least one was also a thane: Macbeth, sheriff of Scone and thane of Strathardle [50] in the 1190s. Thanes probably played a significant part in the development of the Scottish sheriffdoms. But the original concept was so firmly established that in every case except Forfar and Haddington, the thanages themselves survived within the sheriffdoms of the same name until the fourteenth century.

On the other hand, several thanages were alienated by the crown. These alienations, however, were relatively limited. Although much land went to

---

1   *ESC*, no. 255. Fochabers subsequently came back into crown hands in exchange for the land of 'Bynin': *APS*, i, p. 110, and cf. *ER*, i, p. 443.
2   *Scone Lib.*, nos. 1, 5; cf. *RRS*, i, nos. 57, 243. The locations of the lands granted by Alexander I can be found in the index to *RRS*, i; apart from Scone itself, none were in Scone parish, and therefore probably lay outwith the thanage.
3   G.W.S. Barrow, *David I of Scotland (1124-1153): The Balance of New and Old* (Reading, 1985), *passim*.
4   *ER*, i, 16; SRO, RH.1/2/55. Cf. *RRS*, ii, nos. 28, 408, for Gilbert and Hervey of Kinross in William I's reign.

the Church, only four thanages were transferred: Malcolm IV gave his 'whole land of Coupar' [51] (though possibly not the entire thanage) to his abbey at Coupar Angus; William I gave Birse [25] to the bishops of Aberdeen; Alexander II gave the bishops of Moray most of Kinmylies [2], and alienated Callendar [70], giving more than two-thirds to Holyrood Abbey. Lay landlords did rather better. Malcolm IV granted Falkland [66] to Earl Duncan of Fife; later earls also possessed Kingskettle [67], and also Cromdale [8] in Moray. Those were probably among a number of thanages granted away by William I. His brother Earl David's endowment included Longforgan [52] and Ecclesgrieg (Morphie) [35]; while his bastard son Robert of London's included Kellie [69]. William's chamberlain Walter of Berkeley received Inverkeilor [43]; Osbert Olifard received Arbuthnott [29]; and Richard de Melville may have had Tannadice [40]. Alexander II was less generous, but gave the Aberdeenshire thanages of Aboyne [24] and Kincardine O'Neil [23] to Walter Bisset and Alan Durward respectively, while the residue of Callendar [70] went to the former thane. Alexander III included Belhelvie [20] in the dowry of his daughter Margaret, queen of Norway;[1] gave Conveth [17] and superiority over Uras [28] to the earl of Buchan; and granted two escheated thanages to R[ichard?] Siward (Kellie) and to the earl of Fife (Kincardine O'Neil). Finally, by the end of the century John of Inchmartin held Strathardle [50] as a barony. That gives a total of 14 grants of thanages (not counting Callendar) during the four kings' reigns: hardly a sweeping alienation.[2] Over half, moreover, went to 'native' Scots who were the kings' close kin: Earl David, Robert of London, Alexander III's daughter, the earls of Fife, and the Inchmartins.[3] And some of the grants were temporary: Aboyne and Tannadice were back in crown hands in the later thirteenth century; Siward only received Kellie on a short-term basis; and the queen of Norway did not have Belhelvie for long. In fact, only Arbuthnott and Inverkeilor provide cases of the permanent acquisition of entire thanages by members of Scotland's 'Norman' families.

It was probably more usual, however, for thanages to be slimmed down, as happened to Glamis [46]: in the 1264 exchequer the thane was allowed 16 merks 'for the lands of Clofer and Cossenys, subtracted from the thanage of Glamis'; and William I had already alienated Ogilvie and Kilmundie, both in the parish and probably the thanage of Glamis.[4] In Alexander III's rental of Aberdeenshire, Aberdeen thanage [22] is followed by the lands of Grandoun and Prestoly, which 'were accustomed to belong to the service of the sheriff of Aberdeen'; since these lands are in the parish of Old Aberdeen, it seems likely that they had been detached from the thanage

1   Menmuir [37] was also to be granted if the other lands did not produce 700 merks a year: Stevenson, *Documents*, i, pp. 313-4.
2   One of the lists of (subsequently lost) Scottish muniments drawn up by English clerks in 1292 (*APS*, i, p. 110) includes references to charters concerning Aberchirder, Aboyne, Fochabers, and Newdosk, but there is no way of telling what transactions were involved.
3   For the Inchmartins, descendants of a bastard son of David earl of Huntingdon, see Barrow, 'Lost Gàidhealtachd', pp. 68-9.
4   *ER*, i, p. 8; *RRS*, no. 140.

when the sheriffdom was established.[1] William I granted four acres in Forteviot [64] to Cambuskenneth Abbey; and gave Humphrey son of Theobald seven davochs in the parish of Fordoun (probably Kincardine thanage [30]), plus four ploughgates in Conveth (= Laurencekirk [34]).[2] Previously alienated lands were excepted from Alexander II's grant of Kinmylies [2].[3] Alexander himself granted Rait and Kinfauns, almost certainly in the thanage of Scone [53], to Scone Abbey; Bamff and other lands in the 'feu' (presumably thanage) of Alyth [49] to his physician Master Ness; and territories in the thanage of Auchterarder [62] to Lindores Abbey.[4] There are no doubt many other examples of this; constituent parts of thanages were probably a common source of royal patronage. In some cases, the result might have been the disappearance of the thanage: that perhaps happened to Haddington [71] and Laurencekirk [34], which had had thanes in the twelfth century but were not subsequently called thanages. Similarly, while most Banffshire thanages kept their status in the fourteenth century, Mumbrie [14] and Netherdale [15] were then only small freeholdings; part of Netherdale, indeed, was held by the thane of Aberchirder in the 1280s,[5] and so its residue had perhaps been incorporated into Aberchirder.

Yet there were probably at most a dozen cases of thanages being whittled away to the point of disappearance during the twelfth and thirteenth centuries;[6] that was no more common than outright alienation. And the main point is that most of the known thanages suffered neither fate, but survived this period more or less intact. A thanage has been counted as having survived if it had a thane in the late thirteenth century, or if it was called a thanage after 1300 and was then demonstrably part of the fourteenth-century royal demesne. No fewer than 41 out of the total of 71 known Scottish thanages meet these criteria;[7] they are shown in Map (B). Thus, although the feudalisation of Scotland obviously diminished the thanages' significance, it did not sweep them away. Well over half the thanages continued to exist, alongside the new feudal tenancies.

Two objections might be raised to that conclusion. First, since the analysis is based, for the most part, on thanages identified from post-1200 records, it might miss many alienated during the twelfth century. Secondly, if north of the Forth the 'Norman' incomers were not endowed from the thanages, then where did their lands come from? Answers to both objections can be derived from Geoffrey Barrow's suggestion about the feudalisation of

1   *Aberdeen Reg.*, i, p. 55; *Aberdeen-Banff Colls.*, i, p. 231.
2   *RRS*, ii, nos. 208, 346, 345.
3   *Moray Reg.*, no. 34.
4   *Scone Lib.*, no. 75, and *RRS*, v, no. 17, comment (also, when Alexander I founded the abbey, he detached its site and immediate surroundings from the thanage); *Bamff Chrs.*, no. 1; *Lindores Cart.*, no. 22.
5   *Moray Reg.*, pp. 319-20.
6   This fate seems to have happened to Mumbrie [14], Netherdale [15], Uras [28], Laurencekirk [34], Idvies [44], Fortingall [57], Findowie [59], Dalmarnock [60], Strowan [61], Dunning [63], Dairsie [68], and Haddington [71].
7   'Surviving thanages', following the above criteria, are nos. 1, 3-7, 9, 10, 12, 13, 16-22, 24, 26, 27, 30-3, 36-42, 45, 46-9, 53, 54, 56, 64, and 65 in the Appendix.

Moray: that, after 1130, the forfeited comital lands were used for feudal grants in the province, while existing royal thanages were kept by the crown.[1] That would explain where the land for the incomers was found; and it also indicates that the places known as thanages after 1200 may indeed be taken as constituting the bulk of the royal thanages of the early twelfth century.

Does this apply elsewhere? North of the Forth, twelfth-century royal grants were concentrated quite remarkably in Fife, eastern Perthshire (the former earldom of Gowrie), Angus, and Mearns.[2] Now in Gowrie, which had come into crown hands under Alexander I, there were both royal and comital demesnes: Malcolm IV referred to 'all my manors of Gowrie, both of the earldom and of the crown'.[3] But other royal charters mentioned 'these my manors of Gowrie, namely Scone, Coupar, Longforgan and Strathardle'.[4] Those (almost certainly) were thanages, and the implication is that they constituted the ancient royal demesnes in Gowrie, while the rest of the crown's possessions there were former earldom lands. In Angus, no earls are recorded before about 1150 and thereafter they were poorly endowed, which might indicate that Angus had previously been in royal hands and that much of the old mormaers' territory had been kept by the crown.[5] In Mearns, the last record of a mormaer is of his killing Duncan II in 1094;[6] after Duncan's death was in effect avenged by his half-brother Edgar, Mearns was probably forfeited to the crown. Thus Gowrie, Angus and Mearns may all have been taken over by the early twelfth-century kings — which raises interesting questions about the politics of that period. As for Fife, judging by the relatively limited possessions of the early twelfth-century earls and their family's close kinship with the kings, something similar perhaps happened there some generations earlier.[7] Be that as it may, it does seem likely that in all four provinces the twelfth century kings had both royal thanages and former mormaer/earldom lands — and that, as Barrow suggested for Moray, the latter were chiefly used for the new ecclesiastical endowments and grants to 'Norman' incomers, while the thanages mostly stayed in royal hands.

Admittedly the pattern is not exact: during the twelfth century, for instance, the Fife thanages and at least two of the four royal *maneria* in Gowrie were granted away.[8] On the other hand, it has been seen that most

1   Barrow, 'Badenoch and Strathspey: 1', pp. 2-3.
2   See *RRS*, i and ii, *passim*. There are also some grants of land in Moray, a few in Banff, and, of course, Earl David's lordship of Garioch in Aberdeenshire.
3   *RRS*, i, no. 245: 'tam de comitatu quam de regali'.
4   Ibid., i, nos. 57, 243: 'de hiis quatuor maneriis meis de Gouerin, scilicet de Scon et de Cubert et de Forgrund et de Straderdel'.
5   Duncan, *Making of the Kingdom*, p. 165.
6   *RRS*, p. 125; Anderson, *Early Sources*, ii, p. 90.
7   Not only Falkland [66], Kingskettle [67] and Dairsie [68], but also the territories of Strathmiglo and Strathleven were all outside the early 12th-century earldom: *RRS*, i, no. 190; ii, no. 472; and see Bannerman, 'MacDuff of Fife', above, pp. 24-6.
8   I.e. Falkland [66], Kingskettle [67], Dairsie [68] and Kellie [69] in Fife; and Coupar Angus [51] and Longforgan [52] in Gowrie.

grants of thanages were to close relatives of the kings, not typical run-of-the-mill crown patronage. Broadly speaking, the suggestions made here would answer the objections raised above, and also help account for the survival of so many thanages at the time of the feudalisation of Scotland. If the twelfth-century kings were reluctant to alienate their thanages, then the 71 of which we have record probably do give a good indication of the royal thanages at the beginning of the feudal era. Moreover, if the feudalisation of northern Scotland, especially beyond the Tay, was mostly at the expense of former mormaer/earldom lands, then that might illuminate the relationship between thanage and shire. The former mormaer/earldom lands — 'my manors ... of the earldom', as Malcolm IV put it for Gowrie — were no doubt shires; if these were extensively alienated in the twelfth century, they would not have been known as thanages when the latter term evolved around 1200. This would explain the fact that, while shires and thanages were obviously very similar and sometimes identical, many of the Scottish shires were never known as thanages.[1]

Whatever the truth of the above hypotheses, the salient point is that well over half the known thanages of medieval Scotland were still in existence at the end of the thirteenth century. What, however, of the thanes? Does the survival of the thanages imply the survival of thanes in the same measure? The answer is, not necessarily; despite the word-association, by the thirteenth century it was not automatic for thanages to be run by thanes. As was remarked earlier, where a sheriffdom was based on a thanage, the sheriff probably took over the thane's function (though Aberdeen is an exception). There are, moreover, cases of magnates' replacing thanes without having had feudal grants of the thanages: in 1266, for instance, Reginald Cheyne of Inverugie was fermer of Formartine [19] and Alexander Comyn earl of Buchan was bailie of Dingwall [1]; later, Dingwall was fermed by the earl of Ross, while Dull [56] was fermed by Alan Durward in 1264 and by Earl Duncan of Fife in 1288-9. On the other hand, where we have evidence of how surviving thanages were run during the thirteenth century, more often than not it was by thanes rather than by any other agents. There are references to individual thanes in 21 of the known thanages after 1200, of which 14 date from the second half of the thirteenth century or later.[2] The thanes' survival is highlighted by the case of

1 E.g. when William I founded Arbroath Abbey, he gave it the shires of Arbroath, Dunnichen and Kingoldrum, all in Angus (*RRS*, ii, no. 197). Dunnichen and Kingoldrum had belonged to the bishop of Caithness (*RRS*, ii, no. 223), but that endowment can hardly be earlier than *c*.1150 (cf. Duncan, *Making of the Kingdom*, pp. 258, 266).
2 I.e. the thanes of: Strowan [61], in 1200; Uras [28], in 1214 × 24; Dull [57] and Fortingall [56], in 1214 × 49; Idvies [44], in 1219; Callendar [70] (former), in 1233; Dunning [63], in 1247; Kintore [21], in 1253; 'Molen' [12], in 1261; Aberdeen [22], Glamis [46] and 'Rathenech' [10], in 1264; Forteviot [64], in 1266; Montrose [42], in 1268; Aberchirder [16] (plus Conveth [17]), in *c*. 1280; Cowie [27], in 1281; Cawdor [4], and Moyness [5], in 1295; Glentilt [55], in 1341 × 6; Cromdale [8], in 1367; and Brodie [6], in 1492.

Callendar [70]. In 1154 Ness of Callendar, presumably the thane, was killed for treason against Malcolm IV.[1] That gave an ideal opportunity for his replacement by a feudal lord; in fact, Callendar was still held by thanes in the early thirteenth century. Then, in about 1233, it was surrendered to Alexander II, who gave most of it to Holyrood Abbey in feu-ferme, for 160 merks a year — but left 40 librates (subsequently the barony of Callendar) with Malcolm, 'the former thane'. In effect, therefore, Holyrood took over the thane's function for about 70 per cent of the thanage, in return for a fixed ferme. But there was no question of simply removing the thane, who actually did very nicely, gaining the equivalent of a fair-sized knight's feu or barony.[2]

Clearly, therefore, the feudalisation of Scotland did not involve any deliberate anti-thane policy. What probably happened to the thanes was more prosaic. In general, their importance must have declined as a new upper class, distinguished by knightly rank and tenure, closely connected with the central royal establishment, and in the localities soon monopolising the new offices of sheriff, appeared in Scotland. But it does not follow that the thanes became obsolete. Twelfth- and thirteenth-century legislation shows them helping to supervise the payment of teind, swearing never to harbour or assist criminals, and performing and recruiting common army service; while Edward I's list of Scottish muniments in 1292 included '23 rolls of accounts of sheriffs, bailies, fermers, thanes, burghs, etc.'[3] Thus although they were no longer the crown's main local agents, thanes did retain a function in the localities.

There, they were certainly not insignificant. Consider the example of Ewen thane of 'Rathenech' [10] (to whom Geoffrey Barrow first drew attention). He was an active royal agent, who in 1264 supervised the construction of a new hall at 'Rathenech' and a boat to cross the Spey, in preparation for a visit by Alexander III, and provided a large granary and a malt-house for the king's household. He was also a local landowner, the fourth of his line to possess the land of Meft (in Urquhart parish) and fishing-rights in the Spey. He is found witnessing charters of Malcolm de Murray and Eva Murthac, lady of Rothes, in the company of the bishop of Moray, various lords of Anglo-Norman descent, and (in Eva Murthac's charter) his neighbour Dugal thane of 'Molen' [12] and another thane called Hugh. Also, in 1261, he and Dugal of 'Molen' headed a list of jurors in the Elgin sheriff court.[4] These two thanes were clearly well integrated into Moray's landowning society, of both native and Anglo-Norman stock. Their

---

1   He died in judicial combat following the accusation: *RRS*, i, p. 8.
2   *Holyrood Lib.*, no. 65. For thanes and barons of Callendar, see Appendix entry [70].
3   *RRS*, ii, no. 281; *APS*, i, p. 378, c. 20; 398, c. 2 (a particularly interesting illumination of their military role, from 1221); *APS*, i, pp.113-4 (and they no doubt featured in the further 185 account rolls of sheriffs and other ministers).
4   Barrow, 'Shires and thanes', p. 49; *ER*, i, pp. 10, 14; *APS*, i, p. 101; *Moray Reg.*, p. 461, and no. 125; *APS*, i, p. 99. In Eva Murtach's charter the name of Thane Hugh's thanage is, unfortunately, illegible.

appearance on the jury is interesting, since several other thanes occur in that context: for instance Macbeath thane of Falkland [66] in Robert of Burgundy's lawsuit in the 1120s, Gilys thane of Idvies [44] in an Angus lawsuit in 1219, John thane of (Old) Montrose [42] in a case before the justiciar at Forfar in 1250, and both Donald thane of Cawdor [4] and William thane of Moyness [5] on a jury in Nairn in 1295.[1] On each occasion, the thanes were in the company of leading nobles and of at least one *judex*, that other example of a pre-feudal office to survive into the thirteenth century and beyond.[2] Moreover, in the 1219 Angus lawsuit the thane of Idvies helped to supervise a perambulation conducted by a number of local Gaelic worthies, including the thane's brother Malcolm, who seem rather lesser men. The implication is that the thane still had a relatively high status. That was probably the case in general: in twelfth- and thirteenth-century Scotland from Perthshire to Moray, it would appear that thanes continued to enjoy quite a prominent role in what could be described as the communities of the sheriffdoms.

This would mean that the thanes were part of a hereditary landowning society — and, as argued earlier, they were presumably hereditary landowners in their own rights. Now one feature of office-holding in medieval Scotland was its tendency to become hereditary — witness the sheriffdoms. Although the Arbuthnott evidence indicates that thanages were not hereditary in the twelfth century,[3] there were no doubt strong pressures in that direction. There was also, perhaps, a desire for the same social standing as incoming 'Norman' lords. That was the case with at least one native Scots landlord in the later twelfth century: Swain son of Thor gave land in Perthshire to Scone Abbey 'as freely as any religious house in the whole kingdom holds of any baron', thereby asserting baronial status.[4] Other twelfth-century native lords such as Merleswain of Ardross (a 'king's knight') or Orm of Abernethy doubtless did the same.[5] At least one thane thought likewise: Malcolm thane of Callendar [70], for whom, as we have seen, Alexander II converted part of the thanage into an apparent knight's feu or barony. And earlier, Archibald of Longforgan [52] — probably the

---

1   *ESC*, no. 80; *Arbroath Lib.*, i, nos. 228, 250; *Moray Reg.*, pp. 471-2. Also, 'Malise of Idvies' was on a perambulation in 1254: ibid., i, no. 366.
2   G.W.S. Barrow, 'The "Judex"', in *Kingdom of the Scots*. For *judices* in the 14th century, see *RRS*, v, no. 9 (Fergus called Demster, in Nairnshire in 1309), and *RMS*, i, app. I, no. 29 (Duncan, in Angus in 1322). In the former case, Fergus was to pay the ferme of his lands to the thane of Cawdor; he was presumably the Fergus *judex* who was with the thanes of Cawdor and Moyness in 1295 (*Moray Reg.*, pp. 471-2), so it might be that here there was a close relationship between thane and *judex*.
3   *Spalding Misc.*, v, pp. 209-13; and above, p. 41.
4   *Scone Lib.*, no. 21; *Scots Peerage*, iv, p. 254. Swain was progenitor of the Ruthvens.
5   See, e.g., *RRS*, ii, nos. 137, 152. Neil MacYwar of Inverlunan probably had a similar status; he has been called a thane (*RMS*, ii, no. 590, and Barrow, 'Shires and thanes', p. 50), but Inverlunan was never described as a thanage, and his position was probably proprietorial, because in 1247 his widow possessed dower land in Inverlunan (Fraser, *Southesk*, ii, no. 26. Inverlunan is included in a 1322 list of lands owing second teinds to Restenneth Priory, but not as a thanage, whereas most of the other territories were described as such: *RMS*, i, app. I, no. 29).

thane or the thane's son — similarly entered the baronial classes when he married the heiress of William Maule and so gained the feu of Fowlis Easter, in Angus; it is likely that the marriage was given by William I as compensation for granting Longforgan to the earl of Huntingdon.[1]

With Malcolm of Callendar, there was probably a formal hereditary grant, but only of part of the thanage; was that a *quid pro quo* for surrendering the rest? Did something similar happen with other thanages? There is certainly one instance, from the early fourteenth century: in 1309 Robert I gave William thane of Cawdor and his heirs the thanage of Cawdor [4] in feu-ferme for 12 merks a year; the render had already been customary under Alexander III, so the purpose of Robert's charter must have been to grant secure hereditary tenure, probably in return for political support.[2] Much earlier, in 1238, a reference to feu-fermers in the neighbouring thanages of Moyness [5], Brodie [6] and Dyke [7] suggests that these thanages may already have been heritable then[3] — though feu-ferme tenure was not always hereditary. On the other hand, consider the case of Simon thane of Aberchirder [16] and Conveth [17] in the 1280s. It was recorded by an inquest held years later, in 1369, that Simon had offended King Alexander III, so that the king seized both thanages: 'and the thane, seeing that he could not recover [anything?] from the king, granted six davochs in Conveth to the earl of Buchan, in order that he could recover the other thanage of Aberchirder.'[4] The wording surely implies that Simon had less than full feudal ownership protected by the Scottish common law over Aberchirder and Conveth. He did, however, have a brother called William of Aberchirder;[5] thus he and his family must have had a close association with that thanage. Perhaps the relationship was something like that later known as kindly tenure: hereditary possession which was tacitly accepted but not guaranteed in law. That might well have been the case with John of Kinalty, who in 1305, during the English occupation, complained to the Westminster parliament that he had been denied sasine of the thanage of Kinalty [39], which (according to him) his late father had held hereditarily of the Scottish king.[6] Other instances of the names of thanages being used as surnames — such as Malise of Idvies [44] in 1250, Duncan of Downie [47] in 1254, Macbeth of Dyke [7] in 1262, Adam of Alyth in 1266 [49], and William of 'Rathenech' [10] in 1296 — perhaps denote the same.

At any rate, if, by the thirteenth century, thanes had generally established full or tacitly accepted hereditary rights over their thanages, then they were vulnerable to the great problem of hereditary tenure: failure to produce sons. Since thane was an office, it could only be held by men; but

1   *RRS*, ii, nos. 16, 205, 302-3; cf. ibid., i, p. 42, and K.J. Stringer, *Earl David of Huntingdon, 1152-1219: A Study in Anglo-Scottish History* (Edinburgh, 1985), p. 289 (n. 84).
2   *RRS*, v, no. 9.
3   *Moray Reg.*, no. 40.
4   Ibid., pp. 319-20.
5   Ibid., p. 280.
6   *Memoranda de Parliamento, 1305*, ed. F.W. Maitland (Rolls Ser., 1895), p. 191.

in the Middle Ages it was never easy to ensure succession by male heirs. Studies of late medieval noble families in England and France show that their direct male lines died out on average once every four generations, and in Scotland, although (especially in the fifteenth century) male extinction rates were not so high, the turnover in noble families was still significant.[1] For the thanes, the best comparison might be with the Scottish earls. In the course of the thirteenth and fourteenth centuries, all but two of the thirteen comital families either died out altogether or descended to females: five in the thirteenth, six in the fourteenth.[2] If something similar happened with the thanages where hereditary succession had been fully or tacitly established, then that succession would have been broken in almost half the cases during the thirteenth century, and in most of the others by the end of the fourteenth. The point may be illustrated by the examples of Archibald of Longforgan [52], who probably died childless in the 1190s, and Simon thane of Aberchirder [16], who seems only to have left a daughter in the 1290s;[3] they are unlikely to have been untypical. Now, if a thane left no sons, the crown could always have appointed another thane, or transferred the thanage to a collateral (Simon of Aberchirder had a brother). But it is equally — and perhaps, judging by the history of the earldoms, much more — likely that in those circumstances the thanage would be used for crown patronage and be given to a feudal landlord, either outright or in ferme, or (if the thane left a daughter) through marriage to the 'heiress'. In such cases (in contrast to the earldoms) the recipients of the thanages would not have called themselves thanes, and so fewer and fewer thanes would have remained. As with so many aspects of the history of the thanages, the argument is conjectural; but it is probably what happened. The steady yet far from absolute disappearance of the thanes is more likely to have been caused by natural wastage and succession by females than by any deliberate policy of sweeping them away on the part of the twelfth and thirteenth-century kings.

Although, when the term thanage came into contemporary use at the end of the twelfth century, it must have meant the territory of a thane, it is now clear that during the thirteenth century the two concepts diverged. Thirteenth-century thanages can be found being run by sheriffs, feu-fermers (including lay magnates and ecclesiastical corporations), and fermers, as well as by thanes. From the crown's point of view, of course, it would hardly have mattered who ran them, so long as they continued to provide

---

1   E. Perroy, 'Social mobility among the French *noblesse* in the later Middle Ages', *Past and Present*, xxi (1962); K.B. McFarlane, *The Nobility of Later Medieval England* (Oxford, 1973), pp. 141-70; A. Grant, 'Extinction of Direct Male Lines among Scottish Noble Families in the Fourteenth and Fifteenth Centuries', in *Essays on the Nobility of Medieval Scotland*, ed. K.J. Stringer (Edinburgh, 1985).
2   Details are in *Handbook of British Chronology* (3rd edn, Royal Historical Soc., 1986), pp. 499-521.
3   *RRS*, ii, nos. 302, 338, show that Archibald of Longforgan's marriage to the heiress of William Maule of Fowlis Easter was childless and had ended by 1194 at the latest. *RRS*, v, no. 339, shows that Simon of Aberchirder's heiress was his daughter, Sibilla.

royal revenue and bases in the localities. That was probably as important to the twelfth- and thirteenth-century kings as to their pre-feudal predecessors.

The almost complete loss of account rolls for the period means that a full analysis of the thanages' contribution to royal finance in the later twelfth and the thirteenth centuries is impossible. But the survival of conveth, or waiting, from Forfar, Glamis, Kinross, Fettercairn and Aberchirder has already been seen.[1] By the later thirteenth century, however, the levy of waiting does not appear common. The surviving accounts, scrappy as they are, make it clear that waitings were levied as a matter of course only on one or two thanages per sheriffdom, and that by no means all the produce was used in the royal household: most of the cheese, for instance, was sold for cash. Indeed the 'royal service' on which the cattle and about half the grain from Forfar and Glamis was expended in the mid-1260s is as likely to have been the provisioning of castles and support of local agents, especially sheriffs, as the feeding of the court. Perhaps, therefore, in the thirteenth century certain thanages were earmarked for maintaining the crown's establishments in the sheriffdoms.

In most thanages, however, the revenue owed to the crown was no longer known as cain and conveth. The terminology was already changing in the mid-twelfth century: whereas David I referred to his cain from the *maneria* of Gowrie, Malcolm IV gave Scone a tenth of his *redditus* of wheat from one of them, Longforgan, or, if that estate were set at a different kind of ferme, then a tenth of that ferme.[2] From then on, *redditus*, or renders, and fermes became the usual terms for crown revenues from thanages. Cain and conveth had probably been generally consolidated into a single rent — which increasingly, as with most Scottish estates of the period, was to be paid in cash.[3] That was, no doubt, particularly the case with thanages which were held at ferme, either temporarily or permanently (feu-ferme). It has been suggested that, in thirteenth-century Moray at least, thanes still paid the thanage dues in kind whereas the fermers of thanages did so in cash;[4] but that does not seem to be generally applicable, for thanes can be found paying cash, too.[5] Perhaps the distinction came to be that whereas fermers of thanages only had to pay in cash, a thane might also be required to provide hospitality, or waiting, if the king actually visited the thanage. That is suggested by the fact that the thane of 'Rathenech' handed over a large grange and malthouse to the royal clerks of the liverance and provender in 1264, without any payment apparently having been made.[6]

---

1   Above, p. 48.
2   *RRS*, i, nos. 57, 243, 248.
3   Duncan, *Making of the Kingdom*, pp. 392-9. When the renders owed by thanages are recorded, they are mostly single sums of money.
4   Ibid., p. 393.
5   E.g., the thanes of Aberdeen and Kintore in 1264 (*ER*, i, p. 11); and although William of Cawdor feued his thanage from Robert I for the accustomed ferme of 12mk, he and his descendants were always known as thanes. In general, thanages with thanes are recorded as owing lump sums in cash just as those in the hands of fermers did.
6   *ER*, i, p. 14. On other hand, the sheriffs footed the bill for repairs at thanages after royal visits: *ER*, i, pp. 3, 8, 14.

Whether the revenues produced by thanages were in kind or in cash, they were certainly substantial. Under Alexander III the value of the waitings from Forfar and Glamis (together, some £75-85) and Aberchirder (£40, plus an additional £13 6s.8d. in cash) was probably equivalent to the rents from a fair-sized knight's fee or barony; in 1264 proceeds from the earldom of Angus, then in crown hands, were only £80.[1] During the 1260s, the ferme of Tannadice thanage [40] was £60; the receipts of the bailie of Inverquiech castle, almost certainly levied from Alyth thanage [49], were £53 6s.8d; and even when most of the four royal *maneria* of Gowrie had been alienated, the remnant, little more than a part of Scone thanage [53], still seems to have been valued at some £75 a year.[2] Earlier in the century, Callendar thanage [70] had probably been worth well over £100 a year.[3] Later, revenue from the *manerium* of Dull [56] (perhaps incorporating the thanages of Dull and Fortingall [57] and the *abthen* of Dull) was assessed at no less than £503 13s.4d. for 1288-9;[4] that is over £250 a year. But, most interestingly, in 1290 the ferme owed from Menmuir thanage [37] was reduced by £33 6s.8d. a year for the previous two years, 'because the ferme can in no way be levied because of the poverty of the bondmen living on the said land, as the chamberlain and all the neighbourhood testify; the sum of fifty merks a year was added in hatred of the said bondmen by Sir Hugh of Abernethy'. Too much was obviously being demanded from a fairly small thanage; nevertheless, the £40 customarily paid from Menmuir was still a sizeable sum.[5]

These are isolated examples, but we also have a more general picture for the sheriffdoms of Aberdeen and Banff. Some time before 1157 the bishops of Aberdeen were granted a tenth of all the royal dues 'which are between the waters of Dee and Spey';[6] because of subsequent disputes over these 'second teinds', various statements about royal finances in the region were preserved by the bishops, in particular the summary of a rent roll of Alexander III.[7] That gives a good indication of the value of crown lands in the two sheriffdoms during the later thirteenth century. And the figures may be put in perspective by comparing them with a statement of the value of all the Scottish sheriffdoms, which immediately precedes the treatise on 'The Scottish King's Household' in an English manuscript of material concerning Scotland; the treatise is generally dated to *c*.1292, and the 'Valuation of the

1   *ER*, i, p. 9.
2   Scone Abbey's teind of the royal revenues from what was left of the four *maneria* was £7 9s. in 1263-6: *ER*, i, pp. 2, 18. See also *ER*, i, pp. 3-4, 9.
3   In 1234 it was divided into £40-worth of land which was kept by the former thane, and the rest which was feued to Holyrood Abbey for 160mk, making a total of £146 13s.4d.
4   *ER*, i, p. 48.
5   *ER*, i, p. 50; *RMS*, i, no. 214 (second teinds on the crown's revenue from Menmuir were £4).
6   *Aberdeen Reg.*, i, pp. 5-7; cf. *RRS*, i, no. 116. Under Alexander II this was confirmed as a tenth of the fermes of all thanages and proceeds of royal courts in the two sheriffdoms: *Aberdeen Reg.*, i, pp. 17-18.
7   Ibid., i, pp. 55-68; see also i, pp. 50, 54, 57-8, 69-70, 72-4, 86-7, 138-41, 155-70.

Sheriffdoms' probably comes from the same date.[1] The revenues from thanages and other crown lands in the two sheriffdoms, according to the 'Alexander III Rental', and the totals noted in the 'Valuation of the Sheriffdoms', are as follows:

		Aberdeen			Banff	
'Alexander III Rental':						
	Aberdeen:	£33	6s.8d.	Aberchirder:	£53	6s.8d.
	Aboyne:	£134	13s.4d.	Boyne:	£113	
	Belhelvie:	£62	13s.4d.	Conveth:	£53	6s.8d.
	Formartine:	£106	13s.4d.	Glendowachy:	£20	
	Kintore:	£67	6s.8d.	Mumbrie:	£34	8s.8d.
				Netherdale:	£49	1s.4d.
	Totals:	£456	13s.4d.		£323	3s.4d.
	(Other land:	£28	0s.4d.	Other land:	£174	13s.4d.)

'Valuation of Sheriffdoms':

£707 16s.0d.   £414   2s.0d.

How reliable are these figures? According to what survives of the 1264 sheriff's account for Aberdeen, the thanes of Aberdeen and Kintore paid £12 and £17 13s.4d. respectively; but since the king had been at both places that year, other amounts (perhaps in kind) had presumably been delivered directly to the chamberlain or household clerks.[2] More significantly, in 1266 Reginald Cheyne, the fermer of Formartine, paid 14½ merks as second teind from the thanage,[3] which means that the ferme was then 145 merks or £96 13s.4d., only £10 less than in the 'Alexander III Rental'; and later in the century the earls of Buchan paid 80 merks for Conveth, the same as the amount given in the 'Rental'.[4] That suggests that it is reasonably accurate. In general, what emerges strikingly is how high a proportion of crown revenue was derived from thanages (the 'other land' total for Banff was made up chiefly of £120 from the barony of Rothiemay, probably temporarily in crown hands, and £40 from the burgh of Banff itself). When the thanage totals are set against those from the 'Valuation of the Sheriffdoms', they work out at 65 per cent and 78 per cent respectively of the crown revenues from the two sheriffdoms — which probably included the profits of justice. All the sums are, of course, misleadingly precise, and probably indicate gross rather than net amounts. Nevertheless, it seems safe

1   *SHS Misc.*, ii, pp. 24-6. The inclusion of the sheriffdoms of Dingwall, Cromarty, Nairn, Forres, Elgin and Auchterarder as contributing to the royal revenues indicates a date before the reign of Robert I.
2   *ER*, i, pp. 11-12.
3   *Aberdeen Reg.*, i, p. 158.
4   *CDS*, ii, no. 1541, taking 'Covenache', said to be held for 80mk and to owe 8mk second teind to the bishops of Aberdeen, to be Conveth.

to conclude that in the sheriffdoms of Aberdeen and Banff, during the later thirteenth century, thanages probably provided well over half, and perhaps over two-thirds, of the crown's regular proceeds from land. And there is no reason to believe that the situation would have been very different in other sheriffdoms north of the Forth. According to the 'Valuation of the Sheriffdoms', at the end of the thirteenth century crown revenue from northern Scotland amounted to £4,591 out of a total £8,100. Thus the north was worth more to the crown than the south; and in the north, the bulk of the regular revenue probably came from the thanages. Little wonder that so many of them stayed in crown hands throughout the century.

As has been emphasised already, however, the value of the royal thanages was not simply in their rents. In the twelfth and thirteenth centuries, several continued to be royal residences, albeit on an occasional basis. The earliest surviving fragments of Exchequer rolls mention, in passing, visits by Alexander III to Scone [53], Kinclaven [54], Forfar [45], Cowie [27], Durris [26], Kintore [21], Aberdeen [22] and 'Rathenech' [10] in 1264.[1] More generally, under William I, Alexander II and Alexander III, a total of 249 *acta* are known to have been issued in the region beyond the River Tay, from 27 different places: 13 of the places were thanages, and 189 of the *acta* (76 per cent) were issued from them. Admittedly the most common place-dates, Forfar (73) and Scone (41), were also administrative and ecclesiastical centres, but they were both essentially royal bases, to which royal demesnes were still attached; and even if they are discounted from the calculation, a good majority of the remainder was issued from thanages.[2] It is particularly striking with Alexander III's *acta*, for ten out of 13 place-dates beyond the Tay were thanages, accounting for 43 out of 48 *acta*.[3]

Place-date analysis, of course, is skewed towards the major administrative centres, where formal documents were most likely to be issued. William I and the two Alexanders probably used at least some of their thanages for relaxation, in the favourite royal pastime of hunting. The pre-feudal kings presumably hunted anywhere in their estates, but from David I's reign the best hunting areas were set apart and formalised into royal forests or enclosed deer-parks. In northern Scotland, most of these were associated with thanages; there were forests at Kinross [65], Alyth [49], Cowie [27], Durris [26], Birse [25], Aberdeen [22], Kintore [21],

1   *ER*, i, pp. 3, 9, 12, 14. Cowie and Durris are probable visits; at Aberdeen, the impression is that it was the thanage not the burgh that was visited.
2   Totals from *RRS*, ii, pp. 28-9 (+ Scone); the map of place-dates of Alexander II's charters prepared by Dr Keith Stringer (to whom my thanks); and G.G. Simpson, *Handlist of the Acts of Alexander III, the Guardians and John, 1249-96* (Edinburgh, 1960). The thanages and acts issued from them are: Forfar [45] (73), Scone [53] (41), Montrose [42] (21), Kincardine [30] (12), Aberdeen [22] (10), Alyth [49] (9), Kintore [21] (8), Kinclaven [54] (4), Aboyne [24] (3), Durris [26] (3), Formartine [19] [Fyvie] (2), Kinnaber [36] [Charleton] (2), and 'Rathenech' [10] (1). Acts were also issued from the former thanages of Coupar Angus [51] (2) and Longforgan [52] (1).
3   22 of these were issued at Scone; nevertheless the preponderance of thanages is clear.

Formartine [19], Boyne [13], and perhaps Kinclaven [54], while Kincardine [30] had a large enclosed deer-park.[1] In the fourteenth century these appear as separate from the thanages, but earlier they may have been part of them; it seems, for instance, that at Kinross the forest and thanage lands were separated by a special inquest in 1323.[2] In addition, the large forest of Clunie, adjacent to Strathardle thanage [50], perhaps represented the main hunting area in the royal earldom of Gowrie; the forest of Plater, near Forfar [45] and Glamis [46], probably provided hunting for the Angus thanages;[3] and to the north the forests of Inverness, Forres and Elgin were no doubt similarly connected with at least some of the thanages in Moray. Judging by the place-dates of royal *acta*, Clunie, Alyth, Durris and Kintore were perhaps the kings' most favoured forests, together with Kincardine, where Alexander III had a new park constructed;[4] but most hunting may well have been done from Forfar, where William de Hamyll, the king's falconer, spent 29 weeks in 1263.[5]

At times, however, the kings did not so much go *to* the thanages as pass through them, using them as staging-posts on journeys. The overall impression of the place-dates of *acta* issued beyond the Tay is that when the kings were in northern Scotland they almost invariably lodged in their own burghs or thanages. Elsewhere in this volume Alan Young has pointed to the strategic value to the crown of the Comyn lordships in the north;[6] but these were complemented by the kings' own, equally strategic, estates. That was not only convenient, but politically useful; together, thanages and burghs (which often originated in thanages) provided the essential stepping-stones for routes through the north, and especially into Moray. When Alexander III travelled to Inverness in 1264, his route was stated to have been through Aberdeen [22] and Kintore [21], and thence (probably via Formartine [19] and the Banff thanages) to 'Rathenech' [10] on the Spey, where a new boat was built for his crossing.[7] Thus, in a concrete way, the thanages provided some crucial building blocks for the twelfth- and thirteenth-century kingdom of the Scots. But the most striking evidence of the use of thanages in journeys to the north comes from 1296, when Edward I 'conquered the realm of

---

1   See, in general, J. Gilbert, *Hunting and Hunting Reserves in Medieval Scotland* (Edinburgh, 1979), chs. 1, 2 (though Kinross and Formartine are missed), and pp. 338-41; also pp. 82-5, 215, for Kincardine. It is pointed out on p. 215 that Kincardine park was not created out of forest lands; what probably happened was that areas of royal lands, mostly in thanages, were set aside for hunting either as forests or as parks.
2   *RMS*, i, app. II, no. 697.
3   The second teind of royal revenues from the sheriffdom of Forfar, which went to Restenneth Priory under Alexander III and probably earlier, included a tenth of the hunting from Plater, which indicates that it was the main (only?) royal forest in Angus: *RMS*, i, app. I, no. 29.
4   *ER*, i, p. 21; Gilbert, *Hunting and Hunting Reserves*, pp. 82-5.
5   *ER*, i, pp. 7-8.
6   See 'The Earls and Earldom of Buchan in the Thirteenth Century', below, p. 178.
7   *ER*, i, pp. 12, 14; cf. Duncan, *Making of the Kingdom*, p. 189: 'the main [route to Moray] lay further east from Aberdeen by Kintore, Inverurie and Fyvie [in Formartine] to the coast at Banff'. The building of the boat might suggest that the royal presence at 'Rathenech' was unusual; but Alexander was there in 1261 as well (*APS*, i, p. 99).

Scotland, and searched it ... within 21 weeks, without any more'; the northern part of his itinerary,[1] which no doubt followed the footsteps of successive Scottish kings, is as follows (known thanages are given in italics):

Stirling — *Auchterarder* — Perth — *Kinclaven* — Clunie forest — Inverquiech (=*Alyth*) — *Forfar* — Farnell — *Montrose*[2] — *Kincardine* — Glenbervie — *Durris* — *Aberdeen* — *Kintore* — Fyvie (=*Formartine*) — Banff — Cullen (=*Boyne*[3]) — Enzie — 'Rathenech' — Elgin — Rothes — Inverharroch — Kildrummy — *Kincardine O'Neil* — *Kincardine* — Brechin — Arbroath — Balledgarno — Dundee — Perth.

Having helped to construct the kingdom, the thanages saw its collapse.

After 1296 the kingdom was, eventually, reconstructed — but, in effect, without the thanages. Although Robert I's *acta* stressed continuity with his predecessors,[4] the reality of his land policy was very different: despite the vast forfeitures after 1314, he actually granted away much more than he acquired.[5] Among these grants were many of thanages. The most sweeping alienation of his reign took place when he recreated the earldom of Moray for Thomas Randolph, 'cum omnibus aliis maneriis burgis villis thangiis et omnibus terris nostris dominicis firmis et exitibus'; that included nine thanages, which, judging by a Moray account roll from 1337, produced the bulk of the earldom's revenues from land.[6] As for the one thanage beyond Moray, Dingwall [1], that was given to William earl of Ross, while Glendowachy [18] near Banff went to his son and heir, Hugh. Elsewhere, the king's long-term companion and brother-in-law Alexander Fraser became a north-east magnate through grants of the thanages of Aboyne [24] (initially at ferme but later hereditarily), Cowie [27],[7] and (almost certainly) Durris [26], together with six acres beside the royal *manerium* of Kincardine [30] and pasture rights in that thanage; while his son John Fraser received the thanage of Aberluthnott [33], apart from 31 merklands which went to John Menteith. Grants of thanages also brought another well-known family, the Grahams,

---

1   Stevenson, *Documents*, ii, pp. 28-31; *An Historical Atlas of Scotland, c. 400-c. 1600*, ed. P. McNeill and R. Nicholson (St Andrews, 1975), map 35.
2   Although the burgh of Montrose ('a castle and a good town') was meant, the journey from Farnell to Montrose involved travelling through the thanage of Old Montrose [42] (and then, from Montrose, through Kinnaber thanage [36]).
3   The original thanage (and forest) of Boyne probably covered most, if not all, of the area between Banff and Cullen.
4   See Norman Reid's essay, 'Crown and Community under Robert I', below, pp. 203ff.
5   For discussion of Robert I's land grants, see G.W.S. Barrow, *Robert Bruce and the Community of the Realm of Scotland* (3rd edn, Edinburgh, 1988), pp. 270-86; cf. A. Grant, *Independence and Nationhood: Scotland 1306-1469* (London, 1984), pp. 26-8.
6   *RMS*, i, app. I, no. 31; *ER*, i, pp. 440-4. Thanages (though not actually described as such) produced significantly higher revenues than other lands. In the 'nine thanages' I have counted Fochabers (which was included in the account) and Cawdor, from which the 12mk feu-ferme now went to the earls; but not Kinmylies and Cromdale, which had been alienated by previous kings.
7   Also the forest of Cowie, in return for making a royal park within the thanage.

into Angus: David Graham was given the demesnes and tenancies of Kinnaber [36], and subsequently all the near-by land of Old Montrose [42]. Then there were the following grants of individual thanages:[1] Formartine [19], mostly to John Brown, partly to Mary Comyn; Belhelvie [20], mostly to Hugh Berclay, partly to John Bonneville; Clova [38], to the earl of Mar; Aberlemno [41], to William Blount; Downie [47], mostly to Walter Bickerton, partly to William Dishington; Monifieth [48], to various men, chiefly William Durham; Scone [53], to Scone Abbey; Fortingall [57], to Thomas Menzies; and Auchterarder [62], to William Muschet. In addition, Robert I alienated parts of Alyth [49] and Forteviot [64]; granted out 20 merks from the annual revenues of Netherdale [15] and 10 merks from Aberchirder [16]; and gave the whole 128-merk feu-ferme owed by Holyrood Abbey from Callendar [70] to his grandson Robert the Steward. It should be added that some of these alienations may not have been absolute, and that in several cases the recipient had to pay an annual ferme or feu-ferme: no less than £100 from Aboyne, £20 from Dingwall, probably £10 and four chalders of wheat from Glendowachy,[2] and £20 from Clova. Also, it is just possible that with some of the grants of thanages, the original renders were still payable; certainly revenues were subsequently received from Belhelvie [20], Cowie [27] and Downie [47].[3] Nevertheless, the diminution of the royal possessions in northern Scotland was dramatic. In one way or another, no fewer than 33 thanages contributed to Robert I's grants of patronage.

This continued under Robert's son David II. During the first part of his adult reign, between 1341 and 1346, David included Aberdeen thanage [22] and the feu-ferme from Aboyne [24] in the provision made for his queen; endowed his elder surviving sister Margaret and her husband William earl of Sutherland with Downie [47] in Angus, Kincardine [30], Fettercairn [31] and Aberluthnott [33] in the Mearns, and half of Formartine [19] and Kintore [21] in Aberdeenshire;[4] and gave the rest of Formartine and Kintore to his other sister Matilda and her husband Thomas Isaac. At the same time, Tannadice [40] went to Peter Prendergast; much of Kinross [65] to John Bruce; Glamis [46] (probably) to the countess of Wigtown; and a 20-merk annuity from Morphie [35] to David Fleming. By 1346, therefore, David II had disposed of ten thanages plus revenue from two others; admittedly most of the grants

---

1   See Appendix for details of the grants and for references.
2   The *reddendo* in Hugh Ross's charter of Glendowachy was eight horses for carting per davoch, annually (*RMS*, i, app. I, no. 5); but subsequently £10 + 4 chalders of wheat were paid (*ER*, i, pp. 548, 551). Since, in the 'Alexander III Rental', Glendowachy owed £15 + 5 chalders of wheat, it looks as if most of the original render had been restored (alternatively, it is just possible that when grants of thanages were made, the original renders continued to be payable even if they were not mentioned in the charters).
3   The fact that so many of Robert I's grants are only known from the 16th/17th-century summaries of the Great Seal Register means that in many cases we cannot be sure of the terms of the grant.
4   Downie, Aberluthnott, and Formartine, all the subject of Robert I grants, must have escheated to the crown; that also applies in the other cases, below, of thanages being alienated for a second time.

were for his immediate family, but they still constituted a significant reduction in what was left of the crown lands. And, following his return from English captivity in 1357, despite acts of revocation and prohibitions on the alienation of royal possessions, coupled with apparent efforts by the chamberlain and auditors to exact maximum revenue from the thanages,[1] the process of disposal carried on. After the death of the queen, half of Aberdeen went to John Herries, and William Keith the marischal (husband of the Fraser heiress) seems to have had the payments from Aboyne cancelled.[2] In 1364, after marrying Margaret Logy, David gave Tannadice[3] and the reversion of Glamis to her son John Logy; Queen Margaret herself got the *abthen* of Dull [56], though by 1368 she had given it to her son. In 1366, Alexander Lindsay received Newdosk [32]; in 1369, John Edmonstone received Boyne [13], while William Cunningham got Kinclaven [54]; and in 1370 Walter Leslie was granted both Aberchirder [16] and the reversion of the earl of Sutherland's three thanages in Mearns (which after his wife's death Sutherland only held for life). In all, therefore, 11 thanages were the subject of David's grants in the second part of his reign.

If the thanage system had been sacrificed on the altar of Bruce patronage, the first Stewart king, Robert II, delivered the *coup de grâce*: when previously alienated thanages came back into crown hands, they were swiftly granted out again. Thus when Kintore [21] and Formartine [19] reverted on the death of the earl of Sutherland, the former went to John Dunbar earl of Moray, the latter to the heir to the throne, John earl of Carrick, and subsequently to James Lindsay of Crawford. Lindsay was also granted much of Alyth [49], including the site of what had been the castle of Inverquiech, with previous tenants being compensated elsewhere.[4] Meanwhile, Lindsay's uncle, Alexander, added Downie [47] (another Sutherland reversion) to Newdosk, which he had had from David II. The thanages of Glamis [46] and Tannadice [40] also came back to the crown when John Logy lost what he had been granted, either at the end of David II's reign or at the beginning of Robert II's; both were bestowed on one of the fastest-rising men of the period, John Lyon. And what was in effect much of the thanage of Kinross [65], which probably reverted some time in the 1360s, was granted to one of Robert II's younger sons, David earl of Strathearn. That was a grant to a legitimate son: two of Robert's bastards, John and James Stewart, were endowed with the king's lands in the thanages of Kinclaven [54] and Forteviot [64] respectively; while their mother, Marjorie Cardeny, seems to have been given half of Aberdeen thanage [22].[5]

The fourteenth-century Scottish kings' policy of alienating their property

1   Evident, e.g., from the sheriffs' accounts surviving from 1358-9: *ER*, i, pp. 542-93.
2   *RMS*, i, app. II, no. 1362: 'To William Keith ... and Walter Moygne, of the arrearage and annuels of Oboyn....' Moigne presumably had the remission of the arrears, for which, as sheriff of Aberdeen, he would have had to account.
3   Which, after presumably escheating from Peter Prendergast, had been held in 1359 by John Ramsay.
4   *RMS*, i, no. 630.
5   In the 1370s she shared Aberdeen with John Herries: *Aberdeen Reg.*, i, p. 56.

was not confined to the thanages; almost every piece of crown land, both north and south of the Forth, appears to have been granted away. That is the policy which John of Fordun was, in effect, criticising in his account of Malcolm II:

> Since Malcolm had kept back nothing for himself ... he was compelled by necessity to ask them that a definite income in keeping with his position as king should be provided, either in the form of lands or revenues or at least a sufficient annual subsidy.... So it seems that King Malcolm although great-hearted both in peace and war distributed his possessions recklessly ... because he squandered extravagantly without setting any limit to his generosity not just a part of his possessions but rather the whole of them, keeping nothing for himself. Generosity is rather thoughtlessly employed where it entails the inevitable asking back of the gifts. So Bernard says: 'If the man who makes his own share smaller is stupid, what is he who utterly empties himself and reserves no part of his goods for himself?'[1]

The parallels with the sweeping grants of land and the levying of compensatory taxation to maintain the royal household under Robert I, David II and Robert II are obvious[2] — as is what Fordun thought of them.

There is an obvious explanation for the royal grants, too: the need for the new regime not only to reward support but to ensure that the Scottish nobility had a stake in it. As the century went on, moreover, the customs on wool exports appeared a better basis for financing the crown (though that was to change after 1385). And, particularly with respect to the thanages, economic factors were probably significant as well. The Wars of Independence were accompanied by economic contraction and currency shortage; then the Black Death struck. By 1366 the total valuation of the country's lands had fallen by 48 per cent compared with that under Alexander III[3] — a fall almost exactly reflected in the fermes of Tannadice [40], which were £60 in 1263 and £30 a century later.[4] Now since, in practice, the gross proceeds of thanages were shared by the thanes, fermers or feu-fermers on the one hand and by the crown on the other, an approximate halving of those proceeds would have serious effects; if the crown insisted on maintaining its full payments, ferming a thanage would no longer be worth while economically; but if the thanes or fermers kept their

---

1   *Chron. Fordun*, i, pp. 186-7; translation from *Chron. Bower* (Watt), ii, p. 417.
2   See, in general, Grant, *Independence and Nationhood*, pp. 162-4; R. Nicholson, *Scotland: The Later Middle Ages* (Edinburgh, 1974), pp. 114-5, 174-8, 187-8, 217-8; and Duncan, 'Laws of Malcolm MacKenneth', below, pp. 242-4. The passage from Fordun looks like a summary, verbatim in places, of parts of the indenture for Robert I's 'tenth penny' made at the 1326 parliament: *APS*, i, pp. 475-6.
3   Grant, *Independence and Nationhood*, pp. 26-8, 75-82, 163-4; N. Mayhew, 'Alexander III — A Silver Age? An Essay in Scottish Medieval Economic History', in *Scotland in the Reign of Alexander III, 1249-1286*, ed. N.H. Reid (Edinburgh, 1990).
4   *ER*, i, p. 9; ii, p. 153.

own incomes near the former levels, then the crown would receive very little. The latter seems to have happened; where thanages were still providing royal revenue, the sums were significantly less than in the thirteenth century.[1] Callendar [70] illustrates the process. Since 1234 most of it had been feued by Holyrood Abbey for 160 merks a year; but in 1364, because of the effects of wars and pestilences, Robert Stewart (to whom Robert I had granted the feu-ferme) cancelled the fixed payment and agreed that the monks should simply pay a fair assessment of the estate's value each year.[2] In those economic circumstances, it probably appeared more sensible to gain immediate political benefit by using the thanages and other crown lands as a source of patronage rather than revenue.[3]

At the same time, the geographical focus of royal interests shifted. The trend perhaps started with John Balliol, who alienated the major thanage of Formartine [19], and also probably Belhelvie [20], to John Comyn earl of Buchan, in return for the latter's renunciation of his claim to inherit part of Galloway, Balliol's ancestral lordship;[4] earlier kings might not have considered that a good bargain. King John was the first Scottish king from the south-west, where the Bruce and Stewart heartlands also lay. It was there, to Cardross on the Firth of Clyde (which he acquired from David Graham in exchange for the lands of Old Montrose thanage [42]), that Robert I retired in his last years. David II, admittedly, seems to have been happiest at Edinburgh, Perth and Scone, but Robert II's preferred residence was, if anywhere, probably Dundonald in Ayrshire, while in the early fifteenth century Robert III seems to have withdrawn almost entirely to Rothesay on Bute. And although Robert II did make regular summer hunting trips to the Highlands, these were more commonly to Kindrochit, which belonged to the earldom of Mar, than to forests associated with royal thanages.[5]

The shift in the crown's orientation is highlighted by a charter of Robert II dated 1372: the lands of 'Balmoschenore' and 'Tyrebeg' had been held

---

1   Had full sheriffs' accounts survived, we could be clear about this point; as it is, that is the impression given by the available evidence. Figures, such as they are, for individual thanages are given in the Appendix.
2   *RRS*, vi, no. 294. Robert I had reduced the feu-ferme to 128mk, but that was a rationalisation exercise, because he had also cancelled a former royal grant to Holyrood of 32mk *p.a.* (*RRS*, v, no. 346; vi, no. 294).
3   In other words, in modern jargon, a policy of privatisation.
4   Maitland, *Memoranda de Parliamento, 1305*, p. 190 (also *CDS*, ii, no. 1541). The text, a petition to Edward I, refers to 'terra theinagii de Fermartyn et de Dereleye'. *Aberdeen-Banff Colls.*, p. 504, mentions a 'Derley' near Fyvie in 1503, but a thanage of 'Dereleye' is unknown; had one existed in Aberdeenshire or Banff, it would surely have been recorded in the documents relating to second teind preserved for the bishops of Aberdeen. The likelihood is that the English clerk garbled Buchan's petition, writing 'Dereleye' for a form of Belhelvie, the neighbouring thanage of Belhelvie, and writing the unfamiliar word *theinagium* in the singular.
5   Barrow, *Robert Bruce*, pp. 319-21; *RRS*, vi, at end, 'Map of places of issue of Acts of David II'; my comments on Robert II and Robert III are based on the material collected for the forthcoming volume of their *acta*, *RRS*, vii. For Robert II's hunting, Gilbert, *Hunting and Hunting Reserves*, p. 36; possibly the extension of cultivation earlier in the 14th century had meant that the best hunting was now in more remote areas (cf. ibid., p. 257).

for an annual render of 300 cartloads of peat at the royal *manerium* of
Forfar, but now, 'because in these times we do not make our residence
at Forfar as often as our predecessors', the render was changed simply to the
provision of sufficient fuel whenever a king happened to go there.[1] The
general implication is that the north no longer mattered quite so much to the
Scottish kings. Instead of maintaining its network of royal thanages to
provide a royal presence and balance noble power in the region, overall
power in the north tended to be relaxed into the hands of a succession of
magnates, most notably Thomas Randolph earl of Moray under Robert I,
Alexander Stewart earl of Buchan (the 'Wolf of Badenoch') under Robert II,
and in the fifteenth century Alexander Stewart earl of Mar, who 'ruled with
acceptance nearly all the north of the country beyond the Mounth'.[2] Given
the problems caused by the province of Moray in the later Middle Ages, it
might be asked whether the alienation of the thanages was a contributory
factor. The whole issue is, of course, infinitely more complex than that;[3] but
the contrast between the policies with respect to thanages followed by the
thirteenth-century and the fourteenth-century kings is certainly worth
pondering in that context.

Robert II's statement in 1372 that 'in these times we do not make our
residence at Forfar as often as our predecessors' would make an apt epitaph
for the Scottish thanages — were it not for the fact that in spite of their
disposal by the crown they did not die out altogether. Robert I's alienation
of the Moray thanages actually meant that some, such as Cawdor [4] or
Brodie [6] survived as thanages within the earldom. The thanages of
Crannach [58] and Glentilt [55] similarly survived within the earldom of
Atholl. Also, although many of the former royal thanages eventually became
submerged among the mass of small pieces of land or touns which
constituted the larger estates and baronies, equally many were themselves
erected into baronies and so retained their individual identities, if not their
original functions. In the fourteenth century, the lands of 33 thanages can be
found being held as baronies.[4] And some of these continued to be called

---

1   *RMS*, i, no. 514.
2   *Chron. Bower* (Watt), viii, pp. 292-3.
3   Especially in view of the Moray problem in the 12th and the earlier 13th centuries;
but that was of a different nature, and it is likely that the string of royal thanages was one of
the factors which helped the earlier kings to overcome it.
4   This is noted in the individual Appendix entries. The equivalence between former
thanage and new barony, however, was not necessarily exact (as, e.g., Reid, 'Barony and
Thanage', pp. 179-82, implies). Surviving charters of thanages in free barony generally
show the king granting 'all our lands of the thanage of X'; that would exclude already
alienated parts of the thanages. Many of these new baronies were thus only the remnants of
the original thanages. Sometimes the limitations are even clearer: the barony which Robert I
created for Hugh Berclay consisted of 'the demesne toun of the thanage of Belhelvie', the
lands of Lower and Upper Westertoun, Keir, and Eigie, and the cain of the kirkton; while
Blairton and the davoch of 'Menie' went to John Bonneville: *RMS*, i, app. I, no. 7; app. II,
no. 414. When Kintore became a barony, the land of Thainston was excluded (perhaps this,
the thane's own toun, was left with the descendants of the original thanes): *RMS*, i,
nos. 627, 718; cf. *RRS*, vi, no. 261. Conversely, extra land was sometimes included.
Blairshinnoch (six miles distant) was, for instance, added to Aberchirder: *RRS*, i, no. 339.

thanages even after becoming baronies. Such was the conservatism of the Scottish landowning system, indeed, that 11 of the former thanages were still being described as such at the end of the seventeenth century;[1] Glendowachy, moreover, carried with it the 'Thaynisnet', fishing rights of a salmon net on the River Deveron, as late as 1695.[2] Among them were the two most famous thanages of all, Cawdor [4] and Glamis [58]. When Hector Boece rewrote the story of MacBeth, he presumably used for the weird sisters' prophecy the names of two thanages which were familiar in sixteenth-century Scotland. The sequence of ranks in Boece's story, thane of Cawdor, thane of Glamis, king of Scots, makes no sense in the context of MacBeth's own kin-based society[3] — but it did ensure that thanks to the genius of William Shakespeare those two thanages, at least, would be immortal.

1   As the *Retours* bear witness. The latest reference to each 'thanage' in the *Retours* is: Aberluthnott, 1667 (Kincardine, no. 111); Boyne, 1700 (Banff, no. 171); Cawdor, 1662 (Inverness, no. 89); Crannach, 1640 (Perth, no. 494); Fettercairn, 1693 (Kincardine, no. 161); Glamis, 1695 (Forfar, no. 536); Glendowachy, 1695 (Banff, no. 161); Glentilt, 1677 (Perth, no. 897); Kinclaven, 1681 (Perth, no. 910); Kintore, 1686 (Aberdeen, no. 468); and Tannadice, 1695 (Forfar, no. 536). Caution is needed, however, over the 'comitatum seu thanagium de Murray' (Elgin, nos. 25, 178); there is no medieval evidence for this, and, of course, thanages did not become earldoms. On the other hand, the reference from Fife in 1543 to 'Thainislandis, viz. tertia parte de Strabrune, Fordell et Fotheris' (Fife, no. 2) is more significant, for in the mid-15th century there was a family called Thane of Fordell (see below, p. 81); but the evidence is insufficiently clear for Fordell to be counted as a medieval thanage.
2   Ibid., Banff, no. 171.
3   See Bannerman, 'MacDuff of Fife', above, p. 26.

Appendix

A List of Thanes and Thanages in Medieval Scottish Records

This list is arranged geographically, by modern (pre-1975) counties, and runs roughly from north to south. The numbers correspond with those on the maps. It gives the first reference to a thane or thanage, when and how the thanage was alienated, and any other relevant information.

ROSS AND CROMARTY

1   DINGWALL: thanage, 1382.
1264, sheriff (*ER*, i, p. 26). Earl of Ross had D at ferme from Alexander III, and in feu-ferme from John Balliol (*CDS*, ii, no. 1631). 1321, Robert I gave 'all the land of D with the castle and burgh' to earl of Ross, for £20 *p.a.* (*RRS*, v, no. 196). 1382, thanage of D (*RMS*, i, no. 741).

INVERNESS

2   KINMYLIES: thane, before 1232.
1232, Alexander II granted 'his land of the grieveship of K' to Moray bishops, in feu-ferme for £10 *p.a.*, reserving the fishings which 'the fermers or thanes' used to have' (*Moray Reg.*, no. 34).

3   ESSICH: thanage, 1337.
2nd teinds to Moray, 1238 (*Moray Reg.*, no. 40). After 1312 in earldom of Moray. 1337, thanage in Moray regality (*ER*, i, p. 440). Lands of E went with office of Inverness sheriff, 1509 (*RMS*, ii, no. 3286).

NAIRN

Both thanages in earldom of Moray after 1312, and held of earl of Ross after 1315 × 23.

4   CAWDOR: thane, 1295.
2nd teinds to Moray, 1238. Donald, thane of C, 1295 (*Family of Rose*, p. 30). 1309, Robert I gave 'whole thanage of C' to William, thane of C, for 12mk *p.a.* (*RRS*, v, no. 9). 1430, James I regranted thanage to Donald of C. 1476, other lands added to 'complete thanage of C' (*RMS*, ii, nos. 176, 1241). See also *Cawdor Bk.*, *passim*; the style 'thane of C' continued in use well into the 16th century.

5   MOYNESS: thane, 1295.
2nd teinds to Moray, 1238. 1238, 24mk *p.a.* to Moray bishops 'by hands of [royal] feu-fermers of M' (*Moray Reg.*, no. 40). William, thane of M, 1295 (*Family of Rose*, p. 30).

MORAY

All thanages in earldom of Moray after 1312.

6   BRODIE: thanage, 1311.
2nd teinds to Moray, 1238. 1238, 16mk *p.a.* to Moray bishops 'by hands of [royal] feu-fermers of Dyke and B' (*Moray Reg.*, no. 40). 1311, farmers of [king's] thanages of Dyke and B (*RRS*, v, no. 15). 1337, land of 'Brothi domini' in Moray regality (*ER*, i, p. 441). John, thane of B, 1492 (*Moray Reg.*, no. 199).

7   DYKE: thanage 1311.
2nd teinds to Moray, 1238. 1238, 16mk *p.a.* to Moray bishops 'by hands of [royal] feu-fermers of D and Brodie' (*Moray Reg.*, no. 40). Macbeth of D, 1262 (*APS*, i, p. 101). 1311, farmers of [king's] thanages of D and Brodie (*RRS*, v, no. 15). 1337, in Moray regality accounts, when a third probably went to the countess (*ER*, i, p. 441). Probably combined with Brodie from later 13th century.

8   CROMDALE: thane, 1367.
1222 × 8, held by earl of Fife, perhaps since *c.*1190. John de Dolas, thane of C, 1367 (*Moray Reg.*, nos. 63, 286; cf. Barrow, 'Badenoch and Strathspey, 1', p. 4.)

9   KILMALEMNOCK: thanage, 1426.
1426, earl of Douglas granted 'all his lands in *le Thayndomis* in lordship of K', plus barony of Petty (*RMS*, ii, no. 49). Perhaps Murray property in 13th and 12th(?) centuries, but if so, held in feu-ferme. 2nd teinds to Moray, 1238. 1337, in Moray regality, paid £4 for 1 term (*ER*, i, p. 443). 1498, lands of K in feu-ferme for 50mk *p.a.* (*RMS*, ii, no. 2386).

10   'RATHENECH': thane 1261.
2nd teinds to Moray, 1238. Ewen, thane of R, 1261, 1262, 1263 (*APS*, i, pp. 99, 101; *Moray Reg.*, no. 125). 1263, royal expenditure on R. 1264, Ewen, 'then thane of R' delivered grange to clerk of liverance. 1337, in Moray regality, paid 26s.8d. for 1 term (*ER*, i, pp. 10, 14, 443).

11   FOCHABERS: thane, before 1153.
David I granted F 'by its right marches' to Urquhart Priory, with 'dues of fish belonging to thane', free of royal exactions (*ESC*, no. 255). 2nd teinds to Moray, 1238. By 1292, back in crown hands, in exchange for 'Bynim'. 'Carta de F' in lost 13th-century muniments (*APS*, i, p. 110). 1337, in Moray regality, paid £5 10s.(?) for 1 term (*ER*, i, p. 443).

12   'MOLEN': thane, 1261.
2nd teinds to Moray, 1238. Dugall, thane of M, 1261, 1263 (almost certainly) (*APS*, i, p. 99; *Moray Reg.*, no. 125). 1337, in Moray regality, said to owe by charter 8mk, 1½ chalders meal, 9 chalders malt, and 7 bolls bran *p.a.*, and paid £2 13s.4d. for one term but no grain (*ER*, i, pp. 443, 446). For the maps, 'Molen' has been tentatively identified with Mulben, south of Fochabers to the east of the Spey.

BANFF

13   BOYNE: thanage under Alexander III.
2nd teinds to Aberdeen. Thanage worth £113 *p.a.* in 'Alexander III Rental' of crown lands (*Aberdeen Reg.*, i, p. 55). 1351, land in thanage of B (*RRS*, vi, no. 135). 1358/9, £31 6s.8d. crown fermes of thanage of B and town of Cullen (*ER*, i, p. 548). 1369, David II gave 'all his lands of his thanage of B' to John Edmonstone in barony (*RRS*, vi, no. 430). Also forest of B (*RMS*, i, app. I, no. 67).

14   MUMBRIE: thanage under Alexander III.
2nd teinds to Aberdeen. Thanage worth £34 8s.8d. in 'Alexander III Rental'. Lands in 'tenement' of M, 1327, 1359 (*RRS*, v, no. 312; *ER*, i, p. 548).

15   NETHERDALE: thanage under Alexander III.
2nd teinds to Aberdeen. Thanage worth £49 1s.4d. in 'Alexander III Rental'. Probably in thanage of Aberchirder before 1286. Robert I gave 20mk *p.a.* from N (*RMS* i, app. II, no. 43). 1345, Thomas Lypp's lands of N, etc. (forfeited before 1329 by his grandfather) restored (*RRS*, vi, no. 94A). 1358/9, £3 6s.8d. fixed payment due to crown (*ER*, i, p. 548). *c.*1402, annuity granted from N (*RMS*, i, app. II, no. 1784).

16   ABERCHIRDER: thane under Alexander III.
2nd teinds to Aberdeen. Worth £53 6s.8d. (including 2 nights' waitings, assessed at £40) in 'Alexander III Rental'. Simon, thane of A and Conveth forfeited both to Alexander III; tried to have thanage of A restored, 1286 × 9 (*Aberdeen-Banff Ills.*, ii, pp. 216-7). 'Carta de A' in lost 13th-century muniments (*APS*, i, p. 110). Robert I gave 10mk *p.a.* from fermes of thanage of A, 1313 (*RRS*, v, no. 36). 1358/9, £40 fixed payment due to crown (*ER*, i, p. 548). 1370, David II gave 'all his lands of thanage of A' plus Blairshinnoch (in Banff parish) to Walter Leslie, in barony, for 1 knight's service; if heirs of thanes recovered status, Leslie to keep superiority (*RMS*, i, nos. 316, 339).

17   CONVETH: thanage under Alexander III.
2nd teinds to Aberdeen. Thanage worth 80mk in 'Alexander III Rental'. Having forfeited both C and Aberchirder to Alexander III (above), in 1286 × 9 Simon, thane of Aberchirder and C granted 6 davochs of C to earl of Buchan and founded a chapel, in hope of having Aberchirder restored. But C had been fermed for 80mk by earl of Buchan (*CDS*, ii, no. 1541). 1358/9, £2 crown fermes of half thanage of C, rest having been granted away; but also land 'leased with king's lands of thanage of C' (*ER*, i, p. 548). *c.*1376, 'demesne of C' shared among three lords, and half thanage held by another (*Aberdeen Reg.*, i, p. 58; *RMS*, i, no. 678).

18   GLENDOWACHY: thanage under Alexander III.
2nd teinds to Aberdeen. Thanage worth £20 (£15 plus 5 chalders wheat) in 'Alexander III Rental'. Robert I granted thanage of G to Hugh (later earl of) Ross, for eight horses for carting from each davoch (*RMS*, i, app. I, no. 5). 1358/9, £10 plus 4 chalders wheat fixed

payments to crown (*ER*, i, pp. 548, 551). 1382, 'all lands of thanage of G', forfeited by earl of Ross, given to John Lyon (*RMS*, i, no. 734). Also known as thanage of Doune, which was barony in 1413 (*RMS*, i, nos. 254, 654, 738; ii, no. 109).

ABERDEEN

19   FORMARTINE: thanage, 1266.
Worth 160mk in 'Alexander III Rental'. 2nd teinds to Aberdeen, accounted for 1266 (14½mk paid by Reginald Cheyne, farmer of thanage), and 1328 (*Aberdeen Reg.*, i, pp. 159-60). 1292 × 6, John Balliol granted thanage of F to earl of Buchan (*CDS*, ii, no. 1541). Robert I granted parts (*RMS*, i, app. I, no. 6; app. II, nos. 417, 423). 1345, 1348, 1358, 1359, thanage shared between David II's sisters and their husbands, earl of Sutherland and Thomas Isaac (*RMS*, i, app. I, nos. 120, 149; *ER*, i, pp. 542, 545, 551). 1366, David II gave 'all the half of our thanage of F' to earl of Sutherland and his heirs-male, in barony (*RRS*, vi, no. 357). 1373-7, in crown hands: 1373, £66 fermes; 1374, £68; 1375, £59; 1377, £77 (*ER*, ii, pp. 430, 457, 499, 567). Thereafter held by earl of Carrick, and from 1382 (as barony) by James Lindsay; Carrick compensated with £100 annuity (*Aberdeen Reg.*, i, pp. 138-41; *RMS*, i, no. 801; *ER*, iii, p. 170). Also forest of F (*RMS*, i, app. II, no. 420).

20   BELHELVIE: thanage, 1323.
Worth 94mk in 'Alexander III Rental'. 2nd teinds to Aberdeen, accounted for 1328. 1281, part of dowry of Alexander III's daughter on marriage with king of Norway (*Rot. Scot.*, i, 19). 1292 × 6, John Balliol granted thanage of 'Dereleye' — probably B — to earl of Buchan (*CDS*, ii, no. 1541). 1323, Robert I gave 'demesne toun of thanage of B' plus 4 other lands in thanage, amounting to £40 old extent, to Hugh Berclay in barony (*RMS*, i, app. I, no. 7; cf. app. II, no. 42). c.1324, Robert I gave lands in thanage of B to John Bonneville (*RMS*, i, app. II, no. 414). 1348, 1358, 1359, accounts for payments due to crown from B (*ER*, i, pp. 543, 546, 553). c.1360, 1369, annuity of £20 *p.a.* from B (*RMS*, i, app. II, no. 1354; *RRS*, vi, no. 426).

21   KINTORE: thane 1264.
Worth 101mk in 'Alexander III Rental'. 2nd teinds to Aberdeen, accounted for 1328, 1337 (*Aberdeen Reg.*, i, p. 159). 1345, 1348, 1358, 1359, thanage shared between David II's sisters and their husbands, earl of Sutherland and Thomas Isaac (*RMS*, i, app. I, nos. 120, 149; *ER*, i, pp. 542, 545-6, 551). 1368, David II gave Thainstone in thanage of K to Henry Gothenys, with all cain in cheese and oats from Kinkell and Dyce in the thanage (*RRS*, vi, no. 397). 1375, Robert II gave 'all our lands of our thanage of K' to earl of Moray, in barony but reserving free-tenancies and cain, for 1 knight's service; 1382, free-tenancies and cain added, except for Thainstone (*RMS*, i, nos. 627, 718). Also forest of K (*RRS*, v, no. 261)

22   ABERDEEN: thane 1266.
Sheriff, probably under William I (*RRS*, ii, p. 40). 1266, thane of A, distinct from sheriff (*ER*, i, p. 11). Thanage worth 50mk in 'Alexander III Rental'. 2nd teinds to Aberdeen, accounted for 1328, 1337. 1347, 1358, thanage in possession of queen; 1359, half thanage in king's hands, half given by David II to John Herries (*ER*, i, pp. 543, 546, 551). c.1376, thanage divided between Herries and Marjorie Cardeny, Robert II's mistress (*Aberdeen Reg.*, i, p. 56). Forest of Stocket (*RRS*, v, no. 37) probably forest of thanage.

23   KINCARDINE O'NEIL: thanage(?), 1511.
1249 × 56, Alan Durward gave £1 2s. *p.a.* from his land of Skene to Aberdeen bishops, in exchange for 2nd teinds of O'Neil (*Aberdeen Reg.*, i, p. 17). That shows it had been royal demesne, and explains why it is not in 'Alexander III Rental'. But a late (1511) list of lands from which 2nd teinds were due calls it a thanage; it is correct over other thanages, and so may probably be trusted (ibid., i, p. 357; cf. p. 57). Late 13th century, held as barony by earl of Fife (*RMS*, i, app. I, no. 68).

24   ABOYNE: thanage, 1328.
Worth 202mk in 'Alexander III Rental'. 2nd teinds to Aberdeen, accounted for 1337. c.1230-55, Walter Bisset was lord of A. 'Carta de A' in lost 13th-century muniments (*APS*, i, p. 110). Robert I leased lands of A at ferme to Alexander Fraser, and they remained with Fraser hereditarily. 1328, £50 fixed payment for 1 term from thanage of A, in Alexander Fraser's hereditary possession (*Aberdeen Reg.*, i, p. 157). 1347, 1358, 1359, 1364, £100 fixed payment due from A, but queen held it 1347-58 (*ER*, i, pp. 542, 545, 551, *Aberdeen Reg.*, i, p. 157). c.1376, thanage of A (ibid., i, p. 57). 1407, barony of A, held by William Keith and Margaret Fraser (*RMS*, i, no. 893).

KINCARDINE

25 BIRSE: thane, before 1180 × 4(?).
1180 × 4(?), William I gave 'all my lands of B', plus forest, to Aberdeen bishops, 'with all neyfs, saving only king's thanes'; text is 'doubtful', but 'may be based on authentic act of King William' (*RMS*, ii, no. 251). 1242, Alexander II gave forest rights over bishops' lands of B (*Aberdeen Reg.*, i, pp. 15-16).

26 DURRIS: thanage, under Robert I.
1264, Kincardine sheriff accounted for repairs at D (*ER*, i, p. 12). 1292, earl of Buchan keeper of forest of D (*Rot. Scot.*, i, p. 10). Robert I granted land 'in the thanedom of D.' *c*.1345, David II gave thanage of D to William Fraser, on his father's resignation (*RMS*, i, app. II, nos. 435, 1050). 1358/9, £12 crown fermes of thanage of D (*ER*, i, p. 586). 1369, David II gave 'all lands of our thanage of D' to Alexander Fraser, in barony, for 1 archer's service (*RRS*, vi, no. 448). Also forest of D.

27 COWIE: thane, 1281.
1264, Kincardine sheriff accounted for repairs at C (*ER*, i, p. 12). Thomas son of thane of C, 1281 (*Aberdeen-Banff Colls.*, p. 258). 1292, earl of Buchan keeper of forest of C (*Rot. Scot.*, i, p. 10). Robert I gave Alexander Fraser 'of the thanedom of C' (*RMS*, i, app. II, no. 437); and, 1327, gave him the forest of C, for making a royal park within thanage of C (*RRS*, v, no. 320). *c*.1345, David II gave thanage of C to William Fraser, on his father's resignation (*RMS*, i, app. II, no. 1050). 1358/9, £48 13s.4d. crown fermes of thanage of C. *c*.1360, 1363, £20 payments 'from fermes of [king's] thanage of C', 'by hands of [king's] fermers of C.' 1375, some lands 'formerly in thanage of C' (*ER*, i, p. 586; ii, pp. 78, 115; *RMS*, i, app. I, no. 139; i, nos. 147, 499). 1378, Alexander Fraser 'lord of barony of C' (*Aberdeen-Banff Colls.*, p. 470); 1415, barony of C. (*RMS*, i, app. II, no. 1978). Also forest of C.

28 URAS: thane, 1214 × 24.
1214 × 24, Lorne, thane of U (*Brechin Reg.*, ii, no. 1). Late 13th century, earl of Buchan had superiority of U plus £10 *p.a.* from it. 1331, David II gave William Irvine the £10 plus claim to land (*RRS*, vi, no 5; cf. *RMS*, i, app. II, no. 428). 1370, 1371, Thomas Rate acquired both halves of U from previous owners; Robert II granted them to him in barony (*RMS*, i, nos. 314, 383, 384).

29 ARBUTHNOTT: thane, before 1165 × 78.
1206, 13 thanes in A in living memory. William I gave A to Osbert Olifard, whose son transferred it to what became family of Arbuthnott (*Spalding Misc.*, v, pp. 209-13, and *RRS*, ii, no. 569; cf. *RRS*, vi, no. 358).

30 KINCARDINE: thanage, 1323.
Sheriff, from William I's reign. 1189 × 95, William I added forest rights to previous grant of seven davochs in Fordoun parish (which roughly = K), to Humphrey son of Theobald (*RRS*, ii, p. 64, and no. 346). 1238, 1245, Humphrey's heiress gave all her feu in Fordoun to Arbroath Abbey (*Arbroath Lib.*, i, nos. 261-4). 1266, new park built at K (*ER*, i, p. 21). 1323, Robert I gave 6 acres beside royal *manerium* of K plus pasture rights in 'our thanage of K' to Alexander Fraser. 1345, David II gave earl of Sutherland and Margaret Bruce 'our thanage', castle or manor, and park of K, in barony; and 1360, for earl's life (*RMS*, i, app. I, nos. 72, 120, 149). 1358/9, K paid no crown fermes or cain, because earl of Sutherland held it (*ER*, i, p. 585). 1370, David II gave reversion of thanages of K, Aberluthnott and Fettercairn to Walter Leslie, all as one barony, for 2 knights' service; if heirs of thanes recovered status, Leslie to keep superiority (*RMS*, i, nos. 316, 338.). Barony of K subsequently held by Leslie and heirs (SRO, RH.6/243).

31 FETTERCAIRN: thanage, 1345.
1264, 1266, one night's waitings at F accounted for (*ER*, i, pp. 12, 20). 1345, David II gave earl of Sutherland and Margaret Bruce 'our thanage' of F, in barony; and 1360, for earl's life (*RMS*, i, app. I, nos. 120, 149). 1358/9, F paid no crown fermes or cain, because earl of Sutherland held it. 1370, David II combined thanage of F with thanages of Kincardine (*q.v.*) and Aberluthnott, in one barony.

32 NEWDOSK: thanage, *c*.1345.
'Carta de N' in lost 13th-century muniments (*APS*, i, p. 110). *c*.1345, David II gave thanage of N to Reginald Cheyne (*RMS*, i, app. I, no. 1023). 1358/9, £6 crown fermes of thanage of N, accounted for by sheriff of Forfar. 1364-5, £10 fermes for 3 terms (*ER*, i, p. 586; ii, p. 218). 1366, David II gave 'all our lands in thanage of N' to Alexander Lindsay, in barony (*RMS*, i, no. 226).

33   ABERLUTHNOTT: thanage under Robert I.
Robert I gave John Menteith 31 merklands in thanage of A; and 'of the thanage of A' to
John Fraser. *c*.1346, David II gave lands in A (*RMS*, i, app. II, nos. 585, 436, 1051).
1345, David II gave earl of Sutherland and Margaret Bruce 'our thanage' of A, in barony;
and 1360, for earl's life (*RMS*, i, app. I, nos. 120, 149). 1358/9, A paid no crown fermes
or cain, because earl of Sutherland held it. 1370, David II combined thanage of A with
thanages of Kincardine (*q.v.*) and Fettercairn, in one barony.

34   LAURENCEKIRK (CONVETH): thane, before 1189 × 93.
1189 × 93, William I gave 4 ploughgates in C to Agatha (of Berkeley) and Humphrey son
of Theobald, with 'all the right which the thane used to have in the kirkton of C' (*RRS*, ii,
nos. 344, 345). Before 1242, their heiress gave her lands in L/C to Arbroath Abbey, which
feued them to Aberdeen bishops (*Arbroath Lib.*, i, nos. 89, 272, 279). Rest of territory not
subsequently called a thanage; was it incorporated in thanage of Kincardine?

35   MORPHIE (ECCLESGREIG): thane (E) 1189 × 95; thanage (M), *c*.1345.
1189 × 95, '[king's] thane and tenants throughout parish of E (*RRS*, ii, no. 352). *c*.1195,
Earl David of Huntingdon (to whom William I must, meanwhile, have given E) cancelled St
Andrews Priory's obligation of cain and conveth from its land in E (Stringer, *Earl David of
Huntingdon*, p. 264). Like rest of Earl David's possessions, in crown hands from Robert I's
reign. *c*.1345, David II granted '5 chalder of victual furth of the thanage of M'; and (to
David Fleming) 'ane annuel furth of the thanage of Meikle M' (*RMS*, i, app. II, nos.
1000-1). 1358/9, £13 6s.8d. fixed payment from Meikle M; payment had gone to Fleming
in 1358. 1361, 5 chalders wheat from thanage of M sold for £7 (*ER*, i, p. 586; ii, p. 111).
1364, David II gave David Fleming 20mk *p.a.* due from Meikle M (*RMS*, i, no. 175).
Nether M also owed crown dues; 1366, David II granted it in feu-ferme for 20mk (*ER*, i, p.
585; *RMS*, i, no. 222).

ANGUS

36   KINNABER: thanage, 1326.
Malcolm IV gave Restenneth Priory 10s. *p.a.* from Kinnaber (as 2nd teind?) (*RRS*, i, no.
195), which implies he granted it out for £5 *p.a.* 1326, Robert I gave 3 merklands near K to
David Graham, 'with demesnes and tenancies of said lands and of thanage of K', plus 7mk
p.a. due from the thanage, all in barony, for 1 archer's service (*RRS*, v, no. 293).

37   MENMUIR: thanage, 1347.
2nd teinds to Restenneth. 1263, sheriff of Forfar paid 1mk to gardener of M. 1290,
sheriff's charge for past 2 years reduced by £66 13s.4d. on ferme of lands of M, because of
poverty of bondmen; the ferme had been increased by 50mk *p.a.* 'in hatred of said
bondmen' by Hugh of Abernethy (*ER*, i, pp. 8, 50). 1281, M to be included in dowry of
Alexander III's daughter (see Belhelvie [20]) if other lands granted did not produce 700mk
p.a. (Stevenson, *Documents*, i, p. 314). 1347, justiciar court judged that since Restenneth
Priory was due 2nd teinds from crown thanages and other lands in sheriffdom, this must be
paid from thanages of M and Monifieth as from other thanages; sheriff and 'whoever are
new freetenants in thanages of M and Monifieth' to ensure payments were made (Fraser,
Southesk, ii, 36). 1360, 3 'lords of lands of M' to pay £3 for M's 2nd teinds; if they delay,
amount to be 8mk (*RMS*, i, no. 214). 1358/9, 13s.4d. fixed payment for 1 term from M
(*ER*, i, p. 588).

38   CLOVA: thanage, 1328.
1328, Robert I gave earl of Mar 'all our thanage of C' plus 2 lands in Glen Clova, for £20
p.a. and 'carting and other small services' due under Alexander III (*RRS* v, no. 341). 1331,
earl of Mar had £20 fixed payment from land of C remitted; 1359, sheriff accounted for
none of £42 due from thanage of C, which should pay £42 *p.a.*, because earl of Mar held it
(*ER*, i, p. 588). Barony of C, 1403 (*RMS*, i, app. II, no. 1830).

39   KINALTY: thanage, 1305.
1322 inquest recorded 2nd teinds to Restenneth from thanage of K under Alexander III
(*RMS*, i, app. I, no. 29). 1264, 1289-90, sheriff received 42 stones cheese *p.a.* plus fodder
from K (*ER*, i, pp. 6-7, 50). 1305, John of K petitioned Edward I for restoration of thanage
of K, which his father held hereditarily (Maitland, *Memoranda de Parliamento, 1305*,
p. 191). 1358/9, £43 6s. fixed payments for 3 terms (= £29 4s. *p.a.*) (*ER*, i, p. 588). 1376,
land in tenement of K feued by Malcolm Ramsay. 1385, Robert II gave Walter Ogilvy
annual payment of £29 due from thanage of K. *c*.1395, crown confirmed William
Abernethy's grant of 'lands of K in barony of Rethy' (*RMS*, i, nos. 572, 757; app. II, no.
1706). 1404, Robert III granted William Abernethy 'all lands in barony of Rethy and K
which he had resigned' (*Aberdeen-Banff Ills.*, ii, p. 227).

40   TANNADICE: thanage 1316.
Before 1189 × 95, Richard de Melville gave church of T to St Andrews Cathedral Priory (*RRS*, ii, no. 333; note Humphrey of Berkeley did same with church of Laurencekirk). 1263, £60 crown fermes of T (*ER*, i, p. 9). 1316, Robert I continued an Alexander III grant of 8mk *p.a.* 'from fermes of our thanage of T', which still being paid in 1407 (*RRS*, v, no. 111; *RMS*, i, no. 906). 1322 inquest recorded 20s.10d. *p.a.* to Restenneth from thanage of T under Alexander III. *c.*1345, David II gave Peter Prendergast thanage of T (*RMS*, i, app. I, no. 29; app. II, no. 984). 1359, no account for thanage of T, because John Ramsay held it. 1361, £21 3s.8d. fermes for 1 term; 1362, £15 5s.4d. for 1 term; 1363, £30 for 1 year (*ER*, i, p. 589; ii, pp. 108, 153, 163). 1364, David II gave John Logy 'thanage of T', in blench ferme. 1370, David II freed William son of John from being 'our neyf of our thanage of T' (*RMS*, i, nos. 172, 345). 1375, Robert II gave 'all our land of thanage of T, including tenancies', to John Lyon, in barony and blench ferme (SRO, National Register of Archives, 885: MS Transcripts of Earl of Strathmore's muniments, *s.d.* 4 September 1375).

41   ABERLEMNO: thanage 1322.
1322 inquest recorded 2nd teinds to Restenneth from thanage of A under Alexander III. 1325, Robert I gave William Blount 'all our demesne land in our thanage of A', in barony, for 2 archers' service (*RMS*, i, app. I, nos. 29, 80). Both Robert I and David II (1366) gave William Dishington land 'in our thanage of A', for 1 archer's service (*RMS*, i, no. 217; app. I, no. 78).

42   OLD MONTROSE: thane, 1250.
Malcolm IV gave Restenneth a tenth of his fermes of (land of) Montrose (*RRS*, i, no. 231). John thane of Montrose, 1250 (*Arbroath Lib.*, i, no. 250). 1322 inquest recorded 2nd teinds to Restenneth from thanage of O M under Alexander III (*RMS*, i, app. I, no. 29). 1326, Robert I gave David Graham all land of O M (*RRS*, i, no. 294).

43   INVERKEILOR: thane before 1173 × 8.
1173 × 8: William I gave I to Walter of Berkeley, with sake and soke, etc., for one knight's service. Walter of Berkeley remitted to the parson of I the *grescan* and other services which kirkton of I used to render to thanes of I, and afterwards to himself (*RMS*, ii, nos. 185, 186; *Arbroath Lib.*, i, no. 56). I was subsequently Redcastle barony, held by descendants of Berkeley's son-in-law Ingram Balliol, and later shared between Stewarts of Innermeath and earls of Douglas (ibid., i, no. 293; *RMS*, i, no. 273; app. i, no. 76; app. II, nos. 774, 1127).

44   IDVIES: thane, 1219.
Gilys thane of I, and Malcolm brother of thane of I, 1219. Malise of I, 1254, 1282 (*Arbroath Lib.*, i, nos. 228, 314, 366). Subsequent history unknown; did not pay 2nd teinds to Restenneth.

45   FORFAR: royal *manerium*, owing two nights' waitings, 1264.
Sheriff, from William I's reign on (*RRS*, ii, 64). 1244, annuity from fermes of king's *manerium* of F to be paid by the sheriff and bailies (*Arbroath Lib.*, i, no. 265). 1264, waitings of F and Glamis (*q.v.*) totalled: 37½ cattle (24 from F, 13½ from G.), 75 pigs, 225 cogals of cheese (144 from F, 81 from G.; worth about £34), 291 chickens, 32 chalders malt, 10 chalders barley meal, about 80 chalders fodder, and 960 eels. 1289-90, 'waitings of two nights at F' included 48 cattle, 288 cogals cheese (*ER*, i, pp. 6-7, 49-50). Under Alexander III, according to 1322 inquest, Restenneth Priory received bread and beer from royal kitchen when king was in F, 1mk from mill, 100 eels from lake, and a tenth of rents of '3 bondages' of F. 1372, 2 of 'bondages' owed 300 loads of peat *p.a.*; this changed to provision of a satisfactory fire when king came to F (*RMS*, i, no. 514; app. I, no. 29).

46   GLAMIS: thanage 1264.
1264, 1289-90, thanage of G owed (one-and-a-?) half nights' waitings; for amounts of cattle, pigs, cheese, hens, grain and eels, see Forfar. It also probably owed money, of which 16mk remitted, because 2 territories had been 'subtracted' from thanage of G (*ER*, i, pp. 6-9, 50). 1322 inquest recorded 2nd teinds to Restenneth from thanage of G under Alexander III (*RMS*, i, app. I, no. 29). 1358/9, no account for thanage of G because countess of Wigtown held it; Newton of G paid £8 12s. fermes (*ER*, i, p. 589). 1364, David II gave John Logy reversion of thanage of G. 1372, Robert II gave John Lyon 'all our lands of thanage of G', with tenancies and services, in barony, for 1 archer's service (*RMS*, i, nos. 172, 411).

47   DOWNIE: thanage 1309.
Duncan of D, 1254 (*Arbroath Lib*, i, no. 366). 1309, Robert I gave Walter Bickerton 'all our thanage of D', in barony, for 1 knight's service (*RRS*, v, no. 6). 1322 inquest recorded

2nd teinds to Restenneth from thanage of D under Alexander III (*RMS*, i, app. I, no. 29). 1331, Forfar sheriff received 15 chalders wheat, 1 chalder meal, 30 chalders malted barley, from fermes of D (*ER*, i, pp. 376-7). 1345, 1360, David II gave earl of Sutherland and Margaret Bruce 'our thanage of D', in barony; and 1360, for earl's life. 1372, Robert II gave Alexander Lindsay 'all our lands of thanage of D', in barony, with tenants' services, for 1 knight's service; plus, 1375, 10mk *p.a.* in compensation for the 10mk being paid to Restenneth 'for 2nd teinds of 100 merklands of land of D' (*RMS*, i, nos. 403, 498; app. I, nos. 120, 149).

48   MONIFIETH: thanage 1315.
2nd teinds to Restenneth. 1201 × 5, earl of Angus gave Arbroath Abbey church of M plus some land (in the *abthen*) and common pasture in M. 1247, Alexander II gave Arbroath Abbey 10mk *p.a.* 'from our ferme of M' (*RMS*, ii, nos. 255-6; *Arbroath Lib.*, i, nos. 39, 50, 266). 1315, Robert I instructed that the 10mk be paid, 'from the thanage of M.' 1321, Robert I gave William Durham 'all our land of M called grange of M', plus mill, for 1 archer's service (*RRS*, v, nos. 75, 186). 1324, Robert I granted land in thanage of M (*RMS*, i, app. I, no. 75). 1347, justiciar court judged that since Restenneth Priory was due 2nd teinds from crown thanages and other lands in sheriffdom, this must be paid from thanages of M and Menmuir as from other thanages; sheriff of Forfar and 'whoever are new freetenants in thanages of M and Menmuir' to ensure payments are made (Fraser, *Southesk*, ii, p. 36).

PERTH

49   ALYTH: thanage, 1319.
Elias of A, *c.*1200 (*Coupar Angus Chrs.*, i, no. 9). Alexander II granted Bamff, etc., in feu of A (*Bamff Chrs.*, no. 1). 1264/5, bailie of Inverquiech castle (near A, probably in thanage of A) accounted for 83mk; had land of Inverquiech at ferme. 1262-4, Easter A in ward. 1266, Walter son of Adam of A paid 20mk relief (*ER*, i, pp. 3-4, 18, 27). 1319, Robert I granted lands 'within our thanage of A' (*RRS*, v, no. 145); and gave William Blount 20mk *p.a.* from the lands of A' (*RMS*, i, app. I, no. 470). 1375, Robert II gave James Lindsay the place of Inverquiech castle and lands (including Ochtiralyth) resigned by 6 landowners in thanage of A with services of tenants, in barony. Lindsay's lands included 'thanage of A', 1384. (*RMS*, i, nos. 610, 630, 763). Also forest of A (*RRS*, v, no. 161).

50   STRATHARDLE: royal *manerium*; thane, 1189 × 99.
Under David I and Malcolm IV, one of king's 4 *maneria* of Gowrie (*RRS*, i, nos. 57, 243). Macbeth sheriff of Scone, thane of Strathardle, 1189 × 99 (*Arbroath Lib.*, i, no. 35). 1183 × 95, William I gave church of Strathardle (= Kirkmichael) to Dunfermline Abbey; *c.*1279, Dunfermline surrendered *abthen* of Kirkmichael to John of Inchmartin (*RRS*, ii, no. 242; *Dunfermline Reg.*, no. 227). 1327, John of Inchmartin granted lands 'in his barony of S' (*RRS*, vi, no. 328).

51   COUPAR ANGUS: royal *manerium*.
Under David I and Malcolm IV, one of king's 4 *maneria* of Gowrie (*RRS*, i, nos. 57, 243); cf. Scone, Strathardle. *c.*1161, Malcolm I gave Coupar [Angus] Abbey 'all 'my whole land of C' (*RRS*, i, no. 226). But 1173 × 8 William I gave it a half ploughgate of land as site of abbey, free of waiting due from it, and papal confirmation of abbey's possessions said that Malcolm IV gave the 'grange' of C (*RRS*, ii, no. 154; *Coupar Angus Chrs.*, i, no. 13) — which raises questions about extent of Malcolm's grant.

52   LONGFORGAN: royal *manerium*.
Under David I and Malcolm IV, one of king's 4 *maneria* of Gowrie (*RMS*, i, nos. 57, 243); cf. Scone, Strathardle. 1160, Swain of L and sons Archibald and Hugh (*St Andrews Lib.*, p. 132). 1162 × 4, Malcolm IV gave Scone Abbey a tenth of corn fermes of L, or any other fermes which might be levied from it. 1165 × 74, William I commanded Archibald and Hugh, sons of Swain of F, to pay Scone's corn teinds. 1160 × 1, 1165 × 9, 'whole shire of L'. 1178(?), William I included L in his grants to his brother Earl David (*RRS*, i, nos. 174, 248; ii, nos. 16, 28, 205). L subsequently went to Earl David's heirs, and (in 3 parts) to crown under Robert I, when called a barony (*RRS*, v, nos. 47, 265, 371; vi, no. 193).

53   SCONE: royal *manerium*; thanage, 1234.
Sheriff, David I-Alexander II (*RRS*, i, 47-8; *Scone Lib.*, no. 68). Under David I and Malcolm IV, one of king's 4 *maneria* of Gowrie; tenth of their revenues given to Scone Abbey (*RRS*, i, nos. 57, 243). 1234, Alexander II gave Scone one net in king's fisheries 'in our thanage of S' in compensation for teind from Longforgan (*q.v.*) (*Scone Lib.*, no. 66). By 1263-6, Scone's teind fixed at £7 9s.1d paid by Perth sheriff (*ER*, i, pp. 2, 18). 1312,

Robert I gave Scone 'whole thanage of S', with full jurisdiction over its inhabitants; 1326, this grant cancelled Scone's teind, now £7 8s.10d. *p.a.* (*RRS*, v, nos. 17, 291).

54   KINCLAVEN: thanage, 1382.
Under Alexander III royal estate and residence, paying renders in oats and cash (*ER*, i, pp. 3, 17-18; *RMS*, i, app. I, no. 75). 1314(?), Robert I's 'tenement of K' (*RRS*, v, no. 438). 1359, paid £22 fermes; 1360, £34 13s.4d. 1361, mair of K, receiver of K's fermes, had had nothing from it; taken over by Robert Stewart. 1362, £18 fermes, then given temporarily to Thomas Falside. 1368, £22 13s.4d. fermes for 1 term (*ER*, ii, pp. 48, 73, 81, 111, 117, 297). 1369, David II gave William Cunningham 'all our lands of K', in barony, until £40-worth of lands found elsewhere (*RRS*, vi, no. 438). 1383, Robert II gave his illegitimate son John 'all our lands of K' and five other territories, 'in thanage of K' (*RMS*, i, nos. 729, 745). 1397, Murdoch Stewart (later duke of Albany) was lord of K (*Aberdeen-Banff Ills.*, iii, p. 263).

55   GLENTILT: thane, thanage, 1341 × 6.
In earldom of Atholl. 1341 × 6, Robert Stewart, lord of Atholl, gave Ewen thane of G 'whole thanage of G', for 11mk *p.a.* and provision of 4 horses for hunting in Bencrumby forest; 'or as much as can be raised from the thanage', assessed by assize of local inhabitants (SRO, RH.1/2/111). 1389, thanage of G (HMC, *7th Report*, p. 715). Andrew thane of G, 1461. Finlay, thane of G, 1467. 'Thanage of Abnathie' or 'le thanedome of G', resigned by Finlay Tosache, thane of G, 1502 (SRO, RH.1/2/246, 259, 312-315; *RMS*, ii, no. 2655).

56   DULL: thane, thanage under Alexander II.
Alexander II informed 'his thanes' of D and Fortingall that Scone Abbey could have building material from 'our thanages of D and Fortingall' (*Scone Lib.*, no. 65). 1264, Alan Durward owed 26s.8½d. from fermes of D. 1289, earl of Fife, fermer of *manerium* of D, accounted for past two years, according to extent made after Alexander III's death; receipts were £503 13s.4d. (*ER*, i, pp. 3, 48). *Manerium* of D probably included *abthen* as well as thanage of D (and perhaps Fortingall as well); subsequently, generally called '*abthen* of D.' 1320 × 6, Robert Bruce of Liddesdale (illegitimate son of Robert I) gave Robert Menzies land in *abthen* of D; confirmed by David II, 1343 (HMC, *6th Report*, p. 690, no. 3; *RRS*, vi, no. 64). *c.*1360, David II made John Drummond bailie of 'the Abthain of D' (*RMS*, i, app. II, no. 1273). 1360, £15 12s.8d. fermes of *abthen* of D. 1361, fermed by Robert Stewart; 1362, only £16 10s.10d. paid out of due ferme of £40. By 1368, in hands of queen, who gave it to her son John Logy. 1370, £7 5s. from its fermes. 1373, in hands of Alexander Stewart (of Badenoch?); no payments made. Thereafter, eventually held by Robert Stewart earl of Fife, though from 1397 he had a 204mk annuity instead, 'pro terris abthanie de Dull' (*ER*, ii, pp. 48, 73, 111, 298, 352, 425; iii, p. 427).

57   FORTINGALL: thane, thanage under Alexander II.
Alexander II informed 'his thanes' of Dull and F that Scone Abbey could have building material from 'our thanages of Dull and F' (*Scone Lib.*, no. 65). Perhaps part of *manerium* of Dull, 1289. Barony of F given to Thomas Menzies by Robert I; to Alexander Menzies by David II; to earl of Fife by Robert III (*RMS*, i, app. II, nos. 465, 908, 1744).

58   CRANNACH: thanage, 1334 × 5.
In earldom of Atholl. 1334 × 5, David of Strathbogie, earl of Atholl, gave 'whole thanage of C, in earldom of Atholl', to Robert Menzies, for suit to Rait in Atholl, and 1 archer in royal army. 1382 × 8, earl of Carrick (lord of Atholl) gave thanage of C to John Menzies, on own resignation. Thanage of C, 1441, 1451 (HMC, *6th Report*, pp. 690-1, nos. 6, 11, 23, 24; *RMS*, ii, no. 260).

59   FINDOWIE: thanage 1187 × 1203.
In earldom of Atholl(?). 1187 × 1203, bishop of Dunkeld confirmed earl of Atholl's grant to Scone church of Logierait with its teinds, etc., from Rait and from the whole thanages of Dalmarnock and F (*Scone Lib.*, no. 65; cf. *RRS*, ii, 241). But 1359 David II confirmed grant by earl of Douglas of Little Findowie in sheriffdom of Perth (*RRS*, vi, no. 220; but original in Atholl charters, Blair).

60   DALMARNOCK: thanage 1187 × 1203.
In earldom of Atholl(?). 1187 × 1203, bishop of Dunkeld confirmed earl of Atholl's grant to Scone Abbey of Logierait church with its teinds, etc., from Rait and from the whole thanages of D and Findowie (*Scone Lib.*, no. 65; cf. *RRS*, ii, p. 341). But 1383, Lower Dalmarnock ('Dalmernock minor') was part of thanage of Kinclaven (*RMS*, i, no. 729).

61   STROWAN: thane, 1194 × 8.
In earldom of Strathearn. 1194 × 8, and 1200, Duncan thane of S (G.W.S. Barrow, 'The earls of Fife in the twelfth century', *PSAS* lxxxvii [1952-3], p. 61; *Inchaffray Chrs.*, no. 9).

62   AUCHTERARDER: thanage 1236.
1226, Alexander II gave Inchaffray Abbey a tenth of his rent of A (*Inchaffray Chrs.*, no. 54). 1236, Alexander II gave Lindores Abbey land 'in thanage of A' (*Lindores Cart.*, no. 22). Sheriff under Alexander III: 1290, earl of Strathearn was 'bailie of sheriffdom' of A (*ER*, i, p. 51). 1328, Robert I gave Sir William Muschet 'all our land of A', including rents of burgh, in barony, for ½ knight's service (*RRS*, v, no. 337). 1364, David II gave Malcolm Drummond (Muschet's son-in-law) services of all freetenants of barony of A (*RRS*, vi, no. 318).

63   DUNNING: thane, 1200.
In earldom of Strathearn. 1200, Earl of Strathearn's charter to Inchaffray witnessed by Anechol, thane of D. *c.*1200, earl of Strathearn called him 'Anechol my thane'. 1247, earl of Strathearn gave Inchaffray 20mk *p.a.* from thanage of D; mandate for this to Brice, thane of D (*Inchaffray Chrs.*, nos. 4, 9, 77). 1381, Cristin Clerk, mair of the shire of D, accounted for D before earldom of Strathearn auditors; fermes £55 5s.7d. for 1380 (*ER*, iii, pp. 33-4).

64   Forteviot: thane, 1266.
Gillecrist of F, 1165 × 9. 1178 × 88, William I granted 4 acres of arable in F (*RRS*, ii, nos. 28, 208). 1263, 10mk-worth of land removed from Kinross and added to F. 1266, thane of F had to account for 20mk (*ER*, i, pp. 34, 18). 1314, Robert I granted lands 'in thanage of F' (*RRS*, v, no. 39); and gave 'lands of F' to John Trollop (*RMS*, i, app. II, no. 482; cf. no. 464). 1358/9, £5 crown fermes of F (*ER*, i, pp. 558). 1372, Robert II granted land 'in tenement of F'; and 1383, gave his illegitimate son James 'all our lands and our mill of F' (*RMS*, i, nos. 395, 730).

KINROSS

65   KINROSS: thanage 1323.
Sheriff from 13th century on. 1263, sheriff of K's account showed 10mk-worth of land removed from Kinross and added to Forteviot. 1264, John of K, sheriff of K, accounted for 4 nights' waitings: 40 cows (sold for £10), at least 38 pigs (sold for £2 17s.), cheese, barley meal, malt and fodder (of which 16 chalders sold for £4) (*ER*, i, pp. 34, 16). 1323, inquest into forest lands of K, belonging to John Kinross, separated these from thanage of K (*RMS*, i, app. II, no. 697). Before 1354, Agnes Crombie held land of heirs of John of K, and in feu of K (*RRS*, vi, no. 132). 1358/9, sheriff of K accounted for total £62. 1372, Robert II gave Lochleven castle and various lands in K sheriffdom (including many called demesne lands 1358/9; presumably parts of old thanage) to his queen and their son David; similar lands given by David II to John Bruce *c.*1346; by 1390 these held by Henry Douglas, and then granted in barony (*RMS*, i, nos. 424, 796; app. II, no. 1055). Also forest of K.

FIFE

66   FALKLAND: thane, *c.*1128.
Macbeath thane of F, *c.*1128 (*ESC*, no. 80). 1160 × 2, Malcolm IV gave F to earl of Fife (*RRS*, i, no. 190). Thereafter part of earldom of Fife.

67   KINGSKETTLE: thane, 1160 × 70.
1160 × 2, Malcolm IV gave 'all my ferme of K' to earl of Fife (*RRS*, i, no. 190). Kyneth thane of K, 1160 × 70 (*North Berwick Chrs.*, no. 3). Thereafter part of earldom of Fife; in 1294 extent of Fife, 'de dominico de Catel, per annum, £23 6s.8d.' (Stevenson, *Documents*, i, p. 416).

68   DAIRSIE: thane, 1160 × 2.
Hywan Macmallothen, thane of D, 1160 × 2 (*St Andrews Lib.*, p. 128). 1178 × 86, earl of Fife agreed he owed bishop of St Andrews ½mk for having his mill in bishop's feu of D, namely the kirkton. But earls seem not to have had any other part of D; whole estate probably went to bishops, who had a grange there (ibid., pp. 27, 44-5, 57, 311, 353; *ER*, i, pp. 137, 145-7; Stevenson, *Documents*, i, pp. 415-7).

69   KELLIE: thane, under David I.
Malmure thane of K, under David I (*ESC*, no. 207). Robert of London, illegitimate son of William I, gave land 'in his waste of K' to Priory of May (*May Recs.*, no. 19). Before 1286, R[ichard?] Siward promised to hand over land of K to Alexander III as soon as he

received any marriage with 100mk-worth of land (*APS*, i, p. 115). Robert I gave barony of K to William Siward. Thomas Randolph and Helen, daughter of Richard Siward and wife of Eustace Maxwell, made agreement about K. 1361, David II gave barony of K to Walter Oliphant, on Helen Maxwell's resignation (*RMS*, i, app. II, nos. 638, 701, 1372).

STIRLING

70 CALLENDAR: thane, 1189 × 95.
David I gave lands in C to Newbattle Abbey, and one tenth of profits from Stirling, Stirlingshire and C to Cambuskenneth Abbey (*RRS*, i, nos. 109, 135). Dufoter of C, under David I (*ESC*, no. 109). 1154, Ness of C accused of treason against Malcolm IV and killed in judicial combat (*RRS*, i, 8). Malcolm, thane of C, 1189 × 95, *c.*1200 (Fraser, *Melville*, iii, no. 8; *Cambuskenneth Reg.*, no. 79). Alwyn of C, *c.*1190 (*ESC*, p. 349). 'P. thane of C', 1226 (*Glasgow Reg.*, no. 141). 1233, Alexander II gave lands to 'M., former thane of C' (Fraser, *Carlaverock*, ii, 404). 1234, Alexander II gave 'all our land of C' to Holyrood Abbey, with sake and soke, etc., in feu-ferme, for 160mk *p.a.*; but reserving £40-worth of land in C which he had given to Malcolm, former thane of C (*Holyrood Lib.*, no. 65). 1328, Robert I confirmed above grant, now in blench ferme, for 128mk *p.a.* (cancelling a 32mk payment owed to Holyrood by king), and payable to Robert Stewart; 1364, Robert Stewart reduced amount due, because of war and pestilence, to whatever the monks could pay (*RRS*, v, no. 346; vi, no. 298). 1345, David II gave barony of C to William Livingstone, following forfeiture by late Patrick Callendar (*RRS*, vi, no. 93)

EAST LOTHIAN

71 HADDINGTON: thane, under David I.
David I granted H church and kirkton, etc., 'free of all dues owed to me and to the thane' (*ESC*, nos. 122, 134). Under Malcolm IV, shire of H held by his mother, countess Ada. 1184, sheriff of H (*RRS*, i, nos. 136, 231; ii, no. 252).

ADDENDUM

FORDELL (in Fife)
This may have been a thanage or have had a thane. In 1451 there was a John Thane of Fordell and his son Alexander, and in 1457 an Alexander Thane of Thainisland (*Dunfermline Reg.*, nos. 435-9, 452); these can be linked to the 1543 retour of 'Thainislandis, viz tertia parte de Strabrune, Fordell et Fotheris' (*Retours*, Fife, no. 2). John and Alexander may have been descended from a John 'Thyanus' (i.e. 'Thaynus') who was chamberlain to the abbot of Dunfermline in 1316 (*Dunfermline Reg.*, no. 348). Unfortunately, it is impossible to be certain about it, and therefore Fordell has not been included among the thanages discussed in this essay.

4

Periphery and Core in Thirteenth-Century Scotland: Alan son of Roland, Lord of Galloway and Constable of Scotland

KEITH J. STRINGER

This essay can be said to have begun life many years ago when Geoffrey Barrow passed on to a very junior historian the happy thought that Earl David of Huntingdon, William the Lion's younger brother, would repay detailed study. Alan son of Roland of Galloway, David's son-in-law, became an obvious secondary concern, and my interest was rekindled in 1989 when, by a rare chance, one of Alan's original charters — politically a key document — was repurchased from America and deposited locally.[1] Both these leading members of the Scoto-Norman aristocracy under the Normanised Scottish crown were prominent players on the trans-national stage of 'British history', a concept with which several medievalists have been much occupied of late.[2] Yet their respective careers display some very striking divergences. Earl David successfully amassed territorial riches for himself largely within the more domesticated 'inner' zone of the British Isles, and did so through essentially patient and peaceful strategies. By contrast, Alan, though the son of a Norman mother, a great feudal magnate, and the Constable of Scotland, was also the hereditary chieftain of a semi-independent Celtic province on Scotland's western fringe; he concentrated on expansion in the rugged 'outer' or 'Atlantic' zone; and he pursued his ambitions with a volcanic aggressiveness sufficient to impress even Sturla Thordarsson, the worldly-wise author of *Hakon's Saga*.[3] But Alan's dreams of lording it over east Ulster, the Isle of Man, and English Cumbria were never fully realised, and after his death in 1234 Galloway itself fragmented when Alexander II forcibly intervened in the name of feudal custom to partition the province among Alan's three daughters and their Anglo-Norman husbands.

---

1   Appendix (A), No. 3. For the full story, see R. Hall, 'The Crackanthorpe manuscripts', *Quarto* (Quarterly Bulletin of the Abbot Hall Art Gallery, Kendal, Cumbria), xxviii, no. 4 (1991), pp. 14-17.
2   See especially *The British Isles 1100-1500: Comparisons, Contrasts and Connections*, ed. R.R. Davies (Edinburgh, 1988); R.R. Davies, *Domination and Conquest: The experience of Ireland, Scotland and Wales 1100-1300* (Cambridge, 1990); R. Frame, *The Political Development of the British Isles 1100-1400* (Oxford, 1990).
3   Anderson, *Early Sources*, ii, p. 464 and n. 8.

It has recently been argued that while the higher nobility in thirteenth-century Scotland maintained external interests to a remarkable degree, this phenomenon cannot automatically be regarded as a dangerous structural weakness of the emerging Scottish state.[1] But Alan's quasi-regal position on the outskirts of the kingdom, the breathtaking scale of his expansionism, and his numerous dealings with the Angevin court, have traditionally seemed to mark him out as a mighty frontiersman whose sense of commitment and loyalty to the Scottish crown was minimal, and whose ancestral estates were deservedly broken up after a long career of playing strictly for his own hand. As a leading authority has put it, 'we hear little of support for Alexander II such as his father had enjoyed from Roland of Galloway ... and, while much is known of Alan ... from 1200, it is not as an active participant in Scottish affairs'.[2] Nevertheless, such an emphasis, although not entirely misplaced, in important respects requires qualification. Most of the main sources for a reassessment of Alan's career lie readily to hand. In particular, Joseph Bain calendared many relevant English records, though he confusingly added references to another Alan son of Roland, a tenant of the honour of Wallingford who was dead by 1212.[3] Of the two Appendices included here, one concerns Alan's charters, seven of which are published in special homage to a great charter scholar.

Without question, Alan was one of medieval Scotland's most powerful magnates. A basic distinction must be drawn, however, between his highly privileged position in Galloway (inherited from his father in 1200) and his more restricted role in Lauderdale and Cunningham (inherited from his mother Helen de Morville in 1217). Whereas the crown exercised direct feudal control over the latter lordships, in Galloway proper its superiority was based on the less exacting concept of overlordship. Consequently, Alan enjoyed there a degree of seigneurial authority virtually unsurpassed elsewhere within the realm: domestic government remained independent of regular royal intervention and oversight; Galwegian law was distinctive and autonomous ('Galwydia ... leges suas habet speciales'); and, in sum, although its ancient royal status had been suppressed on Malcolm IV's deposition of Alan's great-grandfather Fergus, the self-styled 'king of the Galwegians', Galloway retained a vestigial royalty as 'something of a kingdom within a kingdom'.[4] For all Alan's familiarity with feudal

1   K.J. Stringer, *Earl David of Huntingdon, 1152-1219: A Study in Anglo-Scottish History* (Edinburgh, 1985), ch. 9 (esp. pp. 207-11). Cf. A. Young, 'Noble Families and Political Factions in the Reign of Alexander III', in *Scotland in the Reign of Alexander III, 1249-1286*, ed. N.H. Reid (Edinburgh, 1990), pp. 16ff.
2   A.A.M. Duncan, *Scotland: The Making of the Kingdom* (Edinburgh, 1975), p. 529.
3   *The Boarstall Cartulary*, ed. H.E. Salter (Oxford Historical Soc., 1930), p. 307. The following entries must be discounted: *CDS*, i, nos. 228, 236, 243, 343, 381.
4   See especially R.D. Oram, 'Fergus, Galloway and the Scots', in *Galloway: Land and Lordship*, ed. R.D. Oram and G.P. Stell (Edinburgh, 1991), and H.L. MacQueen, 'The laws of Galloway: a preliminary survey', ibid. The quotations are from *APS*, i, p. 403, and Duncan, *Making of the Kingdom*, p. 187.

practices, the support of Galloway's native community was crucial to his personal supremacy. Its strongly militarist traditions went together with a fierce sense of regional patriotism, as was most dangerously demonstrated in the great revolt against Scottish overlordship led by Alan's great-uncle, Gilbert son of Fergus, in 1174, and in the uprising provoked by Alexander II's determination to dismember Galloway in 1234. Even in peacetime, Galwegian excesses were justifiably feared. When Alan's younger brother Thomas and his cronies visited England, violence and misery often marked their passage: a breach of the peace in Herefordshire (c.1206); a rape at York (1212); a homicide in Worcestershire (c.1221).[1] As a normal course, Alan's fiery tribesmen expected opportunities for pillage and land-grabbing, and the need to justify his leadership in terms appropriate to the intense acquisitiveness of Gaelic kin-based society no doubt helped to influence his expansionist policies. As hereditary Constable of Scotland, he was also responsible with the earls for leading royal armies under the king.[2] But he nevertheless treated the common army of Galloway (and Cunningham) as a private fighting force, 'exercitus noster',[3] which might form part of the *exercitus Scottorum*, or else might not. His military reserves were formidable, as was his capacity to wage war by both land and sea. Sturla Thordarsson put his strength at 150 to 200 ships,[4] which may indicate an army of 2,000 to 3,000 men — a not implausible suggestion given the well-documented muster of 1,000 select warriors from Galloway in 1212.[5] Few magnates in Britain, let alone Scotland, could command such resources, a point not lost on John of England, who treated Galloway as a recruiting-ground for auxiliaries virtually on a par with Brittany and Flanders. Thomas joined John's military household as a mercenary captain in 1204, took a Galwegian fleet to Poitou in 1206, and then seemed set for a career in the Welsh March as lord of the barony of Richard's Castle — though that ended abruptly on his return to Scotland in c.1209.[6] Nor would it be long before King John beckoned to Alan himself.

Understandably, the Scottish government's attitude towards Alan was ambivalent from the first. His collaboration was highly valued in order to ensure political stability in the south-west and military aid when required. But he concentrated on his regional interests and rarely came to court; his strength gave him an awesome capacity for independent action; and there was no gainsaying that as recently as the 1170s such power had been

1   *Pleas before the King or his Justices, 1198-1212*, ed. D.M. Stenton (Selden Soc., 1953-67), iv, no. 2572; *CDS*, i, no. 531; *Rolls of the Justices in Eyre ... for ... Worcestershire, 1221*, ed. D.M. Stenton (Selden Soc., 1934), nos. 1227, 1402.
2   *RRS*, ii, pp. 37-8.
3   PRO, Special Collections, Ancient Correspondence, S.C.1/3/147; Appendix (A), No. 7.
4   Anderson, *Early Sources*, ii, p. 476 and n. 12.
5   *CDS*, i, no. 529; *Pipe Roll 14 John* (Pipe Roll Soc., 1955), p. 154.
6   *Rotuli Litterarum Clausarum in Turri Londinensi asservati*, ed. T.D. Hardy (Record Comm., 1833-4), i, p. 11b; *CDS*, i, nos. 357, 359-60, 382; *The Cartulary of Worcester Cathedral Priory*, ed. R.R. Darlington (Pipe Roll Soc., 1968), p. xxxvij and n. 3.

harnessed to promote rebellion and provincial autonomy. Moreover, it can only be agreed that Alexander II's assertion of feudal authority over Galloway on Alan's death was a calculated display of political discipline and masterfulness. Yet the king's immediate reasons for acting have still to be properly explained. In reality, a resurgence of Galwegian separatism was *not* the initial problem, and it certainly cannot be inferred from the flexing of royal muscles in 1234-5 that Alan's relations with the crown had been characterised throughout by antagonism and non-cooperation. He kept order in Galloway and never challenged the crown directly; he refrained from making treasonable alliances with its enemies; and he followed in these respects his father's policy of loyalty to the king. That, however, is to make a case negatively and inadequately. The basic question is: how far did the Constable cultivate his expansionist ambitions, dictated as they were primarily by personal and family interests, within the context of both respect for the monarchy and positive support for its strategies? The answer must in large measure determine the verdict on his career, and that in turn will have implications for any assessment of the nature of the contemporary Scottish political community.

King John's speculative grant to Alan of a vast fief in Ulster has hitherto been interpreted largely within the framework of Angevin military-political requirements, and with good reason. Although its terms can be reconstructed from English chancery records, the actual grant does not appear among them — the charter rolls are missing from April 1209 until May 1212 — and historians have yet to agree on its precise date. But, clearly, John's primary motive was to improve the Anglo-Norman position in north-east Ireland after his dispossession in 1210 of the discredited Hugh de Lacy, earl of Ulster, whose fall had created a perilous power vacuum, a situation exacerbated by the enmity of Áed Ó Néill (d. 1230) of Tír Eóghain, the mightiest Gaelic leader in the north.[1] Thus Alan was to serve Angevin interests by imposing his authority in north Antrim and destroying Ó Néill's hegemony west of the River Bann, where he was in effect commissioned to conquer extensive areas.[2] But Alan's involvement with Ulster ought nevertheless to be assessed from other strategic perspectives and, in particular, that of the Scottish government's insecurity in the far north and north-west where, in the twelfth and early thirteenth centuries, dynastic risings and rebellions posed one of the gravest threats to the integrity of the kingdom.

The first task is to reconsider the date of John's grant and establish its immediate political setting. The year 1210 has recently been favoured,[3] but only on the strength of the appearance of 'Alan son of Roland' (the

1   K. Simms, 'The O Hanlons, the O Neills and the Anglo-Normans in thirteenth-century Armagh', *Seanchas Ard Mhacha*, ix (1978-9), pp. 74-7.
2   R. Greeves, 'The Galloway lands in Ulster', *TDGNHAS*, 3rd ser., xxxvi (1957-8), p. 117 and n. 12.
3   Duncan, *Making of the Kingdom*, p. 250; cf. *A New History of Ireland*, viii: *A Chronology of Irish History to 1976*, ed. T.W. Moody *et al.* (Oxford, 1982), p. 89.

Constable's English namesake?) among the knights who served in John's Irish campaign. Curiously, what has hitherto been overlooked is the valuable information preserved by later Scottish chroniclers who drew on earlier annals now lost. John of Fordun states that when John and William the Lion made a new treaty at Norham in February 1212, Alan 'did homage to John ... by his lord king's will and leave for extensive lands in Ireland that John had given him'.[1] Walter Bower repeats this passage, and then — evidently using a source unknown to Fordun — adds that Alan obtained in return for ten knights' service a lordship comprising 160 knights' fees, and that he took the formal oath on William's behalf to uphold the new treaty.[2] The available record evidence inspires much confidence in these accounts. John's charter of confirmation to Alan of 1215 does specify a service quota of ten knights, and although the true number of fees was 140,[3] the error here is trifling. Again, the English exchequer copy of William's 'Norham' letter to John, suspect though it is, contains a corroboration clause mentioning Alan's seal.[4] Finally, the dorse of the sixth membrane of the chancery patent roll for 14 John (1212-13) carries an undated letter from the Irish justiciar announcing that he had made a formal delivery of seisin to Alan's proxies. Bain assigned this text to c.May 1213, but April or May 1212 must be preferred, for the letters enrolled on the face of the membrane belong to 4 May-1 June of the earlier year.[5] It therefore seems reasonable to accept the chronicle accounts of an Angevin grant made during the proceedings subsidiary to the Anglo-Scottish treaty of February 1212, and, if so, Fordun and Bower are surely correct in saying that Alan entered John's service with William's express permission.

Furthermore, there are strong circumstantial grounds for believing that the Constable had obtained more than William's grudging approval. The perceptive 'Barnwell' annalist saw a direct link between William's renewal of his alliance with John and major disturbances in northern Scotland, where in 1211 the pretender Guthred MacWilliam had invaded Ross and Moray to advance his claims to the throne.[6] Heavy fighting was still going on early in 1212, and retaining John's goodwill was a major consideration for William. As has been well said, their new agreement was 'a treaty of mutual security',[7] and a supplementary alliance between Alan and John fitted well with immediate Scottish needs. Nor had William any serious reason to doubt the Constable's loyalty in 1212. Relations had been strengthened in 1209 when Alan was married to the king's niece Margaret, the eldest daughter of

1   Chron. Fordun, i, p. 278.
2   Chron. Bower (Goodall), i, pp. 531 (where the passage is misplaced), 533.
3   CDS, i, nos. 573, 625.
4   RRS, ii, no. 505.
5   Rotuli Litterarum Patentium in Turri Londinensi asservati, ed. T.D. Hardy (Record Comm., 1835), p. 98a-b (= CDS, i, no. 573). C.R. Cheney, The Papacy and England: 12th-14th Centuries (London, 1982), ch. xiv, pp. 297-8, gives pertinent guidance on contemporary English chancery practice.
6   Memoriale Fratris Walteri de Coventria, ed. W. Stubbs (Rolls Ser., 1872-3), ii, p. 206.
7   Duncan, Making of the Kingdom, p. 196.

Earl David, and Thomas of Galloway became earl of Atholl *jure uxoris*; and both men appear as commanders of the royal forces led against Guthred in 1211.[1] William was therefore confident of Galwegian support against the MacWilliams; indeed, this was an old story going back to 1187, when Roland had smashed a MacWilliam army near Inverness. That brings us to the last and most important issue. The Bury St Edmunds annals refer to Guthred's invasion of Scotland 'with a great army'; Fordun and Bower add that he came into Ross from Ireland (possibly by the Great Glen route); and the 'Barnwell' chronicler stresses his ability to count on regular aid from both the Scots and the Irish.[2] Roger of Howden also makes clear that, at any rate in the 1180s, the attacking power of the MacWilliams did not depend solely on disaffected Scottish levies.[3] Almost certainly, valuable support against William the Lion came from Orkney and the Hebrides, though no chronicler actually says so. But it is difficult to resist the conclusion that the MacWilliams traditionally used Gaelic Ulster as a haven and a place where large bodies of troops could readily be recruited, and that their military effectiveness in 1211-12 owed much to the assistance of Áed Ó Néill, who was consequently as much William's enemy as John's.

A conspicuous feature here is the ease with which ambitious leaders could maximise their strength by drawing warriors from one part of the closely knit Norse-Gaelic world to fight in another. Awareness of this problem is clearly reflected in John's Irish Sea diplomacy, which ensnared not only the lord of Galloway but King Ragnvald of Man, Earl Duncan of Carrick, and Harald Maddadson, earl of Orkney and Caithness.[4] Yet it was the earldom of Ulster itself, vividly described by Edmund Curtis as a 'wedge' driven into the 'old Gaelic world of Erin and Alba',[5] which had central strategic importance. If Alan could make John's grant fully operative, he would control a virtually continuous block of territory sweeping round the north-east coast from the Glens of Antrim to Lough Foyle. That in turn would put him in an outstanding position to reduce major threats to William's security, and the Anglo-Galwegian alliance of 1212 can thus be set in the context of a coordinated campaign by the Scottish and English governments to impose control over outlying territories where their respective authority was most seriously disputed. In the same year Earl Thomas, who continued to cooperate closely with Alan, crossed the North Channel with a Galwegian fleet to devastate Derry and Inishowen in conjunction with Ó Néill's rivals, the chiefs of Cenél Conaill; and he

1　　*RRS*, ii, no. 501; *Chron. Bower* (Goodall), i, p. 532.
2　　*Memorials of St Edmund's Abbey*, ed. T. Arnold (Rolls Ser., 1890-6), ii, p. 20; *Chron. Fordun*, i, p. 278; *Chron. Bower* (Goodall), i, p. 532; *Memoriale Walteri de Coventria*, ii, p. 206.
3　　*Gesta Regis Henrici Secundi Benedicti Abbatis*, ed. W. Stubbs (Rolls Ser., 1867), i, p. 277.
4　　*Monumenta de Insula Manniae*, ed. J.R. Oliver (Manx Soc., 1860-2), ii, pp. 25-37; *CDS*, i, nos. 321, 324, 480, 578, 737. Cf. P. Topping, 'Harald Maddadson, earl of Orkney and Caithness, 1139-1206', *SHR*, lxii (1983), pp. 110-12.
5　　Quoted in R. Frame, *English Lordship in Ireland, 1318-1361* (Oxford, 1982), p. 131.

returned in 1214 to fortify Coleraine.[1] These look like determined efforts to strike out bases of MacWilliam support, and it is surely significant that when the MacWilliams again descended on Ross in 1215, this revolt was quashed on the spot, without having to re-mobilise the royal army.[2]

Yet John had now established a direct claim on Alan's loyalty and, characteristically, he intended to make the most of it. On 20 July 1212, following earlier discussions specifically referred to, Alan was formally called on to send 1,000 hand-picked Galwegians to join the English army at Chester for an expedition (later aborted) against the Welsh.[3] If the reports of Henry II's use of Scottish troops in Wales in 1165 are reliable, it is unlikely that they had served with Malcolm IV's blessing.[4] But in 1212 the Galwegian levy had probably been authorised when William, no doubt together with Alan and Thomas, had joined John's court at Carlisle late in June.[5] John had already sent, or promised to send, Brabantine mercenaries against Guthred MacWilliam, and William had an opportunity to reciprocate. Alan refused to supply troops at his own cost, as John had requested.[6] A reason for this (apart from the obvious one) may have been concern to avoid any suggestion that military service was owed from Galloway to the English crown. If so, Alan's stance was very different to that of Edgar son of Donald of Nithsdale, who on 8 July had transferred his allegiance to John by rendering homage for his Scottish lands, an act of astonishing effrontery.[7] Earl Thomas figured in John's schemes for an expedition to Poitou in 1213 until baronial opposition again forced the king to change his plans, and on 24 and 28 July John issued charters granting Thomas his first Ulster fiefs, including Ó Néill lands in Derry.[8]

As fears of civil war in England mounted, so aid from Galloway became even more crucial to King John. From the summer of 1212 his position in northern England was particularly precarious, but he probably anticipated that his ties with Alan would deter open rebellion, or else enable him to crush it in a pincer grip. Yet Alan had to take account of the developing strategies of the Scottish crown over the major unresolved question of the Border shires. Shortly after Alexander II's coronation in December 1214, at which Thomas had been present, Alan was confirmed in the Constableship.[9] Among those duty-bound to pay allegiance to the new king were no fewer than five ringleaders of the growing baronial opposition in England: Robert de Ros, Eustace de Vesci, and the earls of Hereford, Oxford and

1   G.H. Orpen, *Ireland under the Normans, 1169-1333* (Oxford, 1911-20), ii, pp. 291-2.
2   See especially G.W.S. Barrow, 'Macbeth and other mormaers of Moray', in *The Hub of the Highlands*, ed. L. Maclean (Inverness and Edinburgh, 1975), p. 121.
3   *CDS*, i, no. 529.
4   G.W.S. Barrow, 'Wales and Scotland in the Middle Ages', *Welsh History Review*, x (1980-1), p. 309.
5   *RRS*, ii, p. 105; but cf. Duncan, *Making of the Kingdom*, p. 253 (n. 72). John had retired to York by 1 July, apparently accompanied by Thomas (*CDS*, i, no. 531).
6   *CDS*, i, nos. 529, 533, 540; *Pipe Roll 14 John*, p. 154.
7   *CDS*, i, nos. 523, 525.
8   *CDS*, i, nos. 577, 585-6.
9   *Chron. Fordun*, i, p. 283.

Winchester. All these Anglo-Scottish landowners were to figure among the Twenty-Five barons responsible for enforcing Magna Carta, and the first three were Alexander's close kinsmen.[1] They constituted a potentially powerful pressure group at the Scottish court, where a party of 'hawks' apparently did emerge early in 1215.[2] But at this stage the young king and his chief advisers still hoped to capitalise on John's troubles by negotiating concessions, and no one, it may be thought, was better placed than Alan to advance this policy.

On 5 May 1215, the very day when the barons delivered their defiance at Reading, John authorised a payment to Alan of prests to the tune of 300 marks,[3] an obvious indication of the perceived importance of military support from Galloway as the Angevin court steeled itself for all-out war. At the end of June, Alan was confirmed in his Irish lands; Thomas was granted the custody of Antrim castle and a well-defined lordship based on Coleraine.[4] By then Magna Carta had been sealed, and for present purposes the negotiations at Runnymede have two significant elements. First, Alan himself was in attendance on John from an early stage in the talks — they are found together at Windsor on 3 June agreeing to an exchange of gifts (a good Scottish hound for two English geese)[5] — and appears in the preamble of the Charter as one who counselled John to grant it. Second, in Magna Carta (clause 59) John promised that justice should be done to Alexander II concerning, among other things, his 'liberties and rights'. It is hardly rash to suggest that Alan was instrumental in securing this concession, and confirmation that he was acting in a formal capacity may be provided by the inclusion of Enguerrand de Balliol, a leading member of his personal following, in the Scottish embassy sent on 7 July to parley with John.[6]

When, on John's repudiation of Magna Carta, Alexander went to war in alliance with the baronial rebels, it has been supposed that Alan gave only tepid support to the Scottish war effort, or even that (like Duncan of Carrick) he held aloof altogether. In fact, the Constable resolved any conflict of loyalties by standing shoulder to shoulder with the Scots king, and took a considerable part in the war. Early in 1216 an English government memorandum named him as a rebel in arms.[7] Galwegians are said to have served in the army led by Alexander into Northumberland in July 1217, and they had probably fought in the king's campaign against the

---

1   Stringer, *Earl David*, pp. 45, 52, 130, 318 (n. 87), 320 (n. 124).
2   Cf. *Chron. Fordun*, i, p. 283. De Vesci almost certainly attended on Alexander II at Clunie in Stormont on 1 March 1215 (British Library, MS Additional 33,245, fo. 44ʳ); Winchester (Saer de Quincy) was probably with the king at Edinburgh on 3-4 April (*Melrose Lib.*, i, no. 174; ii, no. 366). For a possible visit to Scotland in 1215 by Oxford (Robert de Vere), see ibid., i, no. 256.
3   *CDS*, i, no. 621.
4   *CDS*, i, nos. 625-7, 631.
5   *CDS*, i, no. 628.
6   *CDS*, i, no. 629.
7   *Foedera*, i, p. 144 ('Alanus de Ganneya').

north in February 1216.[1] Even more significantly, when on 23 September 1217 the English government ordered Alexander to surrender Carlisle castle and all his other war gains, Alan was written to 'eodem modo'.[2] That in itself is indicative of his key role on the Scottish side, and precise corroborative details are happily to hand among his surviving *acta*. Although the crucial charter, Alan's confirmation to John of Newbiggin of lands in Kirkby Thore and Hillbeck in Westmorland, was published by F.W. Ragg as long ago as 1917, its true significance has escaped both English and Scottish historians. Ragg was unable to date the charter closely and did not explain how Alan came to exercise lordship in this part of Cumbria, although he shrewdly observed that Alan had a claim to the barony of Appleby or Westmorland proper through his mother.[3] Her grandfather Hugh I de Morville had held this strategically vital district on King David I's grant during the Scottish occupation of northern England from 1141.[4] But the Morville-Galloway claim had lain dormant since 1173, and it must have seemed totally extinguished when in 1203 John gave Westmorland in hereditary right to Robert de Vieuxpont, one of his most energetic and reliable agents. The 1215-17 war, however, presented Alan with an ideal opportunity to pursue his territorial claims and simultaneously serve the Scottish cause.

Twenty-three persons witnessed Alan's charter to John of Newbiggin. Six of this company belonged to the Constable's personal retinue. Of the remainder, three were listed under Cumberland in 1212 as tenants-in-chief by cornage: Adam son of Odard, lord of Wigton; Richard of Levington, lord of Kirklinton; and Richard Gernun, lord of half of the barony of Burgh by Sands in right of his wife, Alan's kinswoman Joan de Morville.[5] Twelve further witnesses can be classed as members of the Cumbrian 'gentry', with tenants of the baronies of Kendal and Westmorland predominating: Thomas son of Ranulf (of Mansergh), Ralph de Aincurt (of Sizergh), Eudo of Carlisle, Robert of Castle Carrock, Ralph de la Ferté (of Bowness on Solway), William of Hillbeck, Roger of Lancaster, Nicholas and Thomas of Morland, Wigan of Sandford, Thomas of Tebay, and William of Warcop. Some of these Cumbrian proprietors had close connections with Scotland. Richard of Levington, for example, had presumably succeeded his father as lord of Hutton in Dryfesdale, Dumfriesshire, in 1210, and four of Richard's sisters eventually married Scottish husbands.[6] Eudo of Carlisle occurs in a

1   *Chron. Melrose*, pp. 63, 68. Cf. G.W.S. Barrow, 'The Army of Alexander III's Scotland', in Reid, *Alexander III*, p. 136.
2   *Patent Rolls of the Reign of Henry III* (London, 1901-3), i, pp. 93-4.
3   F.W. Ragg, 'Five Strathclyde and Galloway charters', *TCWAAS*, new ser., xvii (1917), pp. 228-9, 231; Appendix (A), No. 3.
4   See now G.W.S. Barrow, *The Anglo-Norman Era in Scottish History* (Oxford, 1980), pp. 72-3.
5   *The Book of Fees* (London, 1920-31), i, p. 198.
6   G.W.S. Barrow, 'Some problems in twelfth- and thirteenth-century Scottish history — A genealogical approach', *The Scottish Genealogist*, xxv (1978), pp. 106, 111 (n. 104) (cf. *CDS*, i, no. 704); T.H.B. Graham, 'The de Levingtons of Kirklinton', *TCWAAS*, new ser., xii (1912), pp. 59-75.

southern Scottish context in the entourage of Eustace de Vesci, and probably held the lands of Kinmount in Annandale.[1] It is nevertheless safe to assume that Alan's charter was issued south of the Solway (perhaps at Appleby itself), and that Robert de Vieuxpont had effectively been superseded as lord of Westmorland.

Equally clearly, the occasion was a gathering of rebels against the English crown. John of Newbiggin and ten of the witnesses are definitely known to have been in rebellion in 1217, and they did not begin to enter Henry III's peace until October of that year.[2] Eudo of Carlisle was still regarded by the English administration as in revolt 'cum rege Scottorum' as late as 12 November.[3] Adam son of Odard, Richard Gernun, and Richard of Levington were leaders of the insurrection in Cumbria; indeed, in this regard the only notable absentees from the witness-list are Robert de Brus of Annandale, as lord of Edenhall, Robert de Vaux, lord of Gilsland, and Gilbert Fitz Reinfrid, lord of Kendal.[4] But Roger of Lancaster, who does appear, was Gilbert's illegitimate son,[5] while Ralph de Aincurt, another witness, was among those of Gilbert's knights captured when Rochester castle fell in November 1215 after a seven-weeks' siege. Released from prison early in 1216, he promptly rebelled again.[6] Moreover, the recently rediscovered Lanercost Priory cartulary preserves the full text of a charter by Robert de Vaux whose witness-list opens with the names of Alan of Galloway ('Alanus de Galweth'), the hereditary Stewart (Walter II son of Alan), and Robert de Brus. Then follow twelve further witnesses, six of whom also attested Alan's charter to John of Newbiggin.[7] It can hardly be doubted that both *acta* were issued at about the same time. The latest possible date of these charters is the end of the civil war in the autumn of 1217; the *terminus post quem* can probably be set at June or July 1216. Ralph de Aincurt had been in no position to witness northern charters during the earliest phase of Scottish pressure. In any event, late in 1215 Alexander's main effort was directed against Northumberland, while one important result of John's winter counter-offensive was the strengthening of Robert de Vieuxpont's regional dominance. By February 1216 Carlisle castle and the *comitatus* of Cumberland had been entrusted to his charge, and he now held sway in all but name as Angevin viceroy in the north-west.[8]

---

1   *Kelso Lib.*, i, nos. 207-8, 210; *Scots Peerage*, ii, pp. 374-6.
2   Thus, John of Newbiggin, Adam son of Odard, Ralph de Aincurt, Roger de Beauchamp, Ralph de Campania, Eudo of Carlisle, Robert of Castle Carrock, Ralph de la Ferté, Richard Gernun, Richard of Levington, Wigan of Sandford: *Rot. Litt. Claus.*, i, pp. 244a, 343a, 374b, 375a, 376a.
3   Ibid., p. 343a.
4   J.C. Holt, *The Northerners: A Study in the Reign of King John* (Oxford, 1961), pp. 28, 31, 65, 133.
5   J. Nicolson and R. Burn, *The History and Antiquities of ... Westmorland and Cumberland* (London, 1777), i, p. 64.
6   H. Hornyold, *Genealogical Memoirs of the Family of Strickland of Sizergh* (Kendal, 1928), p. 9; *Rot. Litt. Claus.*, i, pp. 244a, 376a.
7   J.M. Todd, 'The Lanercost Cartulary: An Edition of MS DZ/1 in the Cumbria County Record Office' (Lancaster University Ph.D. thesis, 1991), ii, no. 28.
8   *Rot. Litt. Pat.*, p. 163a-b; *Rot. Litt. Claus.*, i, p. 247b.

Robert, however, had apparently quit the north by the early summer of 1216 when, although Carlisle had yet to submit to Alexander II, John's party was no longer able to control events north of the Ribble.[1]

Since the sources for the Anglo-Scottish war of 1215-17 are notoriously patchy, any evidence that can throw fresh light is specially welcome. Alan must be seen as Alexander II's chief lieutenant in the reassertion of Scottish control over Cumbria, and, to judge from Alan's relations with the provincial nobility, political consolidation went forward with some thoroughness. Indeed, de Vieuxpont was still having difficulty as late as 1223 in recovering the land Alan had confirmed to John of Newbiggin.[2] Nor was the Constable's wartime authority necessarily confined to Westmorland proper. It may well have encompassed Carlisle (as Henry III's government seems to have assumed) and the royal manors centred on and including Penrith — where Alan's clerk Master Adam of Thornton is found drawing revenues unlawfully in c.1219, the under-sheriff of Cumberland having to distrain him by seizing the goods of men coming from Galloway to Carlisle.[3] Alan's ousting of de Vieuxpont consequently appears to have involved nothing less than the substitution of a Scottish for an Angevin viceroyalty in the north-west. This also raises the tantalising possibility that, whatever his English and French allies intended, Alexander had a vision of creating a Scoto-Northumbrian realm such as had motivated David I, a suggestion that receives some additional support from Alexander's exercise of royal rights in imposing his own nominee as bishop of Carlisle.[4] But whether or not the goal was to push the boundary of the kingdom south of the Westmorland fells, Alan's support for Scottish war aims was certainly no less central than that given by his de Morville great-grandfather during King David's struggle with Stephen of Blois.

In an undated letter, written soon after Alexander II's reconciliation with Henry III in December 1217, Alan complained to King Henry that he had gained little advantage from his estates in Ireland, but declared his readiness to serve him faithfully ('quod nos ad voluntatem vestram semper parati sumus per mare et terras in seruicio vestro ire').[5] The Anglo-Norman colony had remained in the Angevin allegiance, and advancement of Alan's Ulster interests depended on his making his own peace with the English government. That this was delayed until 1220 is at first sight perplexing, and may give the false impression that Alan jeopardised the new understanding between the crowns. The real reason for his tardiness was in fact rooted in Angevin politics. At the beginning of Henry III's reign,

1   Holt, *Northerners*, pp. 136ff.
2   *Patent Rolls of Henry III*, i, p. 410.
3   *Royal ... Letters illustrative of the Reign of Henry III*, ed. W.W. Shirley (Rolls Ser., 1862-6), i, no. 142, which supplements the information on Master Adam in D.E.R. Watt, *A Biographical Dictionary of Scottish Graduates to A.D. 1410* (Oxford, 1977), p. 531.
4   *Fasti Ecclesiae Anglicanae, 1066-1300*, ed. D.E. Greenway, ii: *Monastic Cathedrals* (London, 1971), p. 20.
5   PRO, Chancery Miscellanea, C.47/22/9(1); calendared in *CDS*, i, no. 754.

Alan's rival Hugh de Lacy, though he had gone on the Albigensian Crusade, did not lack influential backers in the English council, and no less a personage than the Regent himself, William Marshal, was in principle sympathetic to Hugh's reinstatement as earl of Ulster.[1] In the circumstances, it would have been prudent for Alan to have made some sort of accommodation with the de Lacy interest, and it was possibly in 1219 (rather than 1229, the traditional date) that he married Hugh's daughter Rose.[2] Be that as it may, it was only after the Marshal's death in May 1219 that satisfactory arrangements were finally made to regularise Alan's relations with the English crown.[3] He came back into its allegiance in Alexander II's presence at York in mid-June 1220 when, it is also recorded, Alan was one of the twelve Scottish barons to support Alexander's oath that he would marry Henry III's sister, Joan or Isabella.[4] Alan then seems to have accompanied Walter the Stewart and Robert de Brus on their pilgrimage to Canterbury for the translation of Saint Thomas's remains on 7 July, an appropriate act of piety for the great-nephew of one of the martyr's assassins.[5] Thus the Constable continued to play a responsible part in Scottish affairs, and had given the English administration sufficient assurances that he would cooperate with the government in Ireland.

But the Dublin government had in reality little power, and by 1224 Hugh de Lacy had launched a full-scale war to recover his lands. William Marshal the younger stabilised the position in Meath, but failed to force the passes into Ulster, where Hugh held his ground in alliance with Áed Ó Néill. Earl Thomas's base at Coleraine had already been eliminated.[6] Most of east Ulster must also have been overrun in 1224, and Alan's planned counter-attack was badly mistimed. Writing again to Henry III, he explained how from 9 June to 8 September he had been very active in the king's service, going from island to island and ready at all times to cross with his army to Ireland 'ad honorem vestrum'. But while about to do so, Alan said, he had learned of an agreement between the Marshal and de Lacy, and he petitioned Henry that if Hugh were restored to favour, then his and Thomas's lands should be safeguarded.[7] Little more is heard of Alan in connection with

---

1   J. Lydon, 'The expansion and consolidation of the colony, 1215-54', in *A New History of Ireland*, ii: *Medieval Ireland, 1169-1534*, ed. A. Cosgrove (Oxford, 1987), pp. 156, 158.
2   Cf. K.J. Stringer, 'A new wife for Alan of Galloway', *TDGNHAS*, 3rd ser., xlix (1972), p. 52 (where the argument should turn on the degree of affinity, not consanguinity). This lady, evidently called Rose, had in 1217 been placed in the custody of her uncle, the Regent's ally Walter de Lacy: *Register of the Priory of St Bees*, ed. J. Wilson (Surtees Soc., 1915), p. x and n. 5.
3   *CDS*, i, no. 755 (18 April 1220).
4   *CDS*, i, nos. 762-4.
5   G.W.S. Barrow, 'Early Stewarts at Canterbury', *The Stewarts*, ix (1953), pp. 231-3; Canterbury, Cathedral, City and Diocesan Record Office, MS Register E, fo. 143ʳ (charter of Robert de Brus for Christ Church, Canterbury, attested by Hamund 'clericus domini Alani Galwathie'; another witness, Thomas of Kent, occurs as a clerk in Alan's service in *CDS*, i, no. 754; *Melrose Lib.*, i, nos. 79, 227; cf. Appendix (A), Nos. 3, 7).
6   Orpen, *Ireland under the Normans*, iii, pp. 44-5.
7   PRO, S.C.1/3/147; calendared in *CDS*, i, no. 890.

Ireland. After his restoration in 1226-7, Hugh was intent on controlling all his former earldom lands and had the tacit backing of the Dublin government, despite assurances to both Alan and Thomas from the English council that it still wished to honour King John's grants. In April 1225 Alan had been licensed to colonise his estates, but evidence that he was ever able to entrench his position through systematic settlement is non-existent.[1] Similarly, Thomas failed to put John's grants into full effect. In December 1225 Henry III awarded him an annuity of 100 marks from the Dublin exchequer to set against his military expenses, and no doubt his territorial losses. No payment was made in 1226, but in January 1227 the king instructed the Irish justiciar to make what arrangements he could to sustain Thomas in royal service.[2] In October 1229 Alan and Thomas were named on a list of tenants-in-chief in Ireland summoned to join Henry's expedition to France.[3] Thomas had salvaged enough from the wreck of his Irish ambitions to obey the royal summons.[4] Alan, whom Henry had made no attempt to compensate, did not serve. In any case, he was then much preoccupied by the hornets' nest he had stirred up on Scotland's western seaboard.

While de Lacy's reinstatement goes a long way towards explaining Alan's reverses in Ireland, another consideration is Alan's commitment in the 1220s to warfare in the Isles. Up to a point, this was a sound strategy which might have strengthened him in both areas, but it overstretched his resources, and ultimately had the unforeseen consequences of damaging his relations with the Scottish court and committing Alexander II to a categoric assertion of royal authority over Galloway. Yet this is not to say that throughout his campaigning in the Isles a clear line can be drawn between Alan's interests and the crown's.

The protracted power struggle over the kingdom of Man and the Isles, formally subject to Norway, began as a domestic feud between King Ragnvald and his half-brother Olaf, both of whom were Alan's kinsmen.[5] But there were urgent reasons why the dispute should have attracted the attention of Alexander II, whose early concern for regional security in the west is well attested. Political conflicts among the Islesmen, like freebooting ventures, were unlikely to stop short at the mainland; their feuding could all too easily be exploited by powers hostile to the Scottish crown; and once again the problem of the MacWilliam challenge loomed large. Accordingly, vital Scottish interests were at stake, and there was much to be said for

1   CDS, i, no. 905. Cf. T.E. McNeill, *Anglo-Norman Ulster: The History and Archaeology of an Irish Barony, 1177-1400* (Edinburgh, 1980), p. 21.
2   CDS, i, no. 922; *Rot. Litt. Claus.*, ii, p. 164b.
3   *Royal Letters of Henry III*, i, no. 297.
4   CDS, i, no. 1064.
5   What follows supplements, but scarcely supersedes, the seminal study A.A.M. Duncan and A.L. Brown, 'Argyll and the Isles in the earlier Middle Ages', *PSAS*, xc (1956-7), at pp. 200-2. The main narrative sources consulted are conveniently extracted in Anderson, *Early Sources*, ii, pp. 447, 456ff. See further B. Ó Cuív, 'A poem in praise of Raghnall, king of Man', *Éigse*, viii (1955-7), pp. 283-301.

backing one of the contending brothers in the expectation that he would destroy his rival and restore some measure of stability. Ragnvald had been an adherent of William the Lion during his struggles with Harald Maddadson.[1] As a result of this earlier alliance, Alexander's sympathies are likely to have lain with him, and help could be offered without antagonising the English crown, another interested party. A vassal-king of both Henry III and the pope, Ragnvald was under their joint protection;[2] and the shoring up of his position would reduce the threats to Ireland and Scotland alike. Olaf, by contrast, had been King William's prisoner (1208-14), and naturally looked for support to Hakon IV of Norway, who had few scruples about manipulating regional rivalries for his own ends. In March or April 1224 the possibility of Norwegian intervention in the west caused sufficient alarm at the Scottish court for Queen Joan to alert Henry III that Hakon intended to mount a summer campaign in aid of Hugh de Lacy.[3] And that was probably the main reason why, from June to September, the Galwegian fleet was kept on station in western waters.

When Alan first made common cause with Ragnvald, he therefore probably received Alexander II's active encouragement. Judging from the surviving fragments of his correspondence, that was exactly what Olaf himself believed.[4] But in the long run Alan's involvement in this volatile arena merely increased regional disorder, provoked Norwegian retaliation, and so proved to be entirely counter-productive. In 1221-2 he sent Galwegian levies to assist in Alexander's pacification of Argyll, apparently directed against another of Ragnvald's enemies, Ruaidhri of Garmoran.[5] While approaching Ireland with a Hebridean fleet in 1221, a claimant for the kingship of Connacht (in opposition to Henry III's ally Cathal Crobderg Ó Conchobair) was intercepted and killed by Earl Thomas, who returned to support Alan in 1228 but, perhaps significantly, not any later. On any realistic assessment, it soon became clear that the balance of forces was tilted in Olaf's favour — a conclusion the English government had arrived at by *c*.1228 when it began to cultivate Olaf and, simultaneously, the Irish justiciar negotiated with him to curb piracy.[6] The expedition led by Alan and Ragnvald to Skye and Lewis in 1225 proved fruitless; Galwegian intervention was so resented by the Manxmen that they swiftly transferred their allegiance to Olaf and linked their fortunes with his. Yet Alan would not back down, even after his abortive attempt to conquer Man in 1228 and Ragnvald's defeat and slaying at Tynwald in February 1229.

Alan's persistence suggests, initially, arrant irresponsibility and brazen

1   M. Chesnutt, '*Haralds Saga Maddaðarsonar*', in *Specvlvm Norroenvm: Norse Studies in Memory of Gabriel Turville-Petre*, ed. U. Dronke *et al.* (Odense, 1981), pp. 35, 39.

2   *Monumenta de Insula Manniae*, ii, pp. 43-7, 53-7.

3   *CDS*, i, no. 852 (where the suggested date, March/April 1223, is a year too early). See also *Rot. Litt. Claus.*, i, p. 439b.

4   *CDS*, v, Part II, no. 9.

5   *Chron. Fordun*, i, p. 288; J.G. Dunbar and A.A.M. Duncan, 'Tarbert castle: A contribution to the history of Argyll', *SHR*, 1 (1971), p. 2.

6   *CDS*, i, no. 1001; v, Part II, no. 9. Cf. *Rot. Litt. Claus.*, ii, p. 175b.

personal ambition, but it is possible on reflection to take a rather different view. First and foremost, his behaviour seems to have been dictated by pragmatic dynastic considerations. The key issue here was provision of an alternative power base for his well-known son Thomas, a bastard in the eyes of the Church and therefore barred by feudal rules of inheritance from the Galloway succession. Although in the mid-1220s Alexander II may have had no clear plans for Galloway after Alan's death, it was always a possibility that the crown would insist on treating Galloway like a fief. In any case, Alan must have realised that since there was no doubt about the status of Lauderdale and Cunningham as feudal lordships, the survival of his Scottish patrimony as an integral unit rested, in the first instance, on Galloway's also being dealt with according to feudal law. To be sure, in 1234 only daughters survived from Alan's recognised marriages, and the ultimate result was a threefold partition of the inheritance. Yet Alan's obscure legitimate son, also named Thomas, from his marriage to Earl David's daughter Margaret was possibly still alive in the 1220s, and even if he were already dead, it nevertheless remained feasible for Alan to nurse hopes of a son by his current wife Rose de Lacy who could establish himself in all the three lordships and thereby preserve the status quo (Rose was probably much younger than Alan, and certainly outlived him).[1] But that still left the problem of compensating and placating his bastard Thomas who, under Celtic customary law, was entitled to regard himself as Alan's successor as hereditary lord of Galloway — and in 1235 was indeed recognised as such by the Galwegians (if rather reluctantly and belatedly) in preference to division of the province among Alan's daughters.[2] In 1226, however, Alan had cashed in on Ragnvald's difficulties and married Thomas to Ragnvald's daughter, with the clear intention that Thomas should succeed his father-in-law as king of Man. On this view, it was not so much personal empire-building as desire to protect the future unity and stability of his ancestral Scottish lands that now guided Alan's war policies.

Nor can it be assumed that, in seeking an island kingdom for Thomas, the Constable had overstepped the mark and was engaged purely in freelance activity. He witnessed a charter granted by Alexander II at Stirling on 31 March 1226,[3] and it is conceivable that the king then authorised or condoned Thomas's marriage. The installation of a Scottish client in Man would be a major strategic triumph, and behind such a prospect lay the fact that, as the situation developed, Alexander was especially well placed to control Thomas's loyalty by reserving the ultimate right to decide Galloway's fate

1   On Alan's lawful son Thomas, see *CDS*, ii, no. 169; *Edward I and the Throne of Scotland, 1290-1296: An Edition of the Record Sources for the Great Cause*, ed. E.L.G. Stones and G.G. Simpson (Oxford, 1978), ii, pp. 140, 208. Rose de Lacy was still alive in 1237: *Register of St Bees*, p. x (n. 5).
2   MacQueen, 'Laws of Galloway', pp. 137-8.
3   *Moray Reg.*, no. 29. Earl Thomas attended the royal court at Cadzow in Clydesdale on 31 May 1226: *Lennox Cart.*, p. 92.

according to feudal *or* Celtic norms.[1] If it is acceptable to argue in these ways, it follows that Alexander was not totally opposed to Alan's ambitions; rather, it was the Constable's inability to make a success of them that really caused the trouble.

According to *Hakon's Saga*, news reached the Norwegian court in the summer of 1229 of 'great strife' in the Hebrides and Man, where Alan had returned to the offensive in league with the lords of Argyll and the inner isles. Stung into action, King Hakon sent agents to reinforce his authority in the west. Admittedly, his stratagems failed in their main purpose, due partly to Alan's opportune appearance with his galleys off the Mull of Kintyre (or Galloway?). Yet in 1230-1 the operations of a combined Norwegian, Orcadian and Hebridean fleet presented Alexander II with a military crisis of major proportions. He abandoned other business to be with Alan at Ayr on 28 May 1230.[2] About the same time the Stewart's castle of Rothesay on Bute was successfully stormed by invading Norsemen; on the fleet's return next year from Man, where Olaf was to rule unopposed until his death in 1237, it assaulted Kintyre and severely tested the defending forces before withdrawing. These incursions were no mere plundering raids on Scottish-held territories, but determined attempts to conquer them. Alan's humiliation was complete: nothing had been gained; earlier Scottish advances had been endangered; and no royal sanction for his initial support of Ragnvald or for Thomas's marriage could have saved him from the verdict that he had pursued private ambitions to extremes, regardless of national interests — the more so because Alan could also be held responsible for contributing to the political instability enabling Gilleasbuig MacWilliam, with Hebridean help, to launch a serious revolt in the northern Highlands in *c*.1229-30.[3] The Scottish court's attitude seems to be reflected in the Stewart tradition, recorded by John Barbour, that questioned Alan's commitment and even his loyalty to the kingdom.[4] And, quite understandably, he must have incurred Alexander II's severe displeasure — with decisive consequences for the king's policy towards Galloway and the vigour with which, in 1234-5, he applied the 'feudal noose'.[5] In brief, destruction of the province's unity must be seen as an act of royal domination undertaken specifically to curb expansionist forces that in 1230-1 had brought the realm, however inadvertently, face to face with its gravest national emergency since King John's invasion of Lothian in 1216.

While the importance of magnate enterprise in shaping the political history of the medieval British Isles can scarcely be doubted, sometimes individual

1   Compare royal policy as regards the descent of the Celtic earldoms: Duncan, *Making of the Kingdom*, pp. 199-200.
2   *Paisley Reg.*, pp. 47-8.
3   G.W.S. Barrow, 'Badenoch and Strathspey, 1130-1312, 1: Secular and Political', *Northern Scotland*, viii (1988), pp. 5-6.
4   *Chron. Bower* (Watt), v, p. 262.
5   This graphic phrase is borrowed from Davies, *Domination and Conquest: Ireland, Scotland and Wales 1100-1300*, p. 82.

initiatives faltered or failed. For thirteenth-century frontiersmen, military conquest could still bring dazzling rewards, but, with few easy pickings left, expansion at sword-point was increasingly hazardous. Alan displayed all the arrogance and belligerence of a John de Courcy or a Hugh de Lacy; he gambled for massive stakes, even saw himself as a kingmaker, and yet at every turn was frustrated by forces greater than his own. The majority of contemporary Scottish magnates, most notably members of Anglo-Continental families like the Comyns and the Stewarts, took fewer risks, but emerged more powerful than ever by making the most of their stronger court connections and cautiously expanding outwards from the centre. And their successes, especially when set against Alan's failures, served to confirm the fundamental importance of sustained access to the Scots king's favour and support. Even for the Celtic earls, who were becoming increasingly court-orientated, being a *familiaris regis* was what largely counted now. But this should not conceal the equally basic truth that Galwegian policies were often closely interwoven with Scottish policies. Thus, until he finally overreached himself, Alan's ambitions were more consistent with royal aims than is usually realised; he operated on the periphery, but normally within the ambit of the crown's authority.

There is, therefore, a risk of overstressing the extent to which Alan stood apart from Scotland's aristocratic community. After all, his reverses as a would-be *conquistador* had the effect of recurrently emphasising his metropolitan role as a magnate whose hereditary estates included two first-class lordships, Lauderdale and Cunningham, which had long possessed close ties with the crown. And, as has been seen, it was probably concern for the future of his Scottish power base which ultimately propelled Alan's war machine to the brink of disaster. So, although the marcher emphasis of his career naturally attracts the greatest attention, it should not be allowed to obscure those ties serving to *minimise* the contrasts between Alan and other magnates in the thirteenth-century kingdom.

Alan's dependants as named in charters and other records provide one particularly revealing approach to his social and political orientation. For all his reliance on Galwegian warriors, the impression that emerges from these sources is that of an almost exclusively Anglo-Norman world. This is probably not the outcome of chance or of a built-in quirk of the documentation. Duncan of Carrick's surviving charters, for instance, regularly display Gaelic attestors,[1] but Alan's hardly ever do. As has been argued in a related context, that suggests 'real social exclusivity in the entourage', with members of the native community, rare exceptions apart, figuring 'only as auxiliary troops, labouring bondsmen, or barbarous allies'.[2]

Here, then, was a magnate who was typical of the greater nobility of

1   E.g., *North Berwick Chrs.*, nos. 1, 13, 14; *Melrose Lib.*, i, nos. 29, 30, 32.
2   R. Bartlett, 'Colonial Societies of the High Middle Ages', in *Medieval Frontier Societies*, ed. R. Bartlett and A. MacKay (Oxford, 1989), p. 29.

thirteenth-century Scotland in that his power rested on aspects of both Celtic and feudal lordship, and who, like most 'native' lords, moved easily in the dominant Anglo-French milieu. As his father had done, Alan drew heavily on the former de Morville connection — knights from Cunningham and Largs (e.g., Hugh of Ardrossan, Alexander de Néhou, Alan and Robert de Ros), from Lauderdale (e.g., Peter Haig, Richard Maitland, Vivian de Molyneux), and from English Cumbria (e.g., Roger de Beauchamp, William of Muncaster, Ivo de Vieuxpont).[1] Not least on the political level, especial interest attaches to those dependants recorded in a Galloway context. Basic details on eight of this contingent are supplied in the accompanying table, while brief biographies of Enguerrand de Balliol, Roger de Beauchamp, Ralph de Campania, and Ivo de Vieuxpont are given in Appendix (B).

	Galloway feu	Occurrences with Alan of Galloway	Attestations of Scottish royal acts c. 1200-34
Household Knights			
Sir Roger de Beauchamp	—	3	—
Sir William de la Mare	—	4	—
Tenants			
Sir Adam son of Gilbert	Airdrie	2	5
Gilbert son of Cospatric	Southwick	5	—
(of Workington, Cumberland)			
*Sir Enguerrand de Balliol	Urr	10	30
*Sir Ralph de Campania	Borgue	7	12
Sir Bernard of Ripley	Kirkandrews	1	—
Sir Ivo de Vieuxpont	Sorbie	2	4

* denotes a Scottish royal official.

The names of those who were also tenants of the Scottish crown appear in bold type.

Main sources: Appendices (A) and (B) below; *Wigtownshire Chrs.*; *RRS*, ii; *RRS*, iii (forthcoming).

Eight Members of the Feudal Community in Galloway, 1200-34

Some of the 'Galloway' group were doubtless frequent visitors to the province, resorting to castles such as de Balliol's magnificent motte-and-bailey at Urr or de Campania's impressive residence at Borgue.[2] And all

---

1   Cf. Barrow, *Anglo-Norman Era*, pp. 70ff.
2   C.J. Tabraham, 'Norman settlement in Galloway: Recent fieldwork in the Stewartry', in *Studies in Scottish Antiquity presented to Stewart Cruden*, ed. D.J. Breeze (Edinburgh, 1984), pp. 96-8, 114-17.

were involved in one way or another with maintaining Alan's lordship over the native population. Yet their horizons were by no means limited to those of a wholly frontier-like existence. Their connections with Galloway were restricted because some also had English, even French, estates and interests, or were household knights with no Scottish lands at all. But most significant for our purpose are those who were bound to the Scots crown through ties of tenure and/or formal administrative service. Three can be classed as crown tenants, two of whom carried much weight in Scotland: Enguerrand de Balliol, lord of Inverkeilor in Angus, Adam son of Gilbert, lord of Kilbucho in Peeblesshire, and Ivo de Vieuxpont, lord of Alston in Tynedale. Enguerrand also occurs as sheriff of Berwick in 1226; Ralph de Campania was constable of Roxburgh by c.1228. Both these men thus had a strong stake in the king's government, and their witnessing-patterns in the 1220s suggest regular movement back and forth between Alan's following and the royal court. It can be added that Enguerrand was almost certainly the brother of Alexander II's much-trusted chamberlain, Henry de Balliol; Adam son of Gilbert, who also held Tarbolton in Kyle under the Stewarts and Hutton in Dryfesdale under his Levington kinsmen, was the brother-in-law of William Comyn, earl of Buchan, the father-in-law of Henry de Graham, lord of Dalkeith, and the uncle of both David de la Hay, lord of Errol, and Master Gamelin, future chancellor of Scotland and bishop of St Andrews. Ivo de Vieuxpont was the half-brother of William de Vieuxpont, who held important estates of the crown in Berwickshire and Lothian; Bernard of Ripley was the brother of William of Ripley, on whom William the Lion bestowed Dallas in Moray.[1] It may also be said that Thomas son of Ranulf, a senior clerk of Alexander II from c.1220, sheriff of Dumfries in 1237, and progenitor of the great Randolph dynasty, very probably entered state service through Alan's retinue.[2] Again, the Constable's prominent clerk Thomas of Kent, perhaps of the knightly family of that surname associated with the Stewarts, appears as constable of Edinburgh in the 1230s, and on his death in 1247 received, as would Thomas son of Ranulf in 1262, 'the highest accolade known to thirteenth-century Scotland — an obituary in the Chronicle of Melrose'.[3]

From such evidence it follows that Galloway's separateness from the rest of Scotland, while in some senses very real, must not be overstated. The most influential representatives of the provincial elite identified their interests with those of the Scottish 'establishment' as a whole, and especially

---

1   G.G. Simpson, 'The *Familia* of Roger de Quincy, Earl of Winchester and Constable of Scotland', in *Essays on the Nobility of Medieval Scotland*, ed. K.J. Stringer (Edinburgh, 1985), p. 119 (Bernard of Ripley); on Adam son of Gilbert, see Barrow, 'Some problems' (as cited above, p. 90, n. 6), pp. 106, 112, and, for his relationship with David de la Hay, NLS, MS Advocates 34.6.24, p. 428.

2   Barrow, 'Some problems', pp. 104-5, identifies Thomas son of Ranulf as effective founder of the Randolph family and deals with his career in Scottish royal service; see further Appendix (A), No. 3, comment.

3   See above, p. 93, n. 5; *Newbattle Reg.*, no. 144; *Chron. Melrose*, p. 107. The quotation is from Barrow, 'Some problems', p. 104.

of the king himself. Accordingly, they appear not as members of a closed social and political world, but as members of a wider community who directed its influence on to the affairs of the locality, thus bringing Galloway and its ruler more firmly within the range of royal power than would otherwise have been the case.

As Geoffrey Barrow has taught us, the twelfth and thirteenth centuries were formative periods in the emergence of a united and self-confident medieval kingdom of Scots. Crucial to this process was the attachment of outlying regions more effectively to the royal core, but disruptive forces were likely to prevail when, for whatever reasons, regional concerns clashed with national ones. In that event, the result was to re-focus attention on the forms royal authority needed to assume in dealing with peripheral areas. The crown therefore pounced on the opportunity afforded by the Galloway succession to destroy Alan's provincial supremacy and replace him with more responsible men like John de Balliol, Roger de Quincy, and, later, Alexander Comyn, earl of Buchan. As in Galloway, so in Moray it was the turmoil of *c*.1230 which prompted a bold intensification of royal control, expressed there by the grant to Walter Comyn of the lordships of Badenoch and Lochaber.[1] Frontier warlords had become anachronisms in the kind of kingdom Alexander II wished to rule, and their role was reduced by further redefinition of the relationship between the centre and the localities on terms the crown dictated.

Yet to portray Alan as a champion of the west-coast Gaelic periphery against the Forfar-Edinburgh-Roxburgh core would palpably underestimate the extent to which, even by the early thirteenth century, the concept of a unitary Scottish kingdom had gained real political substance. In its dealings with Alan, the monarchy employed a complex system of regulatory mechanisms: notably, honouring him with high office in the royal household and marriage to a royal bride; stressing his direct tenurial dependence on the crown for the lordships of Lauderdale and Cunningham; enmeshing prominent members of his retinue in a network of royal governance and control. From this standpoint, the interplay between core and periphery was by no means a clear-cut confrontation between two antagonistic entities. Rather, taken in the round, Alan's career exemplifies how far the kings of Scots had progressed in merging the local with the national and the national with the local. Strikingly, his regular charter style, 'Alan son of Roland, Constable of Scotland', stressed not his lordship over Galloway but his links with the crown, and for the most part he worked in harness with it. As for his Anglo-Norman dependants, further evidence of their outlook, if it be needed, is found in the routine manner in which their charters concerning Galloway mention the *regnum Scotie*. A typical instance is Ralph de Campania's grant to Dryburgh Abbey, especially for the soul's weal of

1   Barrow, 'Badenoch and Strathspey, 1', p. 6.

Alan, his lord, of the advowson of the church of Borgue 'as any patron in the kingdom of Scotland can most freely ... transfer to any church or canons any right of patronage or other right in a church'.[1] That still leaves Alan's native followers unaccounted for, but the crisis of 1234-5 does allow us to catch a highly instructive glimpse of their feelings, one that confirms that Galloway was far less detached from the Scottish kingdom than it had been sixty years before. The most remarkable feature of this episode is that the Galwegians then showed sufficient awareness of the strength and prestige of the monarchy, and the reverence owed to it, to ensure that their first response was very different from the demands for independence vociferously voiced in 1174. The Melrose chronicler, who was exceptionally well placed to know the facts, stresses that on Alan's death they begged Alexander II to preserve Galloway's integrity by taking the whole province under his direct lordship and protection.[2] Only when the king spurned them was their loyalty as subjects of the crown found wanting, and even then their rebellion swiftly subsided.

1   *Dryburgh Lib.*, no. 64. See also ibid., nos. 71, 73, 75; *Holyrood Lib.*, no. 70, for similar examples. The political significance of such phrases is well brought out in Barrow, *Anglo-Norman Era*, pp. 154-5.
2   *Chron. Melrose*, p. 83. Adam, abbot of Melrose (1219-46), took a keen interest in the affairs of Galloway, and helped to restore peace there in 1235 (ibid., p. 84).

Appendix (A)

Alan of Galloway's Charters: Seven Texts

The charters edited below, six of which have not been printed previously, relate to Galloway (No. 1), Yorkshire (No. 2), Westmorland (No. 3), Cunningham and Largs (Nos. 4-7). The scribes' punctuation and capitals have been retained in editing the five original texts; elsewhere, punctuation and capitals are editorial. Limitations of space preclude extensive commentary on each act. The political importance of No. 3 justifies a new edition here. No. 1 is a rare example, for pre-1214 Scotland, of a private knight-service charter; Nos. 4 and 5 carry fair impressions of Alan's hitherto unnoticed seal and counterseal (the heraldic decoration, very different to later Galloway arms, merits expert attention); No. 6 is of special interest for early Scots law and No. 7 for military recruitment. The concentrated acquisitions by Hugh of Crawford in west Ayrshire (Nos. 4-6) provide a model instance of the forging of a new estate by a 'rising' younger son. I am most grateful to Geoffrey Barrow for first drawing Nos. 5 and 6 to my attention, and for help in establishing their present whereabouts. I am also indebted to the Marquess of Bute for kindly granting me permission to publish them.

Remarkably few charters of Alan survive. For other texts, see *Calendar of the Charter Rolls* (London, 1903-27), iii, p. 92; *Dryburgh Lib.*, nos. 85, 180, 228; *Holyrood Lib.*, no. 73; *Kelso Lib.*, i, nos. 245-6; *Melrose Lib.*, i, nos. 79, 83-4, 227; *Register and Records of Holm Cultram*, ed. F. Grainger and W.G. Collingwood (CWAAS, Record Ser., 1929), nos. 130, 142; *Register of St Bees*, no. 42; Stringer, *Nobility of Medieval Scotland*, p. 67.

1

Grants to Adam, son of Gilbert the dispenser, Airdrie (in Kirkbean, Kirkcudbrightshire) by those marches by which Gilbert held it of Alan's father Roland; to be held in feu and heritage for the service of the tenth part of one knight. The land of Airdrie is to be free of multure, and Adam's corn shall be ground at the mill of Preston immediately after the corn already in the hopper, unless Alan's own corn or church corn is waiting. (19 December 1200 × 7 January 1210).

Omnibus has liter*as* visuris uel audituris tam presentibus*ᵃ* quam futuris, Alanus filius Roll*andi* Scott' constabill*arius*, salutem. Sciatis me dedisse et concessisse et hac presenti carta mea confirmasse Ade filio Gilberti despensatoris, pro humagio et seruicio suo, Ardarie, scilicet per illas diuisas per quas predictus Gilbertus pater suus predictam terram scilicet Ardarie de patre meo Rollando tenuit. Tenend*am* de me et heredibus meis sibi et heredibus suis in feudo et hereditate, libere et quiete, plenarie et integre, faciendo inde mihi et heredibus meis ille et heredes sui seruicium decime partis vnius militis. Dicta vero terra de Ardarie libera erit et quieta a prestacione multur*e*. Bladum vero predicti Ade et heredum suorum moletur ad molendinum meum de Preston proximo post bladum inuentum in tramuia,*ᵇ* nisi fuerit bladum meum proprium uel bladum ecclesie. Hiis testibus Henr*ico* fil*io* comit*is* D*auid*, Thom*a* fratre meo, Gilbert*o* fil*io*

Cospatricij, Thoma [   ],^{c} Willelmo de Mara, Radulfo^d de Campan*ia*, Alex*andro* de Derwentwater, Willelmo Olifard, Ada de Thore*ntona*, Ric*ardo* clerico presentium^e scriptore.

SOURCE: SRO, Register House Charters, RH.6/308. (Notarial transumpt, made 15 March 1443/4, by Brice Stewart, clerk, of Glasgow diocese. The protocol describes the charter as being sealed on a tag in red wax impressed on white.)

NOTES: ^{a} Source adds scilicet. ^{b} Perhaps traumia. ^{c} Illegible. ^{d} Source has Raduffo. ^{e} Source has presensium.

COMMENT: The earliest possible date is fixed by Roland's death; Alan's brother Thomas, a witness, is not given the title earl of Atholl, which he had obtained by 7 January 1210 (*RRS*, ii, no. 489). It can surely be assumed that the donee was Adam lord of Kilbucho, even though his father Gilbert (son of Richer), a prominent Stewart vassal and a regular witness of William the Lion's *acta*, is not elsewhere styled dispenser (of the king's household?). William Olifard, the eighth witness, became a leading vassal of Earl Thomas in Atholl (*Coupar Angus Chrs.*, i, pp. 49-50).

2

Quitclaims to Roger de Lacy, constable of Chester, the advowson of the church of Kippax (Yorkshire, W.R.). (19 December 1200 × 1 October 1211).

. Sciant omnes presentes et^a futuri has Litteras visuri 7 audituri : Quod ego Alanus filius Roll*andi* . Dominus Galuuath' Scotie Constab*ularius* . quitam clamaui . Rogero de lascy : Cestrie Constab*ulario* 7 heredibus suis . de me 7 heredibus meis : aduocationem ecclesie de kipeis . Hijs Test*ibus* . Eustacio de Vescy . Roberto Walensi . Willelmo de bello monte . hugone despensario . Thoma fratre suo . Gilberto fil*io* Cospatric . Radulfo de Campania . Ricardo clerico de creuequor^b litterarum scriptore . 7 multis aliis .

ENDORSED: de aduoc*atione* ecclesie de kipes (xv cent.?); s' (same hand).

DESCRIPTION: Original, 6.4 × 4.2 in (16.3 × 10.6 cm) as mounted; foot, now open, formerly folded to depth of about 0.9 in (2.4 cm), with slits for tag. Nothing remains of tag or seal.

HAND: Idiosyncratic, backward-sloping charter hand.

SOURCE: PRO, Duchy of Lancaster, Cartae Miscellaneae, D.L.36/1/10. (Copied in D.L.42/1, fo. 167r.)

PRINTED: *CDS*, i, no. 553 (calendared).

NOTES: ^{a} Written 7. ^{b} Sic, for treuequor.

COMMENT: The latest possible date is fixed by Roger de Lacy's death (*Book of Fees*, i, p. 64). One of his daughters, whose name is unknown, was evidently Alan's first wife, and the manor of Kippax formed part of her dowry: *TDGNHAS*, 3rd ser., xlix (1972), pp. 50-3. The first four witnesses after Eustace de Vesci were de Lacy clients. For Richard, clerk of Troqueer (in Dumfries), see also *CDS*, i, no. 596. This act possibly belongs to the same period as Robert Bussel's charter of 1205 to Roger de Lacy, attested by Eustace de Vesci, Robert Walensis, William de Beaumont and Thomas Despenser, in *Lancashire Pipe Rolls and Early Lancashire Charters*, ed. W. Farrer (Liverpool, 1902), p. 381.

3

Confirms to John son of Laurence of Newbiggin property (as specified) in Kirkby Thore (Westmorland) equivalent to one carucate of land; also the moiety of Hillbeck (in Brough), which Thomas of Hillbeck gave to Robert, steward of Appleby, with his sister Dionisia. (*c*.June 1216 × *c*.October 1217).

ALanus filius Rolandi scotie Constabularius . Omnibus Hominibus suis tam Franciis*ᵃ* quam Anglicis saLutem ⁊ sciatis me concessisse et presenti Carta mea confirmasse . Johani*ᵃ* filio Laurancii de Neubiging . et Heredibus suis totas tres partes tocius terre que est Inter trutebec et diuisas de sourebi . et Inter viam Regalem Carleoli et Edene . cum Thoftis et Croftis qui sunt Inter Castellum Welp et molendinum . Cum vno thofto et Crofto Ex altera parte vie pro vna Carucata terre . et commune Aisiamentum de Kirkebi thore . preterea ei concedo et confirmo ⁊ totam medietatem de Hellebec quam thoma*s* de Hellebec dedit Roberto dapifero de Appelbi cum dionisia sorore sua . ad tenendum de successoribus Thome de Hellebec Adheo Libere et quiete sicuti Carta testatur quam habet de predicto thoma*ᵇ* de Hellebec . et Carucatam terre de kirkebithore Ad tenendum de Successoribus Waldeui filii Gamelli . sicuti Carta testatur quam Habet de predicto Waldeuo filio Gamelli . Quare . volo et concedo quod predictus Joh*annes* et heredes s[ui]*ᶜ* teneant et habeant Omnes istas predictas Terras bene et Inpace Libere et quiete cum Omnibus pertinenciis et Aisiamentis . sicuti Carte sue Testantur . His Testibus . .E. de Ballielo Ric*ardo* Gernun . Gilberto fil*io* Cospatricii Ric*ardo* de Leuingt' Rad*ulfo* de Campanio . A. de Wigetona . Rog*ero* de Bello Campo . Rad*ulfo* de feritate Eudone de Carll' . Rob*erto* de castello kairoc . Rad*ulfo* daencurt . thoma fil*io* Rand*ulfi* . Magistro . A. de thorentona Thoma de kent thoma de Morlund . Willelmo de Warh*er*cop . N. de morlund . Rog*ero* de Loncastre . Wigano de saunford . Joh*anne* d[e]*ᶜ*mebi . Thoma de tibai . W. fil*io* Hamonis de [H]*ᶜ*ellebec . Simone de uenacione et Aliis .*ᵈ*

ENDORSED: Kirkebythore et Helleb' (xiii cent.); for the Castell of Whelp and iij partes of Down more (xvi cent.?).

DESCRIPTION: Original, 7.5 × 5.9 in (19.1 × 15.0 cm); foot folded to depth of about 0.8 in (1.9 cm), with holes for laces. Central fragment of seal, in green wax, survives; both sides much defaced.

HAND: Highly decorative charter hand, using crossed forms of tironian 'et'.

SOURCE: Kendal, Cumbria Record Office, Crackanthorpe Muniments, WD/Crk/A1428/M-48.

PRINTED: *TCWAAS*, new ser., xvii (1917), pp. 228-9, with facs.; reprinted, with errors uncorrected, in *TDGNHAS*, 3rd ser., v (1916-18), pp. 258-9.

NOTES: *ᵃ* Sic. *ᵇ* Altered from thome. *ᶜ* Parchment holed. *ᵈ* Full point placed at end of line, with series of flourishes between it and Aliis.

COMMENT: Robert, steward of Appleby - John of Newbiggin's grandfather - was the steward of Alan's maternal great-uncle Hugh II de Morville (*Pipe Roll 22 Henry II*, p. 119). Of the witnesses, Thomas of Morland and W[illiam] of Hillbeck had attested for Alan's uncle William de Morville (*CDS*, i, no. 265; *Melrose Lib.*, i, no. 99; cf. *APS*, i, p. 92). Both Thomas and N[icholas] of Morland were influential in the affairs of Carlisle bishopric,

Thomas as the archdeacon's official (*Register of the Priory of Wetherhal*, ed. J.E. Prescott [CWAAS, Record Ser., 1897], nos. 19, 56, 151; *Abstracts of the Charters ... of the Cistercian Abbey of Fountains*, ed. W.T. Lancaster [Leeds, 1915], i, p. 200), and they no doubt helped to secure the see for Alexander II's nominee during the Scottish occupation.

This charter may also illuminate the vexed question of the origins of the Randolph family, which have yet to be traced beyond Alexander II's worldly clerk, Thomas son of Ranulf, and his first appearance in Scottish royal service (*c*.1220). He was most probably one and the same as the twelfth witness above. This Thomas also attested Robert de Vaux's charter (see above, p. 91), with the Constable and the Stewart. He held land in Mansergh in Kendal barony (*Chartulary of Cockersand Abbey*, ed. W. Farrer [Chetham Soc., 1898-1909], III, ii, p. 1035; cf. Carlisle, Dean and Chapter Muniments, MS Machel, iv, pp. 13-14; v, p. 509), and was presumably a younger son of Ranulf son of Walter (d. *c*.1190), lord of Greystoke, Cumberland. Ranulf's son and heir William, who came of age in 1194 and died in 1209, married Helewise de Stuteville, widow of Alan's kinsmen William II of Lancaster and Hugh III de Morville of Burgh by Sands. By *c*.1280, and possibly much earlier, the Randolphs controlled Redpath in Earlston, Berwickshire (*Melrose Lib.*, ii, pp. 685-7); for Alan's lordship there, see Stringer, *Nobility of Medieval Scotland*, p. 68.

4

Grants to Hugh of Crawford the land of Munnoch (in Dalry, Ayrshire) by specified marches, to be held heritably, saving stags, hinds, roe-deer and young hawks, for an annual *reddendo* of one unmewed sparrowhawk. (*c*.1225).

Omnibus ad quos presens scriptum peruenerit Alanus filius Rollandi scoc' constabul*arius* . Salutem . Nouerit uniuersitas vestra nos concessisse dedisse 7 hac presenti carta nostra confirmasse hugoni de crauford' pro homagio 7 Seruitio suo terram de Monoch per istas diuisas . scilicet ab illo loco In quo albi lapides iacent Qui locus uocatur swynissete . Et sic asscendendo[a] versus partem orientalem per rectas diuisas que sunt inter cunigham 7 larges . vsque In Riuulum qui Riuulus cadit de kalderunh . 7 Ita per Riuulum qui cadit de kalderunh descendendo usque illum locum In quo idem Riuulus cadit In kalf . 7 sic per kalf descendendo usque diuisas terre quam Norman equiciarius tenuit . 7 Ita per rectas diuisas terre Norman extranuerso[a] uersus partem occidentalem usque illum locum in quo ille diuise cadunt In Riuulum qui vocatur Monoch 7 per Monoch asscendendo[a] versus partem borealem vsque predictum locum qui vocatur swynissete . Habendam 7 tenendam de nobis 7 heredibus nostris Sibi 7 heredibus suis . libere . Quiete Integre . 7 plenarie . In pace 7 honorifice . In pratis 7 pascuis . In moris 7 mariscis . In stagnis . 7 molendinis . In boscis 7 planis . In pannagijs 7 assartis . 7 In omnibus alijs aysiamentis 7 libertatibus ad predictam terram pertinentibus . Exceptis ceruis . 7 Bisis[a]. 7 capriolis 7 nido ancipitris . Reddendo inde nobis annuatim 7 heredibus nostris ipse 7 heredes sui vnum nisum sor . ad festum assumptionis beate marie pro omnibus seruitijs 7 consuetudinibus 7 demandis . Nos uero 7 heredes nostri predicto hugoni 7 heredibus suis predictam terram contra omnes warantizabimus tam masculos quam feminas

Charter of Alan of Galloway to John of Newbiggin (No. 3)

Size reduced; reproduced by permission of the Cumbria Record Office, Kendal.

. Et vt hec donatio nostra 7 concessio perpetuum Robur firmitatis obtineant
huic scripto sigillum nostrum apposuimus . Hijs testibus . Domino Waltero
fil*io* alani . Domino . InGelRamo de balliolo . Domino . dunechano de
karrich . Domino . Willelmo de brus . Andr*ea* . de loudun . Rob*erto* de Ros
. alex*andro* . de neuh' . hug*one* . de . ardrossan . ada . de hormishoc .
Ricardo capellano . Symone . clerico . Radulfo . de clift' . 7 aliiS .[b]

ENDORSED: No medieval endorsement.

DESCRIPTION: Original, 7.0 × 5.7 in (17.8 × 14.5 cm); foot folded to depth of about 1.3 in
(3.4 cm), with holes for laces. Seal and counterseal in green wax. On obverse, diameter
about 2.8 in (7.1 cm), knight on horseback to sinister wearing hauberk, long surcoat with
tasselled fringe, flat-topped helmet; sword in right hand and shield on left arm bearing
heraldic device (a saltire?); horse with caparisons. Legend: + SIGIL[ ]V[ ] ALANI
FIL[   ]DI. On reverse, the central device, possibly a saltire, is rubbed and indistinct.
Legend: + SECRETVM ALANI DE GALWEIA CONSTABŁ.

HAND: Handsome, regular charter hand.

SOURCE: Glasgow, Strathclyde Regional Archives, Floyer-Acland Muniments, TD.826/1.

NOTES: [a] Sic. [b] 7 aliiS . spaced out to fill entire last line.

COMMENT: Alexander II confirmed this charter on 31 May 1226: *Selectus Diplomatum et
Numismatum Scotiae Thesaurus*, ed. J. Anderson (Edinburgh, 1739), pl. 32. The marches are
from 'Swynissete' (near Glenton Hill) eastwards by the marches between (the districts of)
Cunningham and Largs to the burn that falls from Caldron Hill, thence downstream to the
Caaf Water, thence by the Caaf Water downstream to the marches of the land of Norman the
stud-keeper, thence by those marches westwards to the Munnoch Burn, and so upstream to
'Swynissete'.

Nos. 4-7 seem from the witnesses alone to be closely related in date. Hugh of Crawford,
the beneficiary in all save No. 7, was a cadet of the prominent Clydesdale family based at
Crawford John. His father Reginald was sheriff of Ayr *c*.1219-*c*.1227 but, contrary to *Scots
Peerage*, v, p. 489, there is no evidence that Hugh ever served as such. In *c*.1228 Hugh
occurs with other 'knights of Cunningham' as a witness to Enguerrand de Balliol's charter
anent churches in Galloway (*Holyrood Lib.*, no. 70).

5

Grants to Hugh of Crawford the land of Crosbie (in West Kilbride,
Ayrshire) by those marches by which Ivolotus Parvus held it; to be held
heritably with all the easements and liberties which any knight has in Alan's
whole *tenementum*, saving stags, hinds and young hawks, for an annual
reddendo of one unmewed sparrowhawk, and performance of the forinsec
service due from one ploughgate of land. (*c*.1225).

Omnibus has literas visuris uel audituris Alanus filius Rollandi scoc'
constabularius Salutem . Nouerit uniuersitas uestra nos dedisse concessisse[a]
7 hac presenti carta nostra confirmasse . hug*oni* de crauford' pro homagio 7
seruitio suo terram de crossebi per illas diuisas per quas yuolotus paruus eam
tenuit illa die . qua hug*oni* de crauford' illam terram concessimus habendam
7 tenendam de nobis 7 heredibus nostris sibi 7 heredibus suis . libere .
Quiete . Integre . In pace . 7 honorifice . In boscis 7 planis . In pratis 7
pascuis . In Moris 7 mariscis . In stangnis . 7 molendinis . In vijs 7 semitis

Charter of Alan of Galloway to Hugh of Crawford (No. 4)

Size reduced; reproduced by permission of the Strathclyde Regional Archives, Glasgow.

. In pannagijs 7 assartis . volumus etiam quod habeat omnia aisiamenta 7
omnes libertates infra predictam terram quas aliquis miles habet in toto
tenemento nostro . Exceptis . ceruis . 7 bissis . 7 nido ancipitris . Reddendo
nobis 7 heredibus nostris ille 7 heredes sui annuatim pro predicta terra vnum
nisum . sor . ad fest*um* assumptionis beate marie virginis . Preterea faciet
forinsecum seruitium quantum pertinet ad vnam carucatam terre . 7 nos 7
heredes nostri dicto hug*oni* 7 heredibus suis prenominatam terram contra
omnes homines warentizabimus . Et ut hec donatio nostra perpetuum Robur
firmitatis optineat presens scriptum sigilli nostri munimine corroborauimus .
Hijs testibus domino . Ingelramo de balliol' . Domino Willelmo de brus .
Domino erchebaut de duueg*las* . Domino Willelmo de Geuelist' . Domino
patricio de suthayc . Domino andr*ea* de loud*un* . domino Rob*erto* de log*an* .
Domino Rob*erto* de Ros . hug*one* de ardrossan . Al*ano* de ros . Ric*ardo*
capellano . Rob*erto* de stubhil . yuoloto paruuo*b* . Willelmo fil*io* cospat*ricii* .
Willelmo de Mauecestria . Et AlijS .

ENDORSED: No endorsement.
DESCRIPTION: Original, 6.4 × 4.1 in (16.3 × 10.4 cm); foot folded to depth of about 0.7 in
(1.8 cm), with holes for laces. Same seal, with counterseal, as that on No. 4, in green wax;
less well preserved.
HAND: Same as that of No. 4.
SOURCE: Mount Stuart, Isle of Bute, Marquess of Bute's Muniments, Loudoun Charters,
no. 3.
NOTES: *a* Altered from concessesse. *b* Sic.

6

Confirms to Hugh of Crawford the third part of the toun of Stevenston
(Ayrshire) sold to him by Margaret, daughter of Adam Loccard, for forty
merks, Margaret having resigned her right *per lingnum et baculum* in Alan's
court. Margaret has also renounced her right to the church of Irvine and, as
a woman of full age and in lawful ownership, has for greater security given
Hugh her charter of resignation in the court of Alexander II. Hugh is to hold
the land of Alan for the third part of the service of one knight, as Margaret's
charter bears witness. (*c*.1225).

Omnibus Has litteras visuris uel audituris presentibus et futuris Alanus fil*ius*
Roll*andi* Scott' Constab*ularius* . Salutem . Sciatis nos concessisse et hac
presenti Carta confirmasse Hugoni de Crauford' terciam partem ville de
steuentun' . quam magareta*a* filia ade loccard uendidit eidem Hug*oni* . pro
quadraginta Marcis argenti . et eidem Hug*oni* et heredibus suis totum Jus
quod habuit in predicta tercia parte sua quietum Clamauit de se et heredibus
suis In perpetuum . Et Illud predicto Hugoni et heredibus suis In curia
nostra per lingnum et baculum reddidit . Et omne Jus suum Inde In
perpetuum In Ecclesia de yrewin abiurauit . Et ad Istud tenendum . et In
perpetuum . firmiter obseruandum ⁙ Cartam suam quiete Clamationis . Inde
in Curia Domini nostri Alex*andri* . Regis scocie ⁙ predicto Hugoni dedit ⁙

tamquam mulier legittime etatis . et ligie potestatis . Hanc autem terram concessimus et confirmauimus predicto Hug*oni* et heredibus suis tenendam de nobis et heredibus nostris . libere . solute . honorifice . faciendo tamen nobis et heredibus nostris de predicta tercia parte Ville de Steuentun' : quantum pertinet ad terciam partem seruicij vnius Militis . secundum quod Carta Ipsius Margarete Juste perportat . Hiis Testibus . Ada . de loudun . Roberto de Ros . Arthuro de Ardrossan . Roberto de logan . Alex*andro* . de Neuh' . Andrea de loud*un* . Ricardo de Wintonia Roberto fil*io* ledmeri . Alano de Ros . Hugone de ardrossan . Ricardo de Cuningham . Ada de ormeshoc . et Aliis .

ENDORSED: No medieval endorsement.

DESCRIPTION: Original, 6.7 × 3.2 in (17.0 × 8.1 cm); foot folded to depth of about 0.7 in (1.8 cm), with slits for two tags. One tag remains; no seal survives.

HAND: Similar to that of Nos. 4 and 5, but more cursive and using a crossed form of tironian 'et'.

SOURCE: Mount Stuart, Marquess of Bute's Muniments, Loudoun Charters, no. 4.

NOTE: *a* Sic.

COMMENT: On the Loccard family, see G.W.S. Barrow, *The Kingdom of the Scots* (London, 1973), pp. 289, 350; Barrow, *Anglo-Norman Era*, p. 46. Adam was probably the son and heir of Stephen Loccard, after whom Stevenston is named. *Scots Peerage*, v, p. 489, cites a deed, '*penes* Earl of Loudoun', by John of Samuelston exchanging his rights in Stevenston for Hugh of Crawford's rights in Cousland in the feu of Livingston, West Lothian. This act cannot now be traced.

7

Grants to Alan de Ros all the land of Fairlie (in Largs, Ayrshire) by specified marches, to be held in feu and heritage as anyone most freely holds of the donor for habergeon service, saving stags and hinds, by providing in the expedition of his army one young man with a habergeon, and by paying the aids due from one ploughgate of land. (*c*.1225).

Omnibus hoc scriptum visuris vel audituris, Allanus filius Rollandi Scotie constabularius, salutem. Noveritis nos dedisse, concessisse et hac presenti carta nostra confirmasse Allano de Ros, pro homagio et servitio suo, totam terram de Ferneslie*a* per has divisas, scilicet a loco ubi Kaulburne cadit in mare ascendendo per Kalburn usque ad locum ubi Clererburn cadit in Kaulburne et sic ascendendo per Clererburn versus austrum usque ad capud de Clererburn et sic a capite de Clererburn per quandam vallem sub quadam rupe ex transverso versus austrum usque in*b* caput de Pollroch et sic descendendo per Pollroch usque in mare. Tenendam sibi et heredibus suis de nobis*c* et heredibus nostris in foedo et hereditate adeo libere, quiete, plenarie, honorifice sicut aliquis*d* per servitium haubergelli liberius, quietius, plenarius, honorificentius in tota terra nostra de nobis tenet aut possidet, in boscis et planis, in pratis et pascuis, in moris et maresijs, in panagijs et assartis,*e* in stagnis et molendinis, in ripis et piscarijs et in omnibus alijs aisiamentis et libertatibus ad predictam terram pertinentibus, salvis nobis et heredibus nostris cervis et bissis,*f* inveniendo nobis et heredibus nostris ipse et heredes sui in expeditione exercitus nostri unum juvenem cum haubergello, et reddendo in auxilijs quantum pertinet ad unam carrucatam terre. Et ut hec donatio nostra perpetuum robur firmitatis obtineat, huic*g* scripto sigillum nostrum fecimus apponi. Hijs testibus Domino Ingelramo de Baliolo, Reiginaldo de Crauford, Rogero de Bellocampo, Andrea de Loudun,*h* Richardo de Winchestra,*i* Hugone de Crauford, Will*elmo* fil*io* Lamber*ti*, Thom*a* de Cancia, Joanne de Crauford, Ric*ardo* capellano, Hug*one* de Ardrossan, Simone clerico, Rad*ulfo* de Clifton*j* et alijs.

HEADING: Dowble of a charter granted by Allanus filius Rollandi Scotie Constabular*ius* Allano de Ros terrarum de Ferneslie tenen' per servitium haubergellj.*k*

SOURCE: NLS, MS Advocates 15.1.18, fo. 75*r* (xviii-cent. copy) = *A*. SRO, Deeds Warrants, RD.12/40/942 (of 12 December 1700) = *B*. Text above based on *A*, with some variants of *B* noted.

NOTES: *a* Fernisley, *B*. *b* ad, *B*. *c* *A* omits last five words. *d* *B* adds qui. *e* allartis, *A*; alareis, *B*. *f* billis, *A* and *B*. *g* hoc, *A*. *h* Loride, *A*; Londe, *B*. *i* Wyncestra, *B*. *j* Aitton, *A*; Aillon, *B*. *k* *A* only.

COMMENT: The marches are from the Kel Burn where it falls into the sea upstream to the confluence of the Clea Burn and the Kel Burn, thence upstream to the head of the Clea Burn, thence by a valley beneath a certain craig southwards to the head of the 'Pollroch' (Fairlie Burn), and so downstream to the sea.

The donee's family had been established by the de Morvilles as tenants of Ardneil (in West Kilbride) and Stewarton. On the form of military service due, suggesting 'a tentative feudalization of "common army"', see *RRS*, ii, p. 56, and Barrow, *Anglo-Norman Era*, p. 140 (commenting specifically on this charter). Ralph of Clifton, whose surname is garbled in the charter copies, was one of Alan's clerks (*Melrose Lib.*, i, nos. 79, 227).

Appendix (B)

Alan of Galloway's Dependants: Four Brief Biographies

(CLD = Cumberland; WLD = Westmorland)

SIR ENGUERRAND DE BALLIOL (tenant)

Main source: G. Stell in Stringer, *Nobility of Medieval Scotland*, pp. 153-4.

Almost certainly younger son of Eustace de B. (d. *c*.1208) of Barnard Castle; bro. of Henry de B., chamberlain of Alexander II (1223-30, 1241-6); uncle of John I de B., husband of Alan's dau. Dervorguilla. Probable lord of Tours-en-Vimeu, Picardy. Married dau. of Walter of Berkeley, chamberlain of William I. Held *jure uxoris* Urr (Galloway) and barony of Inverkeilor (Angus); also held Dalton in Hartness and at Bolam (Co. Durham) (*Victoria County History, Durham*, iii, p. 255; *Rot. Litt. Claus.*, ii, p. 202a). First occ. in Scotland *c*.1200 (*Glasgow Reg.*, i, no. 90); supported Alan in Cumbria 1216-17. Often at Alexander II's court to 1236; sheriff of Berwick 1226 (*Moray Reg.*, no. 29).

SIR ROGER DE BEAUCHAMP (household knight)
    Main source: *TCWAAS*, new ser., xix (1919), pp. 128-9, 137.
De Morville kinsman (Barrow, *Anglo-Norman Era*, p. 71); bro. of Richard de B., constable
of Rouen 1203 (*TCWAAS*, new ser., xi [1911], p. 318; *Rotuli Normanniae* ..., ed. T.D.
Hardy [Record Comm., 1835], pp. 107, 115). Married widow of Thomas son of Cospatric
of Colvend (Galloway) and Workington (CLD) by 1209. Held *jure uxoris* Knipe Patrick
(WLD); also held Staffield (CLD). Regular witness for Hugh III de Morville (d. 1202) of
Burgh by Sands. Occ. as Alan's knight by 1212 (*CDS*, i, no. 533); supported him in
Cumbria 1216-17. Under-sheriff of WLD 1199; juror and justice in CLD or WLD 1219-24.
Benefactor of Cockersand Abbey (*Chartulary of Cockersand*, II, ii, p. 599), St Bees Priory,
Shap Abbey (British Library, MS Harleian 294, fos. 207ᵛ-8ʳ, 209ᵛ), Wetheral Priory; gave
body for burial at Wetheral 1223 × 1229.

SIR RALPH DE CAMPANIA (tenant)

Son of Alice of Croft, *alias* of Thurlaston (Leics); bro. of Robert de C., who held 3½ fees
of Leicester honour and (after 1219) land at Great Doddington (Northants) of Huntingdon
honour (British Library, Additional Charter 21,198; *Curia Regis Rolls*, i, p. 417; *Pipe Roll
16 John*, p. 114; W. Farrer, *Honors and Knights' Fees* [London and Manchester, 1923-5], ii,
p. 345). Held with Henry de C., another bro., 1 fee of Leicester honour at Shapwick
(Dorset) (*Book of Fees*, i, p. 92); also held Borgue (Galloway). Occ. *c.*1200 in retinue of
Roger 'de Beaumont', bishop of St Andrews, who probably brought him to Scotland
(Raine, *North Durham*, no. 469). 'Cum Alano de Gawee' 1213 (*Praestita Roll 14-18 John*, p.
103); supported him in Cumbria 1216-17. Fairly often at Alexander II's court to 1233;
constable of Roxburgh *c.*1228 (*Melrose Lib.*, i, no. 282; cf. *Glasgow Reg.*, i, no. 148).
Benefactor of Dryburgh Abbey. Roger Waspail, step-father, became seneschal of Ulster
1224 (*Rot. Litt. Claus.*, i, pp. 278a, 588a). Robert de C. served as steward of John of
Scotland, earl of Chester and Huntingdon (d. 1237), and was buried at Lindores Abbey
(*Arbroath Lib.*, i, no. 84; *Lindores Cart.*, no. 87).

SIR IVO DE VIEUXPONT (tenant)
    Main sources: *TDGNHAS*, 3rd ser., xxxiii (1954-5), pp. 94ff, with Barrow, *Anglo-
    Norman Era*, pp. 73 (n. 73), 76 (n. 88); *TCWAAS*, new ser., xi (1911), pp. 268, 271-
    81, 316-19.
Son of William II de V. and Maud, dau. of Hugh I de Morville; bro. of Robert de V., lord
of Westmorland; step-bro. of William III de V. of Bolton (E. Lothian), Carriden (W.
Lothian), and Langton (Berwickshire), and of Hardingstone (Northants) in Huntingdon
honour. Held, *inter alia*, Sorbie (Galloway), Glengelt (Lauderdale) (*Dryburgh Lib.*, no.
186), Alston (Tynedale), Maulds Meaburn (WLD), and 4 fees of Leicester honour (*Pipe
Roll 13 John*, p. 195). Participated in 3rd Crusade and fought against Turkish pirates
(*Chronicles ... of the Reign of Richard I*, ed. W. Stubbs [Rolls Ser., 1864-5], i, pp. 93, 104).
Employed by King John as treasury official (PRO, S.C.1/1/4; *Rot. Litt. Claus.*, i, p. 98b),
custodian of Alston silver-mines, etc. Occ. with Robert de V. in Wendover's list of John's
'evil advisers', but took Scottish side 1217. Benefactor of Dryburgh Abbey, Hexham
Priory, St Peter's Hospital, York. Nicholas de V., son, was bailiff of Tynedale for
Alexander III.

5

The March Laws Reconsidered

WILLIAM W. SCOTT

The earliest surviving text of the March laws — the code believed to have regulated the affairs of the Anglo-Scottish Border in the twelfth and thirteenth centuries — states that there gathered 'in the year of grace 1249 on the feast of Saints Tiburtius and Valerian [14 April] at the March the sheriff of Northumberland, on behalf of the king of England, and the sheriffs of Berwick, Edinburgh and Roxburgh, on behalf of the king of Scotland, to recognosce the laws and customs of the March by twelve knights of England and twelve of Scotland'. The record continues with the names of the knights who took part — seven from the sheriffdom of Berwick, five from the sheriffdom of Edinburgh, and twelve English knights — and then follow the fourteen articles of the laws.[1]

That gathering had been preceded by an inquiry on 16 November 1248. George Neilson, in comments on a translation of the laws prepared about 1900 but not printed until 1971, proposed that 'these proceedings of 1248 ... must have been the principal inducing cause of the recognisance of 1249'. The importance of the 1248 meeting as a preparation for the formal promulgation of the laws in 1249 was not developed further by Neilson, although he did remark that of the twelve jurors of 1248 no fewer than seven reappear on the jury of 1249.[2] Clearly, there was some continuity between the meetings. But can anything more be said about Neilson's proposal? Did the 1248 meeting have any special significance?

The 1248 inquiry was held by a group of knights from England and Scotland, who had gathered to investigate certain points about the March laws. A writ of Henry III of England, which had set the arrangements in motion, recalled that he had heard from the king of Scotland's envoys that the laws and customs of the Marches of the kingdoms were now less well observed than formerly, and that a particular injury had been done to Nicholas de Soules against the laws. A meeting was therefore convened at the March on the River Tweed for the purpose of correcting offences against the March laws and customs and looking into the particular case.[3]

Both the 1248 and the 1249 meetings can be put into a longer

---

1   Latin text of the laws in *APS*, i, pp. 413-16; translation in G. Neilson, 'The March Laws', in *Stair Society Miscellany*, i (1971), pp. 15-24.
2   Ibid., p. 15.
3   *CDS*, i, no. 1749 (calendar), pp. 559-60 (Latin text).

perspective, because they were not the first times when men had gathered from both sides of the Border to consider disputed points. In 1245 an attempt to define the line of the March over Wark common had ended in deadlock; another attempt was made just over a year later on 1 December 1246. In 1245 the Scots had been led by David of Lindsay, the justiciar of Lothian, and Patrick II earl of Dunbar. No other names from the Scottish side are recorded.[1] In 1246 the Scottish team was once again led by David of Lindsay, accompanied by David de Graham (probably then sheriff of Berwick), David the clerk (probably David of Machan, the justiciar's clerk), and Nicholas de Soules, then sheriff of Roxburgh. No other Scottish names are known, although there is a full list of the English delegation.[2] In 1248, in contrast, we do not know the leaders of the Scottish delegation, but for the first time we know the names of the Scottish knights. They were Ranulf of Bunkle, Robert son of the earl, Robert de Bernham, Robert of Durham, William of Mordington, and Simon of Grubbit.[3] What contribution could these men make to the purpose of the meeting?

The writ convening the meeting makes it clear that those gathered at the March were expected to ensure that the customary laws were to be observed fully. This in turn implies that together they should have some knowledge of the substance of the laws or be in a position to draw on the knowledge of others, and their retour claimed that this was what they had done. The summoning of men to give evidence as to facts was a feature of the laws of both England and Scotland. It is less well attested in Scotland, but one inquiry held at Perth in 1206 into the rights of the bishop of St Andrews in part of the land of Arbuthnott in the Mearns heard evidence from a group of local men who collectively described events over a period of some fifty years previously.[4] Another inquiry, held at Ayr in the 1230s, took the evidence of local men from Old Kilpatrick in the Lennox about the customary services due from some land there. This time the evidence went back to the 1170s, some sixty years before the inquest.[5] In both cases, the men gathered to say what they knew about the past were comparatively humble members of society. But they were all members of a local community speaking from personal knowledge or repeating what their fathers had told them. The knights who gathered in 1248 were higher in the social scale, but they were no less local men.

Ranulf of Bunkle, as his name suggests, was the lord of Bunkle in Berwickshire. His family's connection with this land went back to at least 1203, when an ancestor was sheriff of Berwick. Ranulf first appears on record in *c.*1210 as holding some land in Cumberland and Westmorland. He was still holding English lands in the 1240s, and so at the time of the 1248

1   Stones, *Anglo-Scottish Relations*, no. 8; *CDS*, i, nos. 832 (misdated), 1676.
2   *CDS*, i, no. 1699.
3   *CDS*, i, p. 559.
4   *Spalding Misc.*, v, pp. 209-13.
5   *Paisley Reg.*, pp. 166-8.

inquiry he was clearly able, from considerable family and other local knowledge, to speak about customs on both sides of the Border.[1]

Robert son of the earl was probably a son of Earl Patrick I of Dunbar. If he was a child of the earl's first wife, as has been supposed, he would have been born before 1200. He witnessed acts of his brother Earl Patrick II and of his nephew Earl Patrick III. In August 1247 he held the post of steward to his brother and he may have still been in this position towards the end of 1248.[2] But when the March laws were set out at the larger meeting in 1249, Robert's place had been taken by Alan of Harcarse, who by then was steward of Earl Patrick III.[3] Robert and Alan had probably taken part as representatives of the earl of Dunbar who, as a condition of his tenure of Beanley and other lands in Northumberland, had a special responsibility for enforcing the laws.[4] Earl Patrick II had been involved in the abortive meeting of 1245. In 1248 his brother Robert was probably therefore representing both by his office and also by his blood a family which had long held a leading position in the Borders. He could have had some personal knowledge of the operation of the laws and he was certainly in a position to find out about them through his family.

Although described in the inquiry as a knight, Robert de Bernham is usually known as the first recorded mayor of Berwick upon Tweed, a position he had probably held since 1238.[5] His presence was due perhaps to his position, and perhaps to personal knowledge of local customs, but it also suggests an important question: how did Berwick and its citizens stand in relation to the laws?

The laws do not mention the place. For all they say, it might not have existed. But by 1248 Berwick, an international port since at least the 1170s, could hardly be ignored. Threats to its security had been a cause of tension between England and Scotland earlier in the thirteenth century, and it had been an important target for King John when he raided Scotland in 1216.[6] It was also, of course, an important crossing of the River Tweed. In practice, therefore, it could scarcely be left out of the regulation of March affairs. A practical example occurred in about 1200 when the Tweed bridge was washed away. According to the English chronicler Roger of Howden, Earl Patrick I of Dunbar, then justiciar of Lothian, took the lead in arranging for the bridge to be replaced. After some difficulties with the bishop of Durham, he succeeded in doing so with the help of the sheriff of Northumberland.[7] The earl's position of special responsibility on the March, and his office as justiciar, can explain his interest in the incident.

---

1   Raine, *North Durham*, no. 265 (× 1203); *CDS*, i, nos. 542, 643, 1106, 1296, 1591.
2   *Scots Peerage*, iii, p. 254; *Melrose Lib.*, i, nos. 230, 232, 235-6.
3   Neilson, 'March Laws', p. 16.
4   See below, pp. 123ff.
5   Raine, *North Durham*, no. 72; A.A.M. Duncan, *Scotland: The Making of the Kingdom* (Edinburgh, 1975), pp. 494-5.
6   Lawrie, *Annals*, p. 130; Anderson, *Scottish Annals*, pp. 322, 332-3; *Chron. Fordun*, i, p. 277; Anderson, *Early Sources*, ii, pp. 406-7.
7   Lawrie, *Annals*, p. 320.

But there may be more to it than that. Howden, who was usually well informed on Scottish events, also says that the earl was the 'custos' (keeper or guardian?) of Berwick. There is no contemporary Scottish evidence to show what this position entailed, but it seems a reasonable inference that the *custos* was in a position of some responsibility towards the burgh, and one might say that he was in charge of the place on behalf of the king of Scots. By 1248, however, this would no longer have been true. In 1238 Berwick appears to have freed itself from direct rule by the king and to have achieved a certain amount of self-government under a mayor. The precise powers of the mayor are nowhere defined, but analogies with other contemporary practice would suggest that the mayor became in effect the *custos* of the burgh.[1] In short, just as one keeper of Berwick had been concerned with the replacement of an important cross-Border bridge in *c.*1200, so one of his successors in that position was concerned with the regulation and correction of March affairs in 1248. And so, although Berwick is not itself mentioned in the laws, it looks as though a practice had grown up of involving the chief representative of the community of the burgh in business affecting the March. This, in turn, implies adaptability in the way the laws were administered.

The next person on the list is Robert of Durham, a burgess of Berwick who occurs as its mayor in 1251.[2] Perhaps we should see him in 1248 as second-in-command to Robert de Bernham and as having the same general experience and interests. If his name is to be taken literally, he obviously had links with the north of England.

With William of Mordington, we return to the Berwickshire gentry. He was a tenant of the priory of Coldingham (a daughter house of Durham Cathedral Priory) and can be traced as a witness of charters issued by the prior of Coldingham and other local people from at least the early 1220s and possibly as early as *c.*1214. He was clearly one of the leading men of Coldinghamshire and is regularly found in attendance at the prior's court. His local position probably made him an obvious choice for the inquest. His career had by 1248 reached its peak; at the time of the inquest he either was, or had recently ceased to be, the sheriff of Berwick, for he is known to have held this position in 1247.[3]

The last Scot is Simon, lord of Grubbit, a place-name which survives in Grubbit Law in the parish of Morebattle in Roxburghshire. He must have been a very old man in 1248, for he first appears on record in the 1190s. The inquest was probably his last public appearance, since he was not at the larger meeting of 1249, and by the early 1250s the lord of Grubbit was named John. Simon's father Uhtred was alive in the 1170s and Simon could therefore have drawn from his own experience and from his family a

1   Ibid.; Duncan, *Making of the Kingdom*, pp. 494-5; C. Platt, *The English Medieval Town* (London, 1976), pp. 152-62.
2   *Coldstream Cart.*, no. 52; *Melrose Lib.*, i, no. 314; Raine, *North Durham*, no. 258.
3   Ibid., nos. 183, 239, 302, 331; Fraser, *Douglas*, iii, no. 285.

knowledge of how March affairs were regulated at least as far back as the twelfth century.[1] Grubbit Law is just to the north of Dere Street, the Roman road over the Cheviots, which in this section of the March was known as Gamelspath. Article 1 of the laws provides that this is where the business of the March is due to be done for the men of Redesdale and Coquetdale,[2] and Simon was probably present to ensure that the customs of that part of the March were properly represented.

The six Scottish representatives of 1248 give the impression of being carefully chosen men, attending because of an official position with an interest in March affairs and/or because of personal knowledge, derived from their own experience or from that of their families or communities, which could go back many years. Some certainly, and perhaps all, were conversant with northern England. That they all came from parts so close to the Border line might suggest that they had taken part in the operation of the laws. At the very least, they were well suited, it would seem, to meet a requirement to discover what the laws had been in earlier times. The personal knowledge of Simon of Grubbit certainly went back beyond 1200, and their combined knowledge probably amounted to a very authoritative view of what the Scots believed the laws and customs of the March to have been throughout the first half of the thirteenth century and for some years before then.

Nicholas de Soules, whose complaints had led to the 1248 inquiry, also fits into this pattern. He was the head of a family which had been established in Liddesdale before the mid-twelfth century, and held land in northern England as well. He was also sheriff of Roxburgh.[3] In all likelihood he was a troublesome man in English eyes, but he was described in a Scottish chronicle as 'the wisest and most eloquent man in the whole kingdom'.[4] No doubt he was well aware of his rights under the March laws and in a position to see that they were not forgotten or infringed.

The English representatives in 1248 were Roger son of Ralph, William of Scremerston, Robert Malenfaunt, Robert of Cresswell, Patrick of Goswick, and Gilbert of Beal.[5] The first of these, Roger son of Ralph, succeeded his father in Northumberland as the lord of Ditchburn (three miles from Beanley) in about 1205, and died in 1252. He was also lord of the main part of Togston, near Warkworth, and both estates were held in chief. He had been a member of the English party in the 1245 dispute. By 1248 he had increased his local influence, having been entrusted with the custody of the important royal castle of Bamburgh. In 1246 he and others were ordered to see to the setting out of a certain new town ('ad ordinandam quandam

1   *Melrose Lib.*, i, nos. 116-19, 127, 137, 139-40, 268, 300; *Kelso Lib.*, i, no. 149.
2   Neilson, 'March Laws', p. 16.
3   *RRS*, i, pp. 96, 160, 168; *CDS*, i, nos. 1649, 1654-5, 1699, 1739, 1749, 1765, 1776, 1862.
4   *Chron. Bower* (Watt), v, p. 349.
5   *CDS*, i, p. 559.

villam novam') at Warenmouth.[1] His career appears to have been similar to that of the Scot, William of Mordington — a lord who gradually consolidated a local position and reputation over a long life.

William of Scremerston was lord of Scremerston, just south of Tweedmouth, and represented the second or third generation of a knightly family established in Islandshire, within the lordship of the bishop of Durham. He also held land in Scotland in the 1240s as a tenant of the priory of Coldingham, and occupied during a minority some land usually held by a family of Ridels. He was also the prior's steward and can be found witnessing the prior's *acta* at various dates between 1246 and 1254. At other times he appears in the company of William of Mordington.[2]

Less is known about Robert de Malenfaunt and Robert of Cresswell. As knights of the shire, they can be found taking part in judicial business affecting Northumberland, and they had been members of the English delegation in the dispute of 1245. They therefore participated in the 1248 proceedings with some previous experience of March business, and Malenfaunt appeared again in 1249.[3]

The last two English representatives in 1248, Gilbert of Beal and Patrick of Goswick, were knights whose families held land in Islandshire of the bishop of Durham at least as early as 1208 × 1210. Their names were carried from father to son, and so it is difficult to be clear which generations of the Gilberts and Patricks took part in the 1248 inquest. Men of these names witness documents in the company of William of Scremerston in the 1240s and '50s, and a Gilbert of Beal granted land in both Holy Island and Beal to the monks of Durham.[4] Whatever the precise pedigree, it is clear that they were men of local standing whose roots go back to at least the early thirteenth century, while the families of Roger son of Ralph and William of Scremerston are clearly traceable to the late twelfth century. Here, then, were members of the inquest with access to knowledge and recollection rooted within Islandshire and also, in the person of Roger son of Ralph, with responsibilities at Bamburgh, an old royal and administrative centre of considerable importance in north Northumberland. William of Scremerston was conversant with eastern Berwickshire, and Roger son of Ralph, as a near neighbour of the earls of Dunbar at Beanley, was probably acquainted with their conditions of tenure and responsibilities for enforcing the March laws.

A striking feature is the comparatively small area from which those who

1   W.P. Hedley, *Northumberland Families*, i (Newcastle upon Tyne, 1968), p. 251; *The Book of Fees* (London, 1920-31), i, pp. 203, 598; ii, pp. 1118, 1120; *Close Rolls of the Reign of Henry III, 1242-7* (London, 1916), pp. 410, 467; *Calendar of the Patent Rolls, 1247-58* (London, 1908), p. 13; *CDS*, i, no. 1676.
2   Hedley, *Northumberland Families*, i, p. 276; *Book of Fees*, i, p. 27; Raine, *North Durham*, nos. 210-11, 214, 239, 250, 379.
3   *Book of Fees*, ii, pp. 1120-1; *Northumberland Pleas from the Curia Regis and Assize Rolls, 1198-1272*, ed. A.H. Thompson (Newcastle upon Tyne Records Committee, 1922), pp. 137-8, 158; *CDS*, i, no. 1676.
4   Raine, *North Durham*, nos. 684-5, 693, 702, 756, 776, 778, 784.

took part in the inquest of 1248 were drawn. At York in 1237 Alexander II and Henry III had settled many outstanding Border disputes. The Scots had given up their claims on Cumberland, Westmorland and Northumberland, and old issues had been settled throughout the whole length of the Border.[1] In this new situation, an inquiry into the March laws could draw, one might think, on men from both sides of the Pennines and the Cheviots. Yet in 1248 — and again in 1249 — it was men from the most easterly part of the March who came together. Except for Simon of Grubbit, the Scots in 1248 were all from Berwickshire and the immediate north side of the Tweed, and only Ranulf of Bunkle seems to have had interests on the western March. In 1249 the Scots were represented by men from the sheriffdoms of Berwick and Lothian. On the English side in 1248 and 1249, the choice was from men of Northumberland.[2]

In 1248 the selection of jurors from, in particular, Scremerston and other places within the lands of the bishopric of Durham is a reminder that when a list of the oldest possessions of Saint Cuthbert had been drawn up in or about 975, it included the land from the sea to Norham on the south bank of the Tweed. It also embraced much land now within Scotland: the Bowmont valley, the land between the Whiteadder and Leader Waters, and Old and New Jedburgh.[3] Any settlement of a new Border on or near the Tweed, whether in the tenth or eleventh century, was bound to destroy the integrity of such an estate, and so would be an issue of great importance for Durham and its jurisdiction. The new Border seems to have severed at least one local community — a 'shire' centred on Yetholm[4] — and so it would not be surprising if men from the lands of Saint Cuthbert on the south bank of the Tweed had been involved at the time in attempts to regulate the new Border. Indeed, they probably were, since the place-names in the eastern March listed in Article 1 of the laws — 'Hamisford' (Norham), Reddenburn and the Duddo Burn — are all on the south side of the river.[5] We may therefore be entitled to interpret the presence of men from Northumberland and Islandshire in 1248 as more than a matter of providing local worthies for an inquiry; it might also represent the vestiges of a tradition in which men from these areas had been involved from the time when the Border had been settled on or near the Tweed and the earliest rules for dealing with cross-Border disputes drawn up.

Whether or not that inference is correct, we still have in 1248, on both sides, men of local standing and knowledge who would have been able to deliver an authoritative statement of what they believed to be the correct laws and customs of the March. Their retour after the inquest claimed that

1   Stones, *Anglo-Scottish Relations*, no. 7.
2   Neilson, 'March Laws', p. 16.
3   H.H.E. Craster, 'The Patrimony of St Cuthbert', *English Historical Review*, lxix (1954), esp. pp. 178-80.
4   G.W.S. Barrow, *The Kingdom of the Scots* (London, 1973), pp. 32-5.
5   Neilson, 'March Laws', pp. 16-17; Barrow, *Kingdom of the Scots*, pp. 158-9.

they had made inquiry by old and discreet men of the March in both England and Scotland; they delivered that Nicholas de Soules had been wronged because he had been impleaded before the king of England; and they affirmed that by the laws of the March subjects of the realm of England or of the realm of Scotland, though they may hold land across the Border, ought not to be impleaded other than at the March for trespasses committed by men of one country in the other. Apart from a later order to the sheriff of Cumberland to do right to Nicholas de Soules in accordance with the customs of the March, nothing further seems to have formally resulted from the 1248 meeting.[1] Yet the close coincidence of that inquest and the next meeting in April 1249 is at least suggestive that the one led to the other.

The trigger for the 1249 meeting may have been the unanimously agreed statement about persons being impleaded at the March. This was an important statement of principle, set out again in Article 1 of the laws in 1249. It echoes a much earlier statement made by Malcolm III at Gloucester in 1093, when he declined to accept judgement in the court of William Rufus. According to 'Florence of Worcester', Malcolm 'would by no means do this, unless upon the borders of their realms, where the kings of the Scots were accustomed to do right by the kings of the English, and according to the judgement of the chief men of both kingdoms'. The source for this information and for the words used was compiled early in the twelfth century but probably relied on materials gathered from the mid-1090s onwards.[2] There is no real reason to doubt that it was recording accurately the stand taken by Malcolm III. And if the words are taken literally, the practice of meeting on the Border was in use in the time of Malcolm's predecessors, some of whom indeed do seem to have met individual kings of Wessex, or treated with their representatives, either on 'middle ground' or in border areas.[3]

In any event, the principle behind Malcolm III's statement should not have surprised (although it clearly irritated) Rufus and his advisers, despite the fact of Malcolm's submission at Abernethy — nowhere near the March — in 1072. For the dukes of Normandy, doing business with the kings of France at the march between Normandy and lands under the control of the Carolingians and Capetians had been the normal mode of operation since the treaty of Saint-Clair-sur-Epte of 911. Rufus could hardly have been unaware of that, or of similar precedents affecting Anjou and Brittany.[4] Malcolm III's claim was not only a declaration that Scotland was a separate and independent kingdom, but a statement of custom conforming with acceptable and normal contemporary practice elsewhere in northern Europe. The

---

1   *CDS*, i, pp. 559-60, nos. 1765, 1776.
2   Anderson, *Scottish Annals*, p. 110; A. Gransden, *Historical Writing in England, c.500 to c.1307* (London, 1974), pp. 143-4, 146.
3   Meetings, for example, at Eamont near Penrith and at Chester. See generally A.P. Smyth, *Warlords and Holy Men: Scotland 80-1000* (London, 1984), pp. 199, 201, 228, 232.
4   J.-F. Lemarignier, *Recherches sur l'Hommage en Marche et les Frontières Féodales* (Lille, 1945), pp. 73-85.

statements of 1248 and 1249 that impleading took place at the March are consequently not as unusual and isolated as might appear at first sight. And, even if we did not have the earlier Scottish evidence to support the point, the length of time that some of the Continental arrangements had been in use gives some colour, to put it no more strongly, to Malcolm III's claim that the custom went back to the time of his predecessors, and so at least to the first half of the eleventh century.

The exact precedent of Malcolm III's statement might not have been in anybody's mind in 1248. But the principle, reiterated by the inquest, was drawn from some very long memories. It mattered to the Scots. For a different reason, it might have touched a nerve in England. During the 1240s Henry III's government had begun to take an interest in the content and operation of the laws of the Marches of Wales and also, it has been proposed, in liberties and rights of jurisdiction in general.[1] Against this background, a statement, supported by English jurors, which denied jurisdiction to an English sheriff in his own court and so appeared to deny the pervasiveness of royal justice, could have been seen as a critical issue requiring further attention — hence, perhaps, the attempt early in 1249 to get the March laws properly recorded. Whether the men at the 1248 meeting had had before them an earlier text of the laws, and had then prepared a text which was formally considered and promulgated in 1249, is not known. It may well have been so. At the very least, the fact that seven of the twelve reappear in 1249 gave the necessary knowledge and continuity for full consideration of issues which the 1248 meeting may well have aroused.

How well the jurors of 1249 wrought in delivering the text of the laws is impossible to say, because there are no earlier texts for comparison. But there seems to have been agreement in 1248 and 1249, in contrast to what had happened in 1245. In Scotland the 1249 text was an item in the first known collection of Scots law, a product of the 1270s. It looks as though the text was accepted by contemporaries as an agreed and accurate statement and, on the Scottish side, was later felt to be an important element in the growing body of Scots law. There was a copy of the March laws in the Scottish royal archives in 1292.[2]

Beyond that, there is some scattered evidence which confirms certain aspects of the laws. First, as Neilson pointed out, a letter of Pope Innocent III of 1216 deplored the fact that trial by combat, an important element in the laws, was applied to ecclesiastics, and anathematised the practice.[3]

---

1   R.R. Davies, 'Kings, Lords and Liberties in the Marches of Wales, 1066-1272', *Transactions of the Royal Historical Soc.*, 5th ser., xxix (1979), esp. pp. 56-60; cf. M.T. Clanchy, 'Did Henry III have a Policy?', *History*, liii (1968), pp. 208-10. The provisions of the 1237 treaty of York about jurisdiction in lands to be granted to the king of Scots in northern England (Stones, *Anglo-Scottish Relations*, no. 7) are an earlier attempt at definition in a different but no less sensitive set of circumstances.
2   *APS*, i, p. 114.
3   Neilson, 'March Laws', pp. 13-14.

Second, an unusual surviving fragment reveals arrangements for a legal challenge to be delivered to David earl of Huntingdon (brother of the king of Scots); it was taken to the March.[1] Third, there are a few instances from the twelfth century of the recognised places for March business being used in peace and war.[2] Given these slight but confirmatory pieces of information, the status of the jurors, and the general acceptance of their work, we may surely assume that they faithfully reported within their knowledge. But, that said, it has to be admitted that the result of the inquest of 1249 is a text of some antiquity and complexity.

Neilson's essential view was that the laws were much older than 1249, and he suggested that originally they were a product of the twelfth century. As to which part of that century, he had no firm views: he argued at one point for 'a time rather before than after the reign of Henry II'; at another, for 'an age as early as that of Glanvill and Henry II'; and at yet another, for 'about the close of the twelfth century'.[3] Since his paper was a draft, it would be unfair to emphasise these inconsistencies and, in any case, they are not entirely inappropriate, for different elements of the laws do seem to date from different times. This aspect was well brought out by Geoffrey Barrow, who drew attention to some very archaic (certainly pre-1066) features of Article 1 (the names of the places where March business was to be done) and of Article 5 (the status of the priest of Stow in representing the bishop of St Andrews).[4] Examination of other Articles reinforces the impression that the 1249 text is really a mosaic of provisions from different periods. A full discussion of this aspect of the laws must await another occasion. But a glance now at two features — the responsibilities of the earls of Dunbar under the laws and the provisions for the recovery of fugitive naifs — can throw some useful light on the evolution of the text.

Article 13 of the laws states 'that *inborch* and *utborch* have power to distrain both realms'. In 1212 it was reported that the earl of Dunbar held the lordship of Beanley 'per servicium quod sit inborhe et hutborhe inter regiones Anglie et Scocie' ('by the service that he should be *inborh* and *utborh* between the territories of England and Scotland'), and that 'per eadem servicia tenuerunt omnes antecessores eius post tempus antiqui regis Henrici qui eos feoffavit' ('all his ancestors held by the same service after the time of old King Henry, who enfeoffed them').[5] It was therefore asserted that the tenure was in existence in the time of Henry I. There is no direct evidence from surviving *acta* of Henry I that he made a grant of Beanley in such terms. But the belief that he did so went a long way back, because at York in February 1136, immediately after Henry I's death, King Stephen confirmed Beanley and other lands to Earl Cospatric. The service

1   *CDS*, i, no. 658; K.J. Stringer, *Earl David of Huntingdon, 1152-1219: A Study in Anglo-Scottish History* (Edinburgh, 1985), p. 115.
2   Barrow, *Kingdom of the Scots*, p. 156.
3   Neilson, 'March Laws', pp. 13, 24-5.
4   Barrow, *Kingdom of the Scots*, pp. 158-60.
5   Neilson, 'March Laws', p. 22; *Book of Fees*, i, p. 200.

was not specified. Some of the lands then confirmed, but not Beanley, did appear earlier, however, in an act of Henry I concerning the marriage of Cospatric's daughter Juliana to Ranulf de Merlay, lord of Morpeth.[1]

Nevertheless, there is a very strong probability that a grant of Beanley *was* made by Henry I to Earl Cospatric. He was a son of the Earl Cospatric of Northumbria who had rebelled against William I in 1068 and had ultimately taken refuge in Scotland, where he had been granted lands which formed the basis of the earldom of Dunbar. In 1072 he was formally deprived of his estates in England.[2] His descendants' later ability to hold land again there must have required the agreement of Henry I and some measure of grant or restoration. But Henry I would not have done that without good reasons. What might they have been?

The answer may lie in the unusual service required for Beanley. 'Unique' would be a better word, because service by *inborh* and *utborh* does not appear to have been demanded from anywhere else in the kingdom of England. Article 4 of the laws provides that 'if anyone shall be poinded for his own debt, whether he be of England or of Scotland, he shall repledge his pledges for three terms of fifteen days by *inborch* and *utborch*'.[3] These terms therefore have to do with pledges and it is in that context that *inborh* can be found in Anglo-Saxon and Anglo-Norman law. There is an apposite early reference in the 'Ordinance respecting the *Dun-Seatas*', a set of rules from perhaps the early tenth century regulating arrangements between English and Welsh in the Wye valley. Article 8 of the Ordinance provides that 'if cattle be attached and the party will vouch to warranty over the stream, then let him place an *inborh* or deposit an "under-wed" that the suit may have an end'.[4] *Inborh* is here used in a similar context to that envisaged in Article 4 of the March laws; it is a way to enforce pledges so that disputes about livestock involving folk on both sides of a border can be settled. The notion of *inborh* in relation to pledges was also alive and well in the early part of the twelfth century in the *Leges Henrici Primi*, where it is clear that *inborh* was security to be taken by a lord from his own men so that satisfaction could be given to an injured party.[5]

Utborh is a more elusive word. Its sole appearances seem to be in the March laws and in the conditions of tenure for Beanley. The shape of the word offers no difficulties. Five compounds ending in *-borh* are known,[6] and so there seems to be no inherent linguistic objection to the formation of

1   *Regesta Regum Anglo-Normannorum, 1066-1154*, ed. H.W.C. Davis *et al.* (Oxford, 1913-69), ii, no. 1848; iii, no. 373a.
2   Anderson, *Scottish Annals*, p. 96.
3   Neilson, 'March Laws', p. 18.
4   *Ancient Laws and Institutes of England*, ed. B. Thorpe (Record Comm., 1840), pp. 150-2.
5   *Leges Henrici Primi*, ed. L.J. Downer (Oxford, 1972), pp. 148-9, 178-9, 254-5, 351; *An Anglo-Saxon Dictionary*, ed. J. Bosworth and T.N. Toller (Oxford, 1898-1972), under 'in-borh'.
6   F. Liebermann, *Die Gesetze der Angelsachsen* (Halle, 1898-1916), ii, p. 26, under 'borg'.

utborh as a sixth. But its meaning remains obscure. Attempts to clarify the earls' terms of tenure have included proposals such as their providing escorts for persons travelling to and from the Border, acting as a kind of passport control, and so on. Such ideas still turn up, although their like was roundly condemned many years ago as guesses.[1] A new argument by analogy may help to explore the issues further. *Infangthief*, the right of summary justice on a red-handed thief caught on one's own land, was a usual adjunct of lordship under Anglo-Saxon and early Anglo-Norman law. *Outfangthief*, the right to hang a red-handed thief caught outside one's own land, was a privilege claimed more rarely. *Inborh* was a right to take pledges from one's own men; *utborh*, by analogy, would appear to have been a right to take pledges from men within the lordship of others.[2] An alternative and perhaps no less fruitful approach is to consider the possible significance of *utborh* in light of another compound word, *utware*, known and attested in northern England and elsewhere.[3] *Utware* involved a duty to fight not for one's immediate lord but for the king. It required service not only outside the land of one's lord but also in alien territory, and clearly overrode local obligations. On this analogy, the taking of pledges under *utborh* overrode local jurisdictions and was deemed to take place on behalf of the king. *Utborh*, then, appears to have provided a power to take or enforce pledges (over and above the power of local lords) in certain disputes justiciable under the laws. A system for enforcing pledges across or along a border between two autonomous kingdoms would lack teeth without such a wide power as is proposed here for *utborh* — hence, therefore, the appearance of such a power in this unique context and hence, too, the clarifying statement in Article 13 of the laws that '*inborch* and *utborch* have power to distrain both realms'.[4] In other words, *utborh* was intended to bite on both sides of the Border line and across it.

By a combination of powers in Articles 4 and 13 of the laws, and by the conditions of tenure for Beanley, the earls of Dunbar seem, therefore, to have had a remit to enforce a cross-Border system of pledges. That a subject of the king of Scots should have been empowered to do this in England is an extraordinary situation. But it could be explained and sustained by three factors. The first is that the earls of Dunbar, as descendants of former earls of Northumbria, were not strangers to conditions immediately to the south of the March. Perhaps they still had power to draw on local loyalties and enjoyed some local prestige. The second factor is that by making a grant or restoration of lands an English king could have secured allegiance for those

---

1   Hedley, *Northumberland Families*, i, p. 236; W.E. Kapelle, *The Norman Conquest of the North* (London, 1979), p. 205 — apparently following Hedley, as above; G. Chalmers, *Caledonia* (new edn, Paisley, 1887-1902), iii, pp. 238-9.
2   *Anglo-Saxon Dictionary*, under 'infangenetheof' and 'utfangenetheof'; Liebermann, *Gesetze*, ii, pp. 523-4, under 'infangenetheof'.
3   C.W. Hollister, *Anglo-Saxon Military Institutions* (Oxford, 1962), pp. 31-7, discusses *utware* and provides references.
4   Neilson, 'March Laws', p. 22.

lands and the service which went with them. Beanley is some twenty miles south of the Tweed and not obviously a convenient place from which to attend regular meetings on the March. But this is where the third factor comes in, for the earls of Dunbar also had considerable holdings of land in Berwickshire and, by chance or not, they held lordship at Birgham and Lennel, directly on the north bank of the Tweed and very near Reddenburn, one of the recognised meeting places for March business.[1] An agreement between a king of Scots and a king of England to set up the earl (and no other lord) with a duty to regulate certain Border disputes could therefore have been based on some reasonable expectations of knowledge and reputation, loyalty and availability. From the Scottish point of view, it promised to protect the principle that business should be done at the March. From the English point of view, it could reconcile ancient enmities and would rehabilitate, at least partly, an old Northumbrian family, with the possible added advantage that the representatives of that family would have the local power and prestige necessary to enforce the March laws.

Other evidence may have a bearing on this issue. In 1091 a peace made between Malcolm III and William Rufus provided that Malcolm should have restored to him twelve unspecified vills in England formerly granted by William the Conqueror and, as before, be paid twelve gold marks a year. The source is the Worcester chronicle which has a knack, as here, of including information not found elsewhere. Version E of the Anglo-Saxon Chronicle is in agreement with the Worcester chronicle in main outline: 'King William promised him [Malcolm III] in land and in everything all he had had of his father'.[2] A restoration was envisaged, and to the extent that it was in land, it seems correct to infer from the latter source that that land would have been in England. The Worcester chronicler's information on the location of the vills consequently seems to be correct. His additional information, twelve vills and twelve marks, is precise. It suggests good and trustworthy news.

We are further told by the Worcester source that the agreement did not last. Presumably, therefore, any renewed hold the Scots may have had on the vills was brief and the lands were regranted. These vills have, over the years, been the subject of discussion. One suggestion is that they were in Cumbria, but this has been rejected as unconvincing. An alternative proposal is that the vills were granted for the use of Scottish kings whenever they journeyed south to the English court.[3] That explanation also has its weaknesses. The locations of the vills are unknown, and why, in any case, should vills be given for very infrequent use as bed-and-breakfast halts? There may, however, be another explanation: that the vills were in

1   *Coldstream Cart.*, nos. 8, 11 (both × 1166).
2   Anderson, *Scottish Annals*, p. 108; Gransden, *Historical Writing*, p. 145.
3   A.O. Anderson, 'Anglo-Scottish Relations from Constantine II to William', *SHR*, xlii (1963), p. 12; Duncan, *Making of the Kingdom*, p. 121; R.L.G. Ritchie, *The Normans in Scotland* (Edinburgh, 1954), pp. 386-8.

Northumberland and had some duties attached to them.

The lands confirmed by Stephen's charter of 1136 to Earl Cospatric are an interesting collection. First, there is land said to have been granted to Cospatric's uncle, Edmund. There are no further details of this apparent restoration of family land. Then there are two groups of named properties, coming in all to a little over twelve settlements: six manors formerly held by Winnoch and before that by Hamo, i.e., Brandon, Beanley, Hedgeley, Branton, Titlington and 'Harop'; then there is the land held by Liulf son of Uhtred, i.e., the three Middletons, Roddam, Horsley, Stanton, Wingates, Wotton, Witton and Ritton.[1] This is a list of sixteen named manors or vills in all, and fourteen if the three Middletons are taken as originally one. It is clear from Stephen's charter that before going to Earl Cospatric these named tenements had been held by others, who may therefore have been dispossessed to make way for the earl, and it is noticeable that two separate holdings have been brought together (perhaps deliberately) to form a block of just over twelve settlements. It could be no more than coincidence; but it is difficult to resist the thought that these properties were, or included, the twelve vills mentioned in the 1091 agreement. If that thought is intrinsically correct, or can be taken as a working assumption, these properties had therefore been granted by William the Conqueror to Malcolm III. But when and, above all, why?

By the end of 1072 Malcolm III had become 'the man' of William the Conqueror by his submission at Abernethy, and Earl Cospatric (father of the Cospatric who received the confirmation from Stephen) had been deprived of his lands in Northumbria. This seems the most likely time for Malcolm to have been granted the twelve vills and the twelve annual gold marks. In 1079 he led another expedition into Northumbria 'unmindful of the treaty made between King William and him', as an English chronicle laconically put it. From then until 1091 relations between the two kingdoms were poor. In 1091 Malcolm's submission to William Rufus is commented on by the chroniclers in such a way as to suggest that it was an attempt to restore the terms of his submission to William the Conqueror.[2] If so, Rufus restored in 1091 what his father had given in 1072, and that included twelve vills and twelve marks. From Rufus's point of view, it was a gesture of conciliation.

The original grant of the vills may also have been a conciliating gesture. In 1072 William I needed peace in the North, partly for its own sake, partly to enable him to strengthen his grip on the rest of England and keep control of Normandy. He may have thought that Malcolm III could help, perhaps by keeping the Border quiet through holding in Northumbria lands which already carried the duty of ensuring settlement of disputes at the March. But is it likely that such obligations existed before 1066? First, recall the jurors of 1248 and the thought that their names sparked off: that the establishment

---

1   *Regesta Regum Anglo-Normannorum*, iii, no. 373a; Hedley, *Northumberland Families*, i, p. 236.
2   Anderson, *Scottish Annals*, pp. 95, 100, 108.

of a new boundary on the Tweed broke the integrity of the lands of Saint Cuthbert. Much more significantly, it broke the integrity of the lands of the kingdom of Northumbria. Former subjects of one ruler now came under two. Lordship and jurisdiction had passed into other hands north of the Tweed. Economic links, if not destroyed, now had a major new obstacle, a new frontier, in the way. The 'Ordinance respecting the *Dun-Seatas*' shows that to establish a set of cross-border rules was not beyond the power of the tenth-century English and Welsh. And the slightly earlier treaty (*c*.886) between Alfred and the Danes, which established, however temporarily, a new frontier between Wessex/Mercia and the Danelaw, envisaged some limited cross-border movement for trading and a system of pledges to control that movement.[1] Finally, the March laws, especially in Articles 1 and 5, have some pre-1066 features, and Articles 3 and 14 are framed in terms which derive from Anglo-Saxon usages.[2] It is therefore not inconceivable that arrangements for pledges on and across the Anglo-Scottish Border came into existence during the early eleventh century, or perhaps even earlier than that, and that the forebears of Earl Cospatric, as rulers of the northern part of Northumbria, had carried some responsibility for supervising those arrangements, perhaps holding Beanley and other lands *ex officio* for that purpose. William I, in other words, seems to have been trying in 1072 to stabilise at the March some existing procedures which threatened to founder after the flight of Earl Cospatric in 1068, and to do this he put his trust in Malcolm III.

This is admittedly a speculative scenario. But if Beanley and other lands were already burdened with *inborh* and *utborh* duties, that may explain why the particular properties confirmed in Stephen's charter should have been taken away from other holders and granted to Earl Cospatric, apparently by Henry I. Even though Beanley and its members (or rather the twelve vills) had been granted again to the king of Scots by William Rufus in 1091, a well-established ruler like Henry I could scarcely have contemplated, let alone permitted, another king to hold on such conditions, for that would have conceded jurisdiction on the English side of the Border. A grant, or rather a restoration of lands and responsibilities, to the earl of Dunbar was a neat and potentially effective compromise, satisfying several objectives for both Henry I and David I.

A detailed and ultimately convincing case has been made out for the vigour and thoroughness with which Henry I sought in the 1120s to pacify Northumbria and to bring it and the Scottish Border under better control. Typically this involved castle-building, clearer obligations of service, and the promotion of new men loyal to the king, including some members of pre-Norman northern families.[3] A more recent study has shown how, in

1   *Ancient Laws and Institutes of England*, pp. 150-2; S. Keynes and M. Lapidge, *Alfred the Great* (Harmondsworth, 1983), pp. 171-2.
2   Neilson, 'March Laws', pp. 16-18, 22-3.
3   Kapelle, *Conquest of the North*, pp. 191-202.

other ways, Henry I tried hard in the mid-1120s to put relations with David I on a new footing, especially in church matters.[1] A fresh regulation of March affairs which included a grant to Earl Cospatric of lands with a service of enforcing pledges arising on either side of and across the Border is certainly not in disagreement with the wider picture; rather, it supplements and reinforces it. Articles 4 and 13 of the laws might therefore have taken their present form in the 1120s and probably no later. But their origins could go a long way further back.

With the example of Articles 4 and 13 in mind, how should one interpret Article 3, which provides for the return of fugitive naifs? The Article, which falls into two parts, is an odd mixture of self-help and judicial procedure. Within the first forty days, pursuit is allowed and the naif can be recovered on the lord's personal oath. Thereafter recovery has to be by 'a brieve of the king in whose realm he [the fugitive] shall be'.[2] On the face of it, the Article allows pursuit across the Border without notification. Whether the rule would work thus in practice is unclear, and it is at least worth observing that the 'Ordinance respecting the *Dun-Seatas*' allowed cross-border pursuit of allegedly stolen cattle only in conformity to rules about notification of intent.[3] Pursuit at will looks like a recipe for local trouble, unless it was a response to very unsettled conditions, such as existed in Northumbria after 1068 and up to at least the death of Malcolm III in 1093. Malcolm was remembered by a Durham chronicler for having carried out five slave-collecting expeditions in the north of England, and many of the victims, so we are told, remained in Scotland.[4] Escape attempts after each raid nevertheless seem very likely, as does a Scottish interest in recovery. Recovery was surely also an issue of importance in Northumbria, which had suffered from the slaughters and deportations. It says something about the state of northern England, and the general shortage of manpower there, that William Rufus felt obliged to plant in Cumbria peasants from further south and that a colony of Flemings was temporarily established in Northumberland (Henry I eventually sent them to south Wales).[5] A mutual understanding about the availability of self-help would not be out of place in such a disturbed situation and in the absence of other means of redress.

Recovery by due process under a brieve is the antithesis of self-help by pursuit. It implies that self-help is no longer to be the main recourse, a clear sign that the intention is to pacify the frontier and make it a little more orderly. Writs for recovery of fugitive serfs are known in England from the reign of William Rufus and, significantly, two of the earliest examples (both date before 1103) were issued in favour of Ranulf Flambard, bishop of

---

1   J. Green, 'Anglo-Scottish Relations, 1066-1174', in *England and her Neighbours, 1066-1453: Essays in Honour of Pierre Chaplais*, ed. M. Jones and M. Vale (London, 1989).
2   Neilson, 'March Laws', pp. 17-18.
3   *Ancient Laws and Institutes of England*, p. 150 (Article 1).
4   Anderson, *Scottish Annals*, pp. 108-9, 112.
5   *Symeonis Monachi Opera Omnia*, ed. T. Arnold (Rolls Ser., 1882-5), ii, pp. 192, 245.

Durham.[1] By the 1120s the writ had become well established in English usage, and was probably known to David I from his own experiences of administration in England before he had become king of Scots. Brieves for recovery of fugitive naifs were certainly available in Scotland after 1124.[2] The second part of Article 3 might therefore have come into the March laws during the late 1120s as part of Henry I's general reordering and pacification of the north of England. The survival of the first part of the Article might seem now to be little more than illogical. But could it not, in the context of the 1120s, represent a realistic view? Self-help and pursuit could not perhaps have been eliminated immediately. So, then, why not allow it for the limited period of forty days? There was certainly a risk that it might cause trouble. But against that, it offered a quick and simple solution for essentially local and personal issues.

This essay has examined a few parts of a complex document and some events which predated it. More could be said about inconsistencies in the text of the laws; discussion of the Articles of the laws for trial by combat has been left to one side; and the legal aspects of the laws have scarcely been considered. Such questions as how effective the laws were, and how far they featured in cross-Border relationships in the thirteenth century, have also been ignored. The thoughts presented are perhaps no more than suggestions when taken individually. Yet in total they may be much more. The choice of men at the inquests of 1248 and 1249; the archaic elements of *inborh* and *utborh*; the rehabilitation of the earls of Dunbar in Northumbria; the principle in Article 1 that impleading was to be done at the March within Scotland — all point to the existence of a body of March law and custom before 1066 and perhaps considerably earlier. Whatever that body of law and custom may have been (and more of it may yet be recoverable from the 1249 text), it could scarcely escape the influence of the thoroughly disturbed conditions in the north of England between 1066 and c.1100, and the changing relationships between England and Scotland early in the twelfth century. Henry I made a major attempt to settle Border affairs in the 1120s, and it is to that period that a major review of the laws probably belongs, one which perhaps for the first time saw the inclusion of the provisions for recovery of naifs by brieve, and almost certainly involved the reinstatement of a descendant of former earls of Northumbria in a position of responsibility for operating a revived system of cross-Border pledges in cases of dispute. I hope that this contribution, offered with gratitude for friendship enjoyed and scholarship shared now for over thirty years, may add to the work already done by Geoffrey Barrow to clarify and enhance knowlege of the early history of the Anglo-Scottish Border.

---

1   R.C. van Caenegem, *Royal Writs in England from the Conquest to Glanvill* (Selden Soc., 1959), p. 337 and nos. 104-12.
2   *RRS*, i, pp. 62-4.

6

Kingship in Miniature:
A Seal of Minority of Alexander III, 1249-1257

GRANT G. SIMPSON

In a later chapter, Alan Young presents the Comyns as a prominent magnate family of both local and national significance. The thirteenth century can indeed be viewed as 'the Comyn century'. Yet elsewhere he has also explained the underlying elements of baronial cooperation visible in Scotland during that era. Even in times of political tension, such as the fraught years of Alexander III's minority, a sense of partnership in authority was never wholly lost. 'The minority period as a whole displays a natural inclination to work through the normal channels of government despite the crisis ... the reverence for royal authority is unmistakeable.... The relationship between crown and nobility, under severe strain during the minority, still survived.'[1] The thirteenth-century term which embodies such political ideas is, of course, *communitas regni*; and, as Geoffrey Barrow has taught us over many years, 'the king of Scots was one side, the community of the realm the other side, of a single coin'.[2] That famous phrase emerges into the light of documentation from 1286 onwards, a full generation after Alexander's minority. But even in the troubled times of the child-king, a Scottish political community can be seen to exist; and thorough analysis of the evidence will reveal that it could generate views about kingship and government which were mature and even sophisticated. The purpose of what follows here is to investigate an unusual royal seal employed during the minority, and to try to extract some impression of the concepts which lay behind its creation and its use.

It is essential at the outset to give a basic description of the seal.[3] It is small, only 4.0 centimetres in diameter. The obverse side shows the king crowned and seated on a simple throne which has a footboard but not a

1   A. Young, 'Noble Families and Political Factions in the Reign of Alexander III', in *Scotland in the Reign of Alexander III, 1249-1286*, ed. N.H. Reid (Edinburgh, 1990), p. 7.
2   G.W.S. Barrow, *Robert Bruce and the Community of the Realm of Scotland* (3rd edn, Edinburgh, 1988), p. xi.
3   This description is largely based on that in J.H. Stevenson and M. Wood, *Scottish Heraldic Seals* (Glasgow, 1940), i, p. 25. For illustrations, see the following page. I am much indebted to Professor A.A.M. Duncan's brief study of the seal in his *Scotland: The Making of the Kingdom* (Edinburgh, 1975), pp. 556-9, which is more accurate and perceptive than the comments of previous writers, and to suggestions made by him in later discussions with me. But I have found it necessary to disagree with him on a few points.

Seal of Minority of King Alexander III: *obverse, left, and reverse, right*

From confirmation, dated 8 June 1252: see below, n. 1(ii).
(Enlarged by 25 per cent from actual size).

back. He holds in his right hand a sword laid horizontally across his thighs, and in his left a foliated sceptre. The main legend (in capitals) is round the outer margin, between beaded borders: ESTO PRVDENS VT SERPENS ET SIMPLEX SICVT COLVMBA. An additional legend (in capitals) exists on the background and surrounding the seated figure: DEI GRA REX SCOTT. On the reverse is a heater-shaped shield of the royal arms of Scotland (a lion rampant within a tressure), with a legend (in capitals), again between beaded borders, identical to the outer legend on the obverse.

Only three impressions of the seal have so far been discovered.[1] All are appended to royal charters of confirmation of grants made by crown tenants

---

1   (i) Confirmation, dated at Edinburgh, 3 June 1250, of a gift by Hugh of Abernethy to John Petcarn, his kinsman, of the land of Innernethy, Perthshire (SRO, Yule Muniments, GD.90/1/18).
    (ii) Confirmation, dated at Newbattle, 8 June 1252, of a sale by Richard Burnard of Fairnington to Melrose Abbey of the meadow of Fairnington, Roxburghshire, called 'Estmedu' (SRO, Melrose Charters, GD.55/336; printed in *Melrose Lib.*, i, no. 336; illustrated ibid., ii, plate I, no. 1, and above).
    (iii) Confirmation, dated at Stirling, 24 June 1257, of a gift by Sir Reginald Prat to

to sub-tenants. As these confirmations were issued in 1250, 1252 and 1257, they all fall within the king's minority, which lasted from his accession in 1249 to about 1258.[1]

The small size of the seal has led some previous scholars to mistake its function. It is less than half the size of the two known great seals used successively by Alexander III: the first of these was probably about 9.5 centimetres in diameter — no impression of it now exists entire — and the second measures 9.8 centimetres.[2] Because the seal here discussed is so much smaller, it has been thought to be a privy seal (*sigillum secretum*);[3] but, for a number of reasons, this is highly unlikely. First, all Scottish privy seals are single-sided; this one, like all great seals, is two-sided. Second, the earliest known Scottish privy seals, those of Robert I, were nearly always attached to a tongue, that is, a strip cut from along the foot of the document.[4] But each of the three specimens of this seal is attached on a tag of parchment hanging from a slit in a fold at the foot. Further, Robert I's privy seals were almost wholly employed on documents concerning household and financial matters, and on mandates to the chancellor; the privy seal was virtually never attached to a simple charter or confirmation.[5] But Alexander's small seal survives, as indicated above, on three confirmations. This seal, in other words, was not treated like a privy seal: it has the functions of a great seal, but it has been miniaturised.

To complete the picture it must be added that Alexander III did possess a privy seal. In 1272 he appended it to a letter to Eleanor queen of England, explaining that he did not have his great seal with him at the time. Similarly, in 1278 it was used on a letter to Edward I, with the same excuse being given.[6] No seals now exist on these letters and it is impossible to say what Alexander's privy seal looked like.

The small seal can only be seen in its fullest context when we look into the evidence of chronicles and records and, especially, at the iconography of the seal itself. The politics of Alexander's minority involved, not surprisingly, serious baronial disputes and various struggles over control of

---

William of Swinburne of the lands of Haughton, land in 'Huntland' (now lost, but including Middleburn in Wark), and the manor of Williamston, in the fee of Knaresdale, and land in Slaggyford, west of the River Tyne, in said fee (Newcastle upon Tyne, Northumberland County Record Office, Swinburne of Capheaton Muniments, ZSW/1/16; printed in J. Hodgson, *History of Northumberland*, III, i [Newcastle upon Tyne, 1840], pp. 18-19, with illustration of seal). I am indebted to Geoffrey Barrow for assistance in locating certain places named in this text.
1   D.E.R. Watt, 'The Minority of Alexander III of Scotland', *Transactions of the Royal Historical Soc.*, 5th ser., xxi (1971), pp. 20-1. It should be noted that the minority was unusually administered in one respect. There was no arrangement for a 'tutor', or '*rector* of the king and kingdom', as provided for Henry III of England in 1216; nor for Guardians of the kingdom, as with Margaret, Maid of Norway, from 1286 to 1290.
2   Stevenson and Wood, *Heraldic Seals*, i, p. 5.
3   Ibid., p. 25; H. Laing, *Descriptive Catalogue of Impressions from Ancient Scottish Seals* (Bannatyne and Maitland Clubs, 1850), p. 5.
4   *RRS*, v, p. 193.
5   *RRS*, v, p. 125.
6   *CDS*, ii, nos. 1, 132.

government.[1] Writing in the 1360s, the well-informed chronicler John of Fordun mentioned a small royal seal in a way which connects it with the problems of the 1250s. In the early years of the minority, from 1249, much of the upset arose from a clash between the entrenched power of the Comyn family, especially strong in northern Scotland, and the ambitious attempts 'to break into the forefront of the Scottish nobility' made by Alan Durward.[2] He had been royal justiciar of Scotia (Scotland north of the River Forth) since 1244 and was married to Marjory, illegitimate daughter of Alexander II and step-sister of the new king. Fears of excessive Durward influence in government led to an approach by the Comyns and others in mid-1251 to Henry III of England. The result was the marriage of the young Alexander at York, in December 1251, to Margaret, Henry's eldest daughter. As a consequence, Henry's influence over Scottish affairs was increased and he appears to have put his strength at this point on the side of those, particularly the Comyns, who feared the burgeoning power of Alan Durward and his kin. According to Fordun, the Comyns accused 'certain persons' of attempting treason against the king. But no action was taken at this point against the accused figures, who clearly must have been part of the Durward grouping. The crisis, however, came fully to a head in January 1252, after the return of the king and queen to Scotland. Fordun reports that the chancellor, Robert of Kenleith, abbot of Dunfermline, was accused of attempting to use the great seal to legitimise the king's bastard sister, Durward's wife, so that she might become heiress to Alexander and gain a hope of succession to the throne. The chancellor capitulated: 'he gave up the seal to the king and his magnates and it was immediately broken up in sight of the people, while a smaller seal was given to Gamelin, who became the king's chancellor.' Fordun completes his story by saying that the king's principal councillors were dismissed about this time and new ones, including prominent Comyns, were put in their places.[3]

While the main outlines of Fordun's version of events seem to be sound, the evidence of records does not fully support it at all points. One minor adjustment of his tale has to be made, since the initiative for removal of royal councillors appears to have come from Henry III during the visit to York late in December 1251.[4] Fordun also implies, although he does not precisely state, that the small seal was created in January 1252, as a result of the chancellor's disgrace. In fact, it had existed as early as 3 June 1250.[5]

The story of the making and usage of seals during the minority is a very complex matter. Unfortunately the evidence from surviving seals is not extensive: from the entire period of the minority only eight original acts now exist with their seals intact. Our picture of what was happening in this area

1   Duncan, *Making of the Kingdom*, pp. 552-76; Watt, 'Minority of Alexander III', pp. 20-1; Young, 'Noble Families', pp. 1-8.
2   Ibid., p. 5.
3   *Chron. Fordun*, i, p. 296.
4   Duncan, *Making of the Kingdom*, pp. 560-1.
5   See above, p. 132, n. 1 (i).

of administration is distinctly patchy. As indicated, the small seal is the first to emerge in view, on 3 June 1250. But an act of 3 December 1250 has appended to it the king's first great seal, which is of normal size and appearance: the enthroned monarch on one side, the monarch as a mounted knight on the other.[1] It is impossible to say which of these two seals was made first. It appears to be the case that within the first eighteen months of the reign both seals existed and were in use. Throughout 1251, so far as present documentation permits us to say, the great seal was the one which remained in operation, attached to acts dated 30 April, 26 July, 19 August and 20 October.[2] A.A.M. Duncan rightly sees this great seal as 'a token that the king was now considered of age to grant secure titles and to be represented as exercising government', and considers that 'this change [of seal] ... represents a decisive shift of power into the hands of Alan Durward'.[3] It would be somewhat extreme, however, to describe this simply as 'the Durward great seal'. There has been a faulty tendency among modern historians to talk too much in terms of Durward and Comyn 'factions' or 'parties'; 'the situation ... should not be seen too narrowly as a dispute between Comyns and Durwards'.[4] And yet the two seals do reflect a difference of constitutional attitude. One demonstrates a king allegedly functioning in full royal authority (even although aged only nine), the other, as we shall see, a monarch visibly a minor, whose power, as the lawyers would say, was *in posse* rather than *in esse*.

The first great seal ceased to be used at some point after 20 October 1251, and this fact is consistent with Fordun's statement that its matrix was destroyed when the chancellor was removed from office in January 1252. The small seal was now re-introduced in royal administration and exists on a confirmation of 8 June 1252.[5] For the remainder of the minority, surviving royal acts are very few indeed: from the years 1253-9 inclusive we have only nine texts.[6] Only one of these has a surviving seal affixed to an original: a confirmation dated 24 June 1257, which carries the seal of minority. Seal information thereafter remains very thin, and it is not until 21 July 1264 that we meet the first instance of the second great seal.[7] As Duncan suggests, 'the small seal was brought back into use for about a decade until, about 1260, a second great seal was engraved closely following

---

1   C. Fraser-Mackintosh, *Invernessiana* (Inverness, 1875), p. 31 (in translation; original in Inverness Burgh Records); for description of seal, see Stevenson and Wood, *Heraldic Seals*, i, p. 5.
2   Fraser, *Carlaverock*, ii, pp. 405-6 (with facsimile); *Spalding Misc.*, ii, pp. 307-8; *Laing Chrs.*, no. 7; F. Moncreiff and W. Moncreiffe, *The Moncreiffs and the Moncreiffes* (Edinburgh, 1929), ii, pp. 636-7 (with facsimile in vol. i, facing p. 12).
3   Duncan, *Making of the Kingdom*, p. 559.
4   Young, 'Noble Families', p. 5.
5   See above, p. 132, n. 1 (ii).
6   See G.G. Simpson, *Handlist of the Acts of Alexander III, the Guardians and John, 1249-96* (Edinburgh, 1960), nos. 17-26; and, in addition, a brieve of protection, dated at Roxburgh, 3 February 1256, for William of Swinburne, parson of Fordoun (unpublished original in Northumberland County Record Office, ZSW/1/13).
7   See above, p. 132, n. 1 (iii); *Melrose Lib.*, i, no. 309.

that made for Henry III at the end of 1259 and showing an elaborate throne (like the wooden chair in Westminster Abbey) and a bearded king'.[1] The minority was ended, and the king's new principal seal demonstrated this fact.

The question of who controlled the functioning of a royal seal was of course important. The chancellor was in charge of the great seal, but the activities in that office of Robert abbot of Dunfermline clearly left a residue of suspicion. Fordun is evidently correct in saying that in January 1252 the small seal was put in the custody of Master Gamelin, a royal clerk, but does not make it fully clear that the post of chancellor was in fact left vacant for just over three years. Gamelin is first mentioned as holding that office on 13 February 1254; but a private deed, not dated but certainly earlier than 1254, does refer to him as 'Master Gamelin, the king's clerk, then carrying his seal' ('sigillum eius tunc portante').[2] Gamelin, although illegitimate, was under the patronage of the Comyns, and was very probably a nephew by marriage of William Comyn, earl of Buchan.[3] It would be no more accurate to describe the small seal as 'a Comyn seal' than to call the first great seal 'a Durward seal'. But both the use of the seal of minority and the activities of its custodian are intimately linked to the political upheavals of the time.

Politics and administration can tell us some things, but the iconography and design of the seal are also revealing. The description above has indicated that the seated figure of the king holds his sword across his knees and not in the upright position customary on the great seals of all Scottish monarchs since Edgar (1097-1107). The same iconographical device of a sword on the knees was used on a Scottish royal charter of about a century earlier than Alexander's minority and that document can be used to help explain the significance of this feature. This charter, the great confirmation to Kelso Abbey by King Malcolm IV, which is undated but belongs to 1159, has a splendidly illuminated initial capital 'M', which displays the king seated alongside his grandfather David I, who towards the end of his life had designated the young Malcolm as his successor. David is depicted as a reigning adult monarch, and he holds the sword in the usual fashion; but Malcolm is shown as a beardless youth and he rests his sword on his knees.[4] The sword commonly symbolised justice, and Alexander's seal similarly displays to us a royal figure who could not in person dispense justice,

---

1   Duncan, *Making of the Kingdom*, pp. 556-7.
2   *Melrose Lib.*, i, no. 322. On this part of Master Gamelin's career, and various complications connected with it, see D.E.R. Watt, *A Biographical Dictionary of Scottish Graduates to A.D. 1410* (Oxford, 1977), p. 211.
3   See G.W.S. Barrow, 'Some Problems in Twelfth- and Thirteenth-Century Scottish history — A Genealogical Approach', *The Scottish Genealogist*, xxv (1978), pp. 105-7.
4   *RRS*, i, no. 131 (text); R.L.G. Ritchie, *The Normans in Scotland* (Edinburgh, 1954), frontispiece (illustration of the initial); Duncan, *Making of the Kingdom*, p. 173. It remains somewhat puzzling that Malcolm's charter should display him during his own reign as possessing status inferior to that of David I; but perhaps reverence for his saintly grandfather is implied.

because he was a child.[1] Further, the absence of one other feature from the seal also suggests that it belongs to a minor. The normal great seal shows the king on horseback on the reverse; but since a child-monarch cannot logically be shown in equestrian pose as a knight, a shield of the royal arms has been substituted.[2] The same solution was applied after Alexander's death in 1286, when the kingdom was under the rule of a body of Guardians. Six Guardians on six horses — or even all on one horse! — present problems for a seal-designer, and again on their seal the shield of royal arms was used as the symbol of royal dignity then in commission.[3]

These two visual elements of the seal are therefore telling us about its intention and status. Its motto, presented on both sides, points also in the same direction. In translation this reads 'Be as cautious as a snake and as innocent as a dove'. As Duncan has said, that is an adaptation from the words of Matthew, x.16: 'I send you out like sheep among wolves; *be wary as serpents, innocent as doves.* And be on your guard, for men will hand you over to their courts, they will flog you in the synagogues' (*New English Bible*).[4] Duncan considers that the motto was 'probably chosen because it included the word *columba*' (referring, he implies, to the saint of that name), and he feels that its relevance is obscure.[5] I think not: the context can in fact be shown to be highly apposite. In the New Testament story Christ is sending the apostles out to labour in the world of men and is warning his disciples of its dangers. The seal motto is advice to a young king who has to learn to grow up into the realm of harsh political reality as he reaches maturity. The choice of quotation is brilliantly apt and is evidently that of a shrewd thinker who possessed a good knowledge of the Bible.

Duncan has pointed also to other elements of the design and wording

---

1   The symbolic meaning of the sword changed over periods of time and from region to region, but its presentation as the instrument of justice became very common (Lord Twining, *European Regalia* [London, 1967], p. 230). I am much indebted to Professor Brigitte Bedos-Rezak of the University of Maryland, USA, who tells me (*per litt.* 25:viii:92) that the 13th-century counts of Toulouse used on the obverse of their seals a seated figure with a sword lying on his knees. She comments: 'The near adoption of a majesty type may well be traced back to their ancestor Constance, countess of Toulouse (*c.*1194), daughter of the French king Louis VI, who, the first of this comital lineage, used a majesty type on her seal, a posture that even French queens had not adopted. Constance's successors retained the iconography of royal ancestry, adding the lying sword (L. Douet d'Arcq, *Inventaires et Documents ... Collection des Sceaux* [Paris, 1863-8], i, nos. 741-8)'. The symbolic significance of this usage is not certain, but it might refer to a family which has an association with royal power, through its descent, but cannot exercise the full authority of monarchy.
2   In practice the honour of knighthood could be conferred at a remarkably early age: Alexander was only ten years old when he was knighted by Henry III at York in December 1251 (Duncan, *Making of the Kingdom*, p. 560).
3   See description in Stevenson and Wood, *Heraldic Seals*, i, p. 5, and illustration in W. de Gray Birch, *History of Scottish Seals* (Stirling, 1905-7), i, plate 15.
4   The Scots translation by W.L. Lorimer (1983) effectively catches, as it so often does, certain additional nuances: 'I am sendin ye furth like sheep amang woufs, sae be ye as cannie as ethers [adders] an as ill-less as dous. Tak tent o men. They will gie ye up tae councils an swipe ye i their meetin-housses.'
5   Duncan, *Making of the Kingdom*, p. 556.

which convey deliberate meaning. Several of these features present quite radical departures from the detailing which for about a century or more had been traditional on Scottish royal seals. Perhaps most strikingly of all, the king is shown wearing a crown. The seal matrices in use since at least 1165 had displayed an uncrowned monarch.[1] The Scottish kings made strenuous attempts from the early thirteenth century onwards to acquire the right to coronation and anointing as part of their inauguration ceremony; and the privilege was eventually granted to them by a papal bull of 1329.[2] To depict the king as a crowned figure was to underline his royalty. Further, the king holds in his left hand a foliated sceptre, absent from Scottish royal seals since 1107. Monarchs from then until Alexander III hold instead an orb surmounted by a cross.[3] The foliated sceptre may be seen as a symbolic 'rod of Aaron', which, 'left in the tabernacle where it brought forth blooms, showed God's choice of the Levites as the kin set aside in Israel for the priesthood. So, too, King Alexander was marked by God to rule.'[4] As the crown emphasised his royal status, so also the sceptre indicated his God-given power to govern.

The wording of the king's title is also innovative in two respects. It reads 'By the grace of God, king of Scots', whereas previous seals had said '[Under] God ruling ...' ('Deo rectore'). The use of the phrase *Dei gratia* in royal titles had been common practice among European rulers since the late eighth century and signified the divine favour conferred on a monarch, especially as made visible to all in the sacrament of royal unction.[5] The 'grace of God' formula on the seal is presumably a daring hint pointing towards the desired anointing. Astonishingly, the title also entirely omits the king's Christian name. In relation to the principal seals of all Scottish figures of high authority such as kings, queens and bishops, this omission is unparalleled, and difficult to explain. Duncan concludes that the 'symbols and words depict a king by God's grace: the stress is upon the nature of kingship, to the exclusion even of the king's name'.[6] The seal was small, but through all these features of display it was intended to convey a powerful set of messages.

There is no precise parallel to a small-scale seal of minority such as this

---

1   Stevenson and Wood, *Heraldic Seals*, i, pp. 3-4. The claim in Birch, *Scottish Seals*, i, pp. 23, 26, that the seals of William the Lion and Alexander II show them wearing some kind of 'cap-shaped crown' is not wholly convincing.
2   Duncan, *Making of the Kingdom*, pp. 526, 553-4; *Nat. MSS Scot.*, ii, no. XXX.
3   Stevenson and Wood, *Heraldic Seals*, i, pp. 2-4.
4   Duncan, *Making of the Kingdom*, p. 557.
5   W. Ullmann, *Principles of Government and Politics in the Middle Ages* (London, 1961), pp. 118-21. The phrase *Deo rectore* was standard usage on Scottish royal seals until that of David II (1329-71) (Stevenson and Wood, *Heraldic Seals*, i, pp. 2-7). It is significant that the change to *Dei gratia* occurs just after the papal grant of unction. The only exceptions to the regular use of *Deo rectore* are the wording on this small seal, 1250-7, and in the reign of King John (1292-6), when *Dei gratia* is again employed. It is noteworthy also that these aberrant appearances of that phrase occur in times of political uncertainty and tension, when the meaning of Scottish kingship was under scrutiny.
6   Duncan, *Making of the Kingdom*, p. 556.

in the chancery practice of Scotland or England in the twelfth or thirteenth century. Henry III of England, who succeeded in 1216 at the age of nine, had no seal at all until November 1218. The great seal made for him at that point was subject to certain restrictions in its use, since it had been agreed that no document under it should make any grant in perpetuity until the king came of age, which he did fully in 1227.[1] In that sense it was a 'seal of minority', but its physical form was no different from that of a normal English great seal. In French chancery practice there was a similar habit in the employment of a seal *ante susceptum*, during the period between a king's accession and his actual coronation, when, like a minor, he was not technically in the fullest possible possession of his royal power.[2] Research beyond the British Isles, however, has not so far produced evidence of any example of a specially made seal of minority like that of Alexander III.

Why was such a remarkable and subtly crafted seal created in mid-thirteenth-century Scotland? There can probably never be any fully adequate answer to that question. But the seal's existence is firm proof that the royal councillors and administrators of that day were by no means unsophisticated or out of touch. When in 1320 the principal barons of Scotland sent a collective letter to the pope in support of the monarchy and the rights of the kingdom, they (or their advisers) intelligently recollected and effectively put to use a method which had been employed by other European politicians for more than a century past.[3] The barons and officials of 1249 who were involved in inventing Alexander III's seal of minority were just as widely knowledgeable and as well briefed constitutionally as their successors some seventy years later.

1   D.A. Carpenter, *The Minority of Henry III* (London, 1990), pp. 94-5. It is likely that a similar principle was being applied during the minority of Alexander III: 'the first known major land grant by charter in the reign dates from [1262]' (Watt, 'Minority of Alexander III', p. 20).
2   G. Tessier, *Diplomatique Royale Français* (Paris, 1962), p. 203, which explains that the full phrase concerned was *ante susceptum regni Francie regimen.*
3   G.G. Simpson, 'The Declaration of Arbroath revitalised', *SHR*, lvi (1977), p. 25.

7

The Provincial Council
of the Scottish Church 1215-1472

DONALD E.R. WATT

Of all the lesser communities over which the Scottish crown presided, the Christian clergy as a whole formed by the twelfth century one of the largest, even if they were dispersed throughout the country. As was the case elsewhere in Western Christendom, they regarded themselves as living under two legal systems, the canon law of the Universal Church and the law of Scotland. The former system had encouraged the professional clergy both in the dioceses and parishes and in the monasteries to think of themselves as constituting 'the Scottish Church' in a narrow legal sense — a community which required 'liberty' to perform its functions properly. And as the clergy, in addition to their religious duties, came to offer the country at large a wide variety of services of a social, legal, educational and administrative character, so the crown was cooperative in offering them protection. This left them free to develop their institutions in harmony with general developments in the Church in other countries.

But there were by the end of the twelfth century peculiarities in the organisation of the Scottish Church which made it unique. As a result of a long struggle to resist the ambitions of the archbishops of York to exercise metropolitan powers over the Scottish dioceses, a settlement had been reached by 1192, whereby only the diocese of Galloway was left under York's supervision, whilst the ten other dioceses of the Scottish kingdom at the time (Caithness, Ross, Moray, Aberdeen, Brechin, Dunkeld, Dunblane, St Andrews, Argyll and Glasgow) were recognised by the papacy as exempt from any local metropolitan authority and as equal administrative units each directly under Rome.[1] There were exempt dioceses of this kind in other parts of the Western Church, but usually only in ones and twos;[2] and nowhere else did they form virtually a whole national Church as in the Scottish case. There had been attempts to secure for the bishop of St Andrews an archbishopric with metropolitan authority over the other

---

1   R. Somerville, *Scotia Pontificia* (Oxford, 1982), pp. 4-10.
2   E.g., apart from many such dioceses in Italy, there was one in Germany (Bamberg), one in France (Le Puy), and four in Spain (Compostella, Léon, Orviedo and Burgos); see *Le Liber Censuum de l'église Romaine*, ed. L. Duchesne *et al.* (Paris, 1910-52), i, pp. 243, 249; ii, pp. 105-6.

bishops, but such plans had been foiled by objections from York.[1] And it is likely that King William I, after observing the complications that had arisen in England in the 1160s as a result of the ambitions of Archbishop Thomas Becket to assert the rights of the English Church to liberty in an extreme form, was happy to support this arrangement of ten equal bishops. Scotland in fact was not to have its own archbishop until, in different circumstances, the see of St Andrews was at last so elevated in 1472.

But there were disadvantages in this loose organisation of ten separate dioceses, which were highlighted as a result of the deliberations of the Fourth Lateran Council of the Western Church held in 1215. Four Scottish bishops took part in the deliberations there, when there was support for new definitions of church practices and a general enthusiasm for active reform. In particular the duties of metropolitan archbishops in charge of church provinces were revised and standardised. By canon VI passed by the Council they were ordered to hold provincial councils once a year, which all the bishops of the province were to attend. These were to address 'the correction of excesses and the reform of morals, especially in the clergy, [by] reciting the canonical rules ... to secure their observance, [and] inflicting on transgressors the punishment due'. The individual bishops were then to return to their dioceses and in their own annual synods seek to secure enforcement of the rules decreed in the provincial council.[2] The basic concept was one of the imposition of authority from above, from the pope-in-council down to parish level. At the same time the metropolitan was to appoint an agent in each diocese of his province to report to each meeting of the provincial council on 'what things deserve correction or reform', and there was particular emphasis on the need to review at the annual council the performance of those who had responsibility for making church appointments (canon XXX).[3]

But how could these instructions for better order and discipline be put into effect in a branch of the Church which had no metropolitan archbishop? By the early 1220s some of the Scottish bishops had formed the view that without the exercise of metropolitan authority such as the Lateran Council had envisaged, the Lateran statutes were being neglected in Scotland and many irregularities ('enormia') were being committed which remained unpunished. In 1225 therefore they put a proposal to Pope Honorius III and won his approval for all the bishops of the kingdom of Scotland who did not have a metropolitan (that is all except the bishop of Galloway) to hold a provincial council. Clearly, such a council was to be held on a regular basis in the spirit of Lateran canon VI, which was quoted verbatim, if selectively, by the pope in his bull.[4] Thus was a unique mechanism for oversight devised

1　　In 1125 and 1151 (Anderson, *Scottish Annals*, pp. 160-1; Anderson, *Early Sources*, ii, p. 212).
2　　*Conciliorum Oecumenicorum Decreta*, ed. J. Alberigo *et al.* (3rd edn, Bologna, 1973), pp. 236-7; *English Historical Documents*, iii, ed. H. Rothwell (London, 1975), p. 648.
3　　Alberigo, *Decreta*, p. 249; Rothwell, *Eng. Hist. Doc.*, p. 658.
4　　*SES*, ii, p. 3; Patrick, *Statutes*, p. 1.

for a province of the Church where the dioceses were each directly under papal authority, and where, at least in later years, the pope could state that he himself was in a special way the metropolitan,[1] but which was positioned far from Rome. And it needs to be noted at once that no instruction was given that this Scottish provincial council was to submit reports to Rome or depend on papal confirmation of its acts. The implication is that the council was to be as independent in its acts as was any metropolitan under the general law of the Church.

Any attempt to understand the effectiveness of this new system is necessarily limited by the small quantity and scrappy nature of the sources available. This is a standard problem for historians of the medieval Church in Scotland, who do not have available the rich and detailed episcopal archives which survive, for example, in England. Moreover, the constitution chosen for the provincial council involved a rotating chairmanship, the implication being that there was no permanent secretariat to maintain a continuous record of the council's business. It is known that at the very end of the story in 1471 a record vaguely described as 'the Provincialis Buk' was in existence,[2] and so it is possible that registers with details of the council's business once existed which have now disappeared. But at least some of the council's decisions are known to have been recorded in separate notarial instruments, each authenticated by the seals of the leading participants;[3] and it would be quite usual for records of this kind to be kept in a locked wooden chest somewhere, probably in a church where the council customarily met (which was most often in Perth). But this is only reasonable speculation, for no such records survive in original form. All we have for study are dispersed copies of bits and pieces of council decisions kept by members of the council or by other interested parties for their own reference purposes. Such records may not be complete in themselves, but with the help of a miscellany of documents which mention council meetings for one reason or another, and the various references to council meetings found especially in Walter Bower's *Scotichronicon*, more can be discovered about the character and working of the Scottish provincial council than has previously been thought.[4]

A particular problem attached to the evidence relating to one side of the council's work, namely legislation by 'statutes' or 'canons', has to be addressed. It is evident that this kind of business was of particular concern to the council in its early days, and then from time to time thereafter. But nearly all the record comprises manuscript copies made after, and sometimes

1   *SES*, i, p. lxxxvi (n. 3).
2   *APS*, ii, p. 99; but the contents of this volume may have been limited to a valuation list of the churches of the province.
3   E.g., an act of 1420 (*Brechin Reg.*, i, pp. 38-40; *SES*, ii, p. 77; Patrick, *Statutes*, pp. 80-2).
4   Nearly all the evidence was collected somewhere in *SES*; but the interpretation of this material can be taken further than Joseph Robertson did in his remarkable first volume of that edition.

long after, the event. The problems of interpretation are formidable when it is realised that the copyists usually did not consider it relevant to attach dates to statutes which the council may well have routinely confirmed from time to time over a long period. A few statutes have been preserved in a fragmentary register belonging to Arbroath Abbey, which seems to have been put together in the 1260s or '70s;[1] a fuller collection of statutes was copied for Aberdeen cathedral chapter in the 1380s;[2] and the fullest collection was copied apparently for use in St Andrews diocese as late as 'very soon after 1500'.[3] These copies must have been made because it was still relevant to have detailed knowledge of the council's statutes, even though some of them had originally been drafted years, indeed centuries, before. This continuing regard for earlier legislation was typical of the way in which the general canon law of the Church had been built up, and need generate no surprise. What it also indicates is a continuing respect for the provincial council which authorised it.

It is curious how short and unspecific is the papal bull of 1225 authorising the Scottish bishops as a group to establish a council: 'By apostolic warrant we command you ... to hold a provincial council by our authority'.[4] There is no evidence on how the first meeting of the new council was to be summoned or who was to attend it. Some time before October 1232 a meeting of the bishops of Scotland 'in their council' issued a ruling on the rights of parish priests to have access to common pastures for their animals.[5] The first dated council meeting that is known took place at Dundee some time during 1230-1, when at least six bishops were present, together with two abbots, two archdeacons, and two clerics with the status of *magister*.[6] And in July 1238 a council was held at Perth attended by at least four bishops, four abbots, the archdeacon and dean of Glasgow, and a cleric who was an eminent doctor of theology (Master Peter de Ramsay).[7] Then in a statute about council procedure probably dating from the 1240s, it was anticipated that councils could be attended by bishops, abbots, deans

---

1   Dundee, City Archives, the Ethie MS, GD.130/25/17, fos. 14^v, 16^{r-v} (old foliation); see *SES*, i, pp. cxcii-cxciii, and frontispiece for a facsimile. The three folios of this fragmentary MS appear to have been written earlier than a valuation list of some benefices in St Andrews and other dioceses (written on another part of the same parchment leaf, and datable *c.*1282 because of the inclusion of the church of Garvock as belonging to Arbroath Abbey), since an amendment to one of the statutes has been made in the hand which wrote this list (cf. *SES*, ii, p. ix, at no. 49); hence the suggested dating in the text.
2   NLS, MS Advocates 16.1.10, fos. 25^{v}ff.; this register was probably compiled under the influence of Bishop Adam de Tyningham in the 1380s (cf. G.R.C. Davis, *Medieval Cartularies of Great Britain* [London, 1958], p. 129). See *SES*, i, p. cxciii; for facsimiles of fos. 26 and 27 (formerly 29), see *Aberdeen Reg.*, ii, facing p. 4.
3   London, Lambeth Palace, MS 167, where the provincial statutes are accompanied by synodal statutes of St Andrews diocese; the opinion given here about the date of this MS is that in N.R. Ker, *Medieval British Libraries*, i (Oxford, 1969), p. 93.
4   See above, p. 141, n. 4.
5   *Moray Reg.*, no. 25.
6   *SHS Misc.*, viii, pp. 5-6.
7   *SES*, i, p. iv (n. 2); Fraser, *Menteith*, ii, p. 328; for Peter de Ramsay, see D.E.R. Watt, *A Biographical Dictionary of Scottish Graduates to A.D. 1410* (Oxford, 1977), pp. 460-3.

and archdeacons, and 'other clergy'.[1] The lists of attenders in 1230-1 and 1238 are taken from documents emanating from decisions taken at the council, and do not necessarily represent the full sederunt; and some of the individuals mentioned may well not have been members of the council, but rather parties interested in items of council business, or advisers brought along by bishops or abbots. Other lesser clergy could sometimes be present as proctors for bishops or abbots who could not attend. What is clear is that the council was not even in these early days confined to the bishops, despite the fact that the papal bull of 1225 was addressed to them alone. It is probably correct to say that full membership was available at first only to those who had the status of 'prelates' as defined by canon law, that is, to holders of church offices to which jurisdiction was attached.[2] Such a composition matches the nature of the various kinds of business done in the council: problems had to be solved and practical guidance offered to busy men with administrative and judicial responsibilities. Subsequently the composition of at least some councils was enlarged to include proctors of '[cathedral] chapters, colleges [of secular clergy], and convents [of regular clergy]'.[3] It is likely that larger assemblies of this kind were gathered when the question of a tax on clerical property arose.[4]

By the 1240s at latest we know that arrangements had been made for the chairmanship of the council, to be exercised both during and between council meetings.[5] At each meeting one of the ten bishops was chosen by the rest of them, and by the fifteenth century at any rate his election took place immediately after the opening devotions and sermon.[6] His title is found in varying forms such as 'Conservator Statutorum Concilii', 'Conservator Privilegiorum', 'Conservator Concilii'. During his period of office, the conservator had the duty of 'punishing manifest and notorious offenders against the council and any statute made by it, and of effectively compelling due satisfaction to be made by means of ecclesiastical censure according to the exigency of the law'. It was also his duty to call the next meeting of the

1   SES, ii, p. 4; Patrick, Statutes, pp. 2-3; for dating, see below, p. 148, n. 6.
2   In 1280 a council was described as made up of 'bishops, abbots, priors, deans, archdeacons and other prelates of churches' (Moray Reg., no. 127; SES, i, p. lxxi); a late 14th-century calling-notice for a meeting of the council was addressed primarily to bishops and the prelates (undefined) of their dioceses (Moray Reg., no. 295; SES, ii, pp. 3-4). In a normal provincial council it was customary for the presiding archbishop metropolitan to consult only the bishops when framing statutes, and not the lesser members of the council who attended only variably and irregularly (R.L. Kay, 'The Making of Statutes in French Provincial Councils 1049-1305', summary of a Wisconsin University Ph.D. thesis in Dissertation Abstracts, xx [1959], p. 1004).
3   Moray Reg., no. 295; these proctors probably equated with the clergy 'qui in consilio et synodo generali consuevit congregari' mentioned in 1420 (see above, p. 142, n. 3).
4   Cf. Bower's description of the council in 1275 which discussed the details of a clerical tax levied by the pope as comprising 'prelati et clerus' (Chron. Bower [Watt], v, p. 402), and the mention in c.1328 of a clerical tax authorised this time 'per prelatos et clerum tocius regni nostri in ultimo consilio Scoticano apud Perth celebrato' (Formulary E: Scottish Letters and Brieves 1286-1424, ed. A.A.M. Duncan [Glasgow, 1976], no. 114).
5   SES, ii, pp. 9-10; Patrick, Statutes, p. 9.
6   See above, p. 142, n. 3.

council when, after the initial ceremonial, he handed over his office to his successor.

So much for the rules; but mysteries remain. We know the names of only a handful of the conservators between 1215 and 1472 (see Appendix). It seems unlikely that there was a fixed rota for service among the bishops; yet we do not know whether or not some conservators were re-elected for extensive periods of office. We are ignorant also of the length of intervals between meetings of the council. In terms of the rule laid down by the Lateran Council they were meant to be held annually. This is echoed in the council's own early statutes,[1] and there is reasonable evidence that by the fifteenth century meetings were held regularly at Perth for three days in the middle of July each year. There are also references to meetings being held after 1400 at the accustomed place and time.[2] But the pattern in the thirteenth and fourteenth centuries was probably more variable. Certainly some meetings were then held at Perth, which was not the cathedral city of any of the bishops, and had the advantage that it was often the site of parliaments or exchequer audits of the civil government, some of them in July; but meetings can also be traced (perhaps exceptionally) at Dundee and Aberdeen, and in August, October, November, February and March.[3] We simply do not know whether the conservator had discretion over meeting-dates, and in that connection whether some were inclined to cling to office by postponing meetings, and others were glad to get rid of the office as soon as they could acceptably bring a meeting together. On balance it appears to have been an office that was more of a burden than a pleasure, so that meetings would have been summoned with fair regularity to pass it on. But a judgement of this kind depends really on the observer's view of human nature among aspirants to high administrative office, for there is not sufficient evidence available to prove the point one way or the other.

From the beginning the provincial council was envisaged as a court, with much of its jurisdiction being exercised between meetings by the conservator in its name, but with important business reserved for decision when it met. Without any day-to-day records of its activities in this sphere, we depend on the accidental survival of evidence, without our being able to assess how typical each case might be. The record which mentions the Dundee council of 1230-1 shows how a dispute between two monasteries belonging to different orders (Tironensian and Cistercian) could be resolved at such a meeting, and in 1238 a bishop could secure the backing of the council to a settlement with a leading lay magnate in his diocese.[4] The bishop of Moray in 1280 was to seek help from a council when he in his turn was litigating

---

1   *SES*, ii, p. 9; Patrick, *Statutes*, pp. 8-9; three days were allowed for each meeting.
2   Details in *SES*, i, pp. lxxx-cix; cf. *St Andrews Copiale*, p. 104 (datable 1418 × 43), and *SES*, i, pp. ccxlv-ccxlvi (1465).
3   Perth in 1238, 1242, 1268, 1273, 1275, 1280, 1321; Dundee in 1230-1 and perhaps 1310; Aberdeen in 1359 (*SES*, i, pp. lv-lxxvi).
4   See above, p. 143, nn. 6, 7.

with a prominent layman in his diocese (Sir William de Fenton lord of Beaufort) over some land belonging to the church of Kiltarlity near Inverness; he wanted a procedural sentence of excommunication against his opponent to be repeated in every church, in every bishop's diocese, with the council's authority.[1] This was a right to which he was entitled under the council's own statutes.[2] Sometimes settlements over land were concluded at a council meeting (as in 1321 and 1408);[3] sometimes copies of important documents were authenticated for a cathedral chapter or religious house (as in 1325 and 1359).[4] In such cases, the council was acting as a desirable court of record. The council was asked to play a more active role in 1470, when it interpreted in favour of the University of St Andrews a recent papal privilege obtained by St Salvator's College there, which was ambitious to grant its own degrees.[5] It was a matter which Bishop Graham of St Andrews apparently could not sort out in his capacity of chancellor of the University, and which he arranged to be resolved in the council in his capacity of conservator.

The judicial functions of the council emerge clearly from a meeting held in 1268, when the bishop of St Andrews secured the excommunication of the abbot and some of the convent of Melrose in Glasgow diocese after an armed attack on some of the bishop's property, during which one of his clerks had been killed.[6] It was no doubt advantageous to have the council as a forum for settling disputes between two major prelates without having to have recourse to the royal courts. More significant perhaps is a case in 1388 when a monk of Urquhart Priory in Moray diocese appealed to the conservator against the action of Bishop Alexander Bur in investing a monk of Dunfermline Abbey with the priorship of Urquhart. The case turned on the legal point of whether or not the priorship was technically vacant, and the conservator seems to have found in favour of the appellant. It was not a matter of the interpretation of the council's own statutes, but of the general canon law.[7] It happened that the particular conservator then in office was Bishop Walter Trayl of St Andrews, a lawyer with experience as a judge at the papal court, who had returned to Scotland as bishop in 1386 with a special faculty to hear and decide in Scotland appeals from Scottish bishops to the Holy See;[8] but it was not in that capacity that he was appealed to in 1388. There were therefore at least some categories of case under the canon

---

1   *Moray Reg.*, no. 127: *SES*, i, pp. lxxi-lxxii.
2   *SES*, ii, pp. 27-8; Patrick, *Statutes*, pp. 27-8.
3   *APS*, i, pp. 478-9; *RMS*, i, no. 84; *SES*, i, p. lxxiii (1321); SRO, Calendar of Gordon Castle Writs, GD.44/4; *RMS*, i, no. 905 (1408).
4   *Coupar Angus Chrs.*, i, nos. 105-6 (1325); *Aberdeen Reg.*, i, pp. 84-6; *SES*, i, p. lxxvi (1359).
5   St Andrews, University Muniments, MS Registrum Evidentiarum et Privilegiorum Universitatis Sanctiandree, fos. 70ᵛ-2ᵛ.
6   *Chron. Bower* (Watt), v, pp. 370-1.
7   *Moray Reg.*, nos. 267-8. See also ibid., nos. 269-70, for later stages in the case; the summary in *SES*, i, p. li, is inadequate.
8   Watt, *Scottish Graduates*, pp. 539-42.

law where the provincial council and its conservator provided a level of appellate jurisdiction similar to that of a metropolitan archbishop in other countries; this must have been a welcome convenience. It was also helpful, when the first case of Lollard heresy appeared in Scotland in the early fifteenth century, that so unfamiliar a problem could be dealt with by the prominent theologian Laurence de Lindores as inquisitor in a council of the clergy ('in concilio cleri') at Perth in 1408.[1] This council must surely have been the provincial council that certainly met at Perth in July 1408,[2] for all the prelates of the country would have been alarmed at this unfamiliar phenomenon and glad to be associated with the condemnation of James Resby that followed. It was thus made an act of the whole Scottish Church, and the country was to be very little troubled by heresy thereafter.

Provincial councils everywhere after the Fourth Lateran Council were charged with reform of 'mores', meaning presumably prevailing customs of all kinds.[3] In Scotland, as elsewhere, a consequence was that the decades after 1215 saw local church leaders compiling collections of statutes for approval at both provincial and diocesan levels.[4] In England and Scotland at any rate it was a comparatively short-lived phenomenon, and we must not be misled by the fact that records of such legislation bulk largest in the scrappy evidence for the work of the provincial council that happens to survive. It is characteristic of such collections of statutes in different countries that we know of them only in manuscript copies made in later years where no precise dates are attached to the various sections which would enable us to be sure when they were enacted. The implication may well be that earlier copies of the statutes had been worn out by constant use as handy works of reference; but new copies were needed (even as late as the early sixteenth century in the case of St Andrews diocese[5]) because legislation of the mid-thirteenth century was still regarded as both valid and helpful some hundreds of years afterwards.

It was typical of medieval law-making in general that there was only a very gradual move towards anything new. The usual procedure when problems of doubt about the rules emerged was to search around for old definitions (either by asking experts or by consulting reference books), and then to confirm in written statutes rules which were in danger of falling into desuetude. In some cases the rules had previously been based on unwritten custom, and their reduction to writing was part of the general move that has been observed all over Europe from administration by oral methods to

1   *Chron. Bower* (Watt), viii, pp. 66-72.
2   See above, p. 146, n. 3.
3   I.e., in canon VI (see above, p. 141, n. 2).
4   Cf. C.R. Cheney, *English Synodalia of the Thirteenth Century* (Oxford, 1941 and 1966); *Councils and Synods with other Documents relating to the English Church*, ii, *A.D. 1205-1313*, ed. F.M. Powicke and C.R. Cheney (Oxford, 1964); O. Pontal, *Les statuts synodaux français de treizième siècle* (*Collection de documents inédits sur l'histoire de France*, Octavo Series [Paris, 1971-83]); O. Pontal, *Les statuts synodaux* (*Typologie des sources de moyen âge occidental* [Turnhout, 1975]), pp. 39-51.
5   See above, p. 143.

administration by writing.[1] Amendments to suit changing circumstances were probably introduced unwillingly. There was also a good deal of borrowing of the precise phraseology for whole collections of statutes from one council to another. Scottish bishops found it sensible to copy useful definitions that were circulating also in England and France, not least because they and their advisers by the mid-thirteenth century were familiar with the universities in these countries and their canon law faculties,[2] which would have been consulted by English and French metropolitan archbishops as they planned their reforming councils. It is possible therefore to identify a widespread common interest in certain of the rules which came to be incorporated in the Scottish provincial statutes, and then to separate out the statutes defining rules customary in Scotland alone. But study of this kind in Scotland is peculiarly complicated by the absence of an archive pertaining to the provincial council as such. Its statutes are known to us only through selections made for a particular religious house (Arbroath), cathedral (Aberdeen), or diocese (St Andrews), and they are incorporated in the surviving manuscripts along with (and intermingled with) statutes on related matters approved by the local diocesan synods.[3]

Much work requires to be done before the legislation of the Scottish provincial council can reliably be identified and dated. But preliminary study suggests that following the visit of the legate Otto to Scotland in 1239,[4] the Scottish bishops (led at that time by David de Bernham, the new bishop of St Andrews[5]) agreed to accept a compilation of statutes on about forty topics, which was promulgated at a provincial council meeting held some time between 1242 and 1249, and which can be identified as a basic core of material which came to be copied into all the surviving manuscripts.[6] The legate Otto had himself in 1237 produced for the English Church a major collection of statutes which remained influential as part of the canon law as exercised in England for the rest of the Middle Ages.[7] The Scots borrowed

---

1   Cf. M.T. Clanchy, *From Memory to Written Record* (London, 1979).
2   Scots who studied abroad in the 13th century can be identified in the index of Watt, *Scottish Graduates*, pp. 601-7.
3   See above, p. 143, nn. 1-3.
4   *Chron. Bower* (Watt), v, pp. 164-5.
5   Watt, *Scottish Graduates*, pp. 41-4.
6   The following statutes as numbered in the printed edition (*SES*, ii, pp. 9-24; Patrick, *Statutes*, pp. 8-24) form a series in the same order in the Aberdeen and Lambeth MSS and in such of the Ethie MS as survives: preamble, nos. 1-46, omitting second part of no. 6, nos. 7, 10, 22-5, second part of no. 33, no. 39. No. 46 has the appearance of a final statute of this code. As additional statutes were approved in future years, so they were inserted or added in different ways in each of the MS traditions.
     The early code has the appearance of being carefully drafted in a series of sections. It apparently dates from after the compilation of statutes for St Andrews diocese only, which Bishop Bernham promulgated on 5 May 1242 (*SES*, ii, pp. 53-63; Patrick, *Statutes*, pp. 57-67) — e.g., compare no. 132 of the St Andrews statutes with no. 17 here — and after King Alexander II had been persuaded to show special support in person for the liberties of the Church in c.July 1242 (*Chron. Bower* [Watt], v, pp. 180-1), but before the confused period which followed his death in 1249.
7   Powicke and Cheney, *Councils and Synods*, ii, pp. 238-59.

three items from this code,[1] but did not choose to adopt it as a whole. They compiled their own code. At least eight more statutes were added to the Scottish collection in the 1250s or '60s (before the Arbroath manuscript was written);[2] more were added later in the century, and others in smaller numbers at dates during the fourteenth and fifteenth centuries. But the big effort made between 1242 and 1249 was never repeated; and this pattern of enthusiasm for provincial statutes as a particular phenomenon of the thirteenth century which was not followed up in later centuries is characteristic of the English (though not the French) provinces also.[3]

The content of the earliest 'core collection' of statutes can be summarised as follows: (1) rules for the conduct of the provincial council itself (cc. 1-2); (2) general statements about the faith and the sacraments (cc. 3-4); (3) rules about building churches and chapels (cc. 5-6); (4) rules about the parish clergy and their benefices (cc. 8-9, 11-13); (5) rules for the discipline of the clergy and their duties in looking after church property (cc. 14-21); (6) rules about the privileged position of the clergy in Scottish society (cc. 26-33); (7) detailed rules about the assessment of teind (cc. 34-8, 40-5); (8) a final rule that all these statutes are to be read out in every church of every Scottish diocese three times a year (c. 46). This last provision suggests that the compilers of these statutes regarded them as a self-sufficient code, since it was enough for them alone to be regularly brought to the attention of all concerned. It also illustrates how the provincial council saw itself as legislating for all ten dioceses for which it was responsible. No such code for the whole kingdom (except Galloway) had been attempted before, for visiting legates in the past had contented themselves during brief stays in the country with handling just current problems that were brought to their attention.[4] By exercising its legislative function, the provincial council was providing a service for nearly the whole country, and thus was seeking the convenience of uniformity of practice.

In contrast with this basic code, which was probably drafted as a whole, the comparatively few later additions to the council's legislation were presumably drafted to meet specific problems brought to the council's attention. They are likely to have been the product of debate and refinement of drafting in the course of council meetings, for without a permanent metropolitan there was no single mind at work, devising and imposing necessary reforms at different times. It becomes clear how exceptional the code of the 1240s was. An illustration of the different method of drafting adopted later is found in a record of council proceedings in 1420. The

---

1   Nos. 16-18.
2   The second parts of nos. 6 and 33, nos. 25, 47-51 (but not no. 39) had certainly been added by the time the Ethie MS was written; but since that MS is now incomplete, it is impossible to say whether or not nos. 7, 10, 22-4, 52-5 and 101-7 had been approved by then.
3   C.R. Cheney, *Medieval Texts and Studies* (Oxford, 1973), pp. 149-50; cf. Pontal, *Les statuts synodaux*, pp. 50-1.
4   See P.C. Ferguson, 'Medieval Papal Representatives in Scotland: Legates, Nuncios, and Judges-Delegate, 1125-1286' (Columbia University Ph.D. thesis, 1987).

business for discussion was a review of the customary rights and duties of bishops and their judicial assistants in administering the estates of the dead, whether they had died testate or intestate; the conservator is recorded as having taken sworn testimony about customary practice from senior clergy of the several dioceses, and then an agreed statute was formulated with rules for the whole province.[1] Presumably practices which had to some degree been variable between the different dioceses (which was inconvenient for people with property in more than one diocese) were now rationalised along generally acceptable lines.

Pope Innocent III, who was a lawyer by training, had intended in drafting the canons of the Fourth Lateran Council that reform of church law should flow out and down from the centre, from an increasingly codified *Corpus Juris Canonici* blessed by papal authority[2] through provincial councils to diocesan and even archidiaconal synods; and a hierarchy of church courts, from (in reverse order) deanery and archdeaconry, through diocese and province up to the court of Rome itself, was developed with the aim of enforcing greater uniformity of practice. The papal archives of the later Middle Ages demonstrate clearly how remarkably effective in some ways this policy was. But a study of the legislation of the Scottish provincial council demonstrates how partial was this success. It was not that the central *Corpus* was neglected or unknown in Scotland: indeed, it was the object of detailed study by generations of ambitious Scottish churchmen in the main European universities from the early thirteenth century onwards.[3] Scotland had its share of bishops with canon law degrees and of qualified practitioners of that law, well able to prepare cases for presentation to the judges of the Roman court (whether sitting wherever the papal court happened to be or acting on commission within the bounds of Scotland) according to the internationally understood rules of the canon law, in matters both of substance and of procedure.[4] But the Scottish provincial council chose only very occasionally to frame a statute with the purpose of drawing attention to rules in the *Corpus*. It did so twice in the code of the 1240s — regarding the privileges which churchmen should enjoy everywhere (c. 30), and on the special privileges which should be offered to people going on crusade (c. 31). At some later date emphasis was placed on the canon law rule regarding the treatment for bishops or priests who had sexual relations with women who had come to them for confession (c. 54)! There was occasion also in the code of the 1240s to devise statutes which closely paralleled canons V, VIII and XVI issued by the Legate Otto in England in 1237 on aspects of the discipline of the clergy (cc. 16-18); but it is

1   *SES*, ii, pp. 77-8; Patrick, *Statutes*, pp. 80-1.
2   By Pope Innocent's time various attempts at codification had already been made in preparation for the compilation of the definitive 'Decretals of Gregory IX', which were to be published in 1234 and studied in the canon law faculties of the universities thereafter (*New Catholic Encyclopaedia* [New York, etc., 1967], iii, pp. 42-3).
3   See above, p. 148, n. 2.
4   Ferguson, 'Medieval Papal Representatives', ch. 5.

significant that the text of one of these statutes (c. 17) was lengthened by the insertion of an extra sentence protecting the rights of religious orders. Concern for Scottish conditions was paramount; and apart from these few exceptions, the bulk of the Scottish provincial statutes is concerned with defining matters of local custom, rather than with emphasising the universal law of the Church. This was true of England, France and Italy, as well as of Scotland: 'law, like population, recruits from below'.[1] Whatever a pope like Innocent III might think, the *Corpus* was no monolithic code of law ready to be enforced everywhere throughout the Church: it was a quarry from which church lawyers were constantly excavating rules which they claimed to be the law of the Church, but which were interpreted in widely divergent ways by different schools of lawyers. No doubt there were areas of law where the Scottish council wanted to conform as nearly as possible to standards maintained in the rest of Western Christendom, especially in matters of faith; but for the most part its statutes demonstrate how free it was to take Scottish conditions and traditions into account. It is this that makes these statutes such an important source for studying the Scottish way of doing things.

But how authoritative within Scotland were these statutes of the provincial council? However much the conservator was meant to watch over their implementation throughout every diocese, there clearly remained many areas of activity where variety of practice was tolerated. For one thing, the provincial statutes went into detail only on matters where uniformity throughout the kingdom was important: matters of the definition of teind, or executry work, or the privileges of the clergy are examples. Other matters such as the rules for the administration of the sacrament were expressed at provincial level only in very general terms, and from surviving diocesan statutes for Aberdeen and St Andrews it appears that each bishop was free to interpret these general rules quite differently.[2] Furthermore, it is not at all certain that the council or its conservator was in any position to impose unwelcome rules on any diocese where the bishop chose to disassociate himself from a council decision which he did not like. This must be the implication of the fact that the surviving collections of diocesan statutes for Aberdeen and St Andrews include different selections of items derived from provincial legislation. It would appear that the provincial council in Scotland did not make pronouncements effectively backed with the authority of the pope who had set the council up; it may not even have had the kind of authority enjoyed by a normal metropolitan archbishop. Its members from

1   Cheney, *Texts and Studies*, p. 185; for England and France see above, p. 147, n. 4; for Italy, see R.C. Trexler, *Synodal Law in Florence and Fiesole 1306-1518* (Studi e Testi [Vatican, 1971]). This book deals with diocesan rather than provincial statutes; but since Florence and Fiesole were two exempt dioceses like those in Scotland, their bishops in practice each acted as a metropolitan.
2   Cf. no. 4 of the code of the 1240s with nos. 56-62 of the Aberdeen diocesan statutes (*SES*, ii, pp. 30-5; Patrick, *Statutes*, pp. 30-6) and with nos. 115-21 of the St Andrews diocesan statutes (*SES*, ii, pp. 56-9; Patrick, *Statutes*, pp. 60-3).

the episcopate may well have regarded it as an assembly of equals, producing recommendations for good practice which they were free to accept or reject. The council's authority was seen more as coming from below than from above, and rested more on convenience than on binding law.

If the function of the provincial council as a court was necessarily irregular, depending on cases that happened to come its way, and if the passing of statutes was after the mid-thirteenth century a quite rare event, there still remained its function of providing a useful mechanism for bringing the prelates of the kingdom together to discuss and resolve matters of mutual interest. It is true that from the mid-fourteenth century the higher and lower clergy came to have a place also in the parliaments and general councils of the three estates summoned by the king;[1] but the development of these occasions for meeting did not supplant the traditional opportunities for gathering together provided by the provincial councils — though on at least one occasion in the fifteenth century parliament and provincial council met simultaneously in the same place.[2] At the council meetings the participants enjoyed in the greatest degree possible their 'liberty' as a separate element in the population. They were a mechanism which could be used by the Scottish Church to formulate responses to papal demands on it as a whole. When the pope called a general council to meet at Lyons in 1274, for example, it was in a provincial council that the decision was made about which bishops should attend and which should stay at home;[3] and there too arrangements were presumably made for the collection throughout the province of 'gravamina', that is suggestions about matters requiring reform which might form part of the agenda of the general council.[4] Again it was in a council that in 1268 it was decided to reject a demand from the Legate Ottobono that all the bishops of Scotland should attend his legatine council in London; and they sent just the bishops of Dunkeld and Dunblane instead (along with the abbot of Dunfermline and the prior of Lindores to represent the other clergy). It is not surprising that later they decided not to adopt the reforming canons which Ottobono proclaimed at his council.[5] When responses were needed to papal demands for financial levies from the Scottish clergy in

---

1   The term 'three estates' came into use from 1357 onwards (R. Nicholson, *Scotland: The Later Middle Ages* [Edinburgh, 1974], p. 166).

2   In 1427; see *SES*, i, p. lxxxi (n. 2).

3   *Chron. Bower* (Watt), v, pp. 398-9. For the next general council at Vienne in 1311 the pope decided which Scottish bishops were to attend (*Regestum Clementis Papae V* [Rome, 1885-92], no. 3631).

4   The Scottish replies to this papal demand for 'gravamina' do not survive; for English and continental responses, see Powicke and Cheney, *Councils and Synods*, ii, pp. 804-9, 810 (n. 3); but some twenty-five 'gravamina' were assembled somehow in Scotland and sent to the council at Vienne, of which two are known because they were selected there for further discussion (*Archiv für Literatur- und Kirchengeschichte*, iv [1888], pp. 376, 383; cf. J. Lecler, *Vienne* [Paris, 1964], pp. 60-1); presumably there had been a council meeting in Scotland at some time in 1310-11 in this connection.

5   *Chron. Bower* (Watt), v, pp. 366-9.

1266 and 1268, they were probably worked out in the provincial council, with rejection as the outcome in both cases.[1] A tax levied at the time of the council of Lyons in 1274 could not be avoided, for Scottish bishops had assented to it there; but when the papal collector arrived in Scotland in 1275, it was at a meeting of the provincial council that he was persuaded to go back to Rome to try to have the instructions on how the tax was to be assessed framed in a more favourable way.[2] The ploy was not successful, and the Scottish clergy had to pay up; but the story illustrates how the provincial council provided a mechanism for obtaining the corporate opinion of the Scottish Church regarding what it considered to be its rights and privileges. This mechanism could also be used to put pressure on the king on occasion. Quite probably the publication of the code of statutes in the 1240s should be interpreted as a positive reminder to Alexander II of the liberty which the Church expected to enjoy.[3]

But periods of strain between the crown and the council were not typical. A king such as the usurping Robert I was no doubt positively grateful for any demonstration of support which the Scottish clergy could offer him through the council.[4] That, however, was at a time of emergency and abnormal relationships. More characteristic were the means used by successive kings in normal circumstances to ensure that the provincial council acted in harmony with their wishes. Whatever canonist theorists might claim, the liberty of the Church was always at the mercy of the crown. The king's personal attendance at a council meeting in 1242[5] is not known to have been repeated. Instead, he usually appointed his proctors to attend council meetings and watch over the crown rights, as we know from two form letters of appointment of such proctors dating apparently from the 1290s and from 1321.[6] Such evidence should not be interpreted as necessarily implying a chronic state of suspicion on the crown's part, for the leading clerics who advised the king and devised the language of these form letters are quite likely to have also been members of the provincial council in other capacities — it was not uncommon for the king's chancellor to be a bishop.[7] Nevertheless it is interesting that in the 1290s it was stated that it was the established custom for the king to send 'knights and clerics' (number not stated) as his representatives, who were empowered to threaten

1   Ibid., pp. 356-7, 368-9.
2   Ibid., pp. 402-3.
3   See above, p. 148, n. 6.
4   Both higher and lesser clergy attended Robert's parliament at St Andrews on 17 March 1309, when they supported a manifesto in his favour (*APS*, i, pp. 459-60; cf. G.W.S. Barrow, *Robert Bruce and the Community of the Realm of Scotland* [3rd edn, Edinburgh, 1988], p. 184 and n. 100); and provincial councils appear to have been held twice (once perhaps at Dundee on 24 February 1310, and once at some unknown place June 1314 × November 1316) to confirm their support (ibid., pp. 268-9).
5   See above, p. 148, n. 6.
6   Duncan, *Formulary E*, no. 65; *The Register of Brieves*, ed. Lord Cooper (Stair Soc., 1946), p. 47, no. 67.
7   For lists of chancellors with their clerical status when relevant, see *Handbook of British Chronology*, 3rd edn, ed. E.B. Fryde *et al.* (London, 1986), pp. 181-2.

to appeal to the pope if anything was done against the interests of the crown. In the 1321 example the king was now sending two clerics qualified in Roman civil law as his proctors to the council, and the warning was very specific that nothing must be done that would be to the prejudice of the royal authority. If this was the framework within which the council was accustomed to do its work, it is not surprising that a council in March 1325 confirmed support for regalian rights to patronage in episcopal vacancies against novel papal claims to similar rights.[1] With the king himself nearby, since this was a time when he had summoned parliament to meet at the same time, the council no doubt felt itself encouraged to conform to the crown point of view.

This was to become strikingly obvious a century later when the authoritarian James I was implementing many reforms in a hurry. In 1427 we find him using parliament to promulgate a new law about procedures in church courts, with the aim of making it easier for laymen to bring accusations against clerics, and the provincial council was simply instructed to adopt this reform forthwith.[2] This brisk approach marks a new age, which had been introduced while the Scottish government had tackled the political problems raised by the Great Schism in the papacy from 1378 onward and by the era of general councils of the Church from 1409 onwards. The complexity of the issues raised by these convulsions in the Western Church at large appears to have been regarded as too great for the provincial council as the voice of the Scottish Church to tackle on its own. All the major decisions therefore were taken by the king (or the dukes of Albany during the captivity of James I), sometimes with the advice of the three estates.[3] The provincial council was being kept strictly to its terms of reference with responsibilities for the internal welfare of the Scottish Church. As always, the crown defined the parameters of its activities.

This does not mean that the crown was regarded as hostile to the Church. Indeed the opposite was the normal situation. The crown was expected to support a framework within which the various communities of the kingdom could arrange their affairs as suited them best. Just as the late-twelfth-century papal recognition of the Scottish dioceses, with only one exception, as equals exempt from the unwelcome ambitions of the foreign archbishops of York to exercise metropolitan authority over them must have been arranged with the support of the king at the time, so the request of the bishops in 1225 for powers to set up their own provincial council must have had crown backing. The lack of some guiding authority for the Scottish

---

1   *Glasgow Reg.*, i, no. 270. By the mid-15th century, royal definitions of this and other categories of regalian right could be handled in both parliament and provincial council (G. Donaldson, *Scottish Church History* [Edinburgh, 1985], ch. 4).

2   *APS*, ii, p. 10, c. 12.

3   E.g., in 1418 (*Chron. Bower* [Watt], viii, pp. 86-93; *Acta Facultatis Artium Universitatis Sanctiandree 1413-1588*, ed. A.I. Dunlop [SHS, 1964], pp. 12-13; *St Andrews Copiale*, pp. 23-9); in 1423 (W. Brandmüller, *Das Konzil von Pavia-Siena*, ii [Münster, 1974], pp. 357-8); in 1433 (J.H. Burns, *Scottish Churchmen and the Council of Basle* [Glasgow, 1962], pp. 18-20).

Church as a whole must have been obvious; and the unique solution devised in Scotland on the basis of a very sketchy papal bull of authorisation had a long run ahead of it until 1472. We are entitled to assume that it was an arrangement that suited the Scottish situation, and provided sufficient homogeneity of approach for the Scottish Church as a whole to be able to offer controlled and consistent service to both crown and people. The provincial council may have had its periods of doldrums under ineffective conservators or uncooperative bishops; and certainly the reformist zeal of the era just after the Fourth Lateran Council was to wane in Scotland as it did everywhere else. But the council survived on the Scottish scene as one of the major institutions of government, for neither crown nor people could do without it.

Appendix

A List of Known Conservators

William de Lamberton (St Andrews)?:	9 Jul. 1321	(*APS*, i, pp. 478-9)
William de Sinclair (Dunkeld):	21-22 Mar. 1325	(*Glasgow Reg.*, i, no. 270)
Walter Trayl (St Andrews):	18 Jul. 1388	(*Moray Reg.*, no. 267)
Alexander Bur (Moray):	early 1390s?	(*SES*, ii, pp. 3-4)
William Stephenson (Dunblane):	16 Jul. 1420	(*SES*, ii, no. 166)
John de Crannach (Brechin):	28 Jun. 1445	(*Brechin Reg.*, i, no. 56)
Thomas Spens (Aberdeen):	19 Jul. 1459	(*SES*, ii, pp. 79-80)
Patrick Graham (St Andrews):	17 Jul. 1470	(see above, p. 146, n. 5)

8

An Urban Community:
The Crafts in Thirteenth-Century Aberdeen

ELIZABETH EWAN

Among the many communities which made up medieval Scotland, one of the most confident and self-assertive was 'the community of the burgh'. The burghs played an especially important role in Scotland's economic and political development in the twelfth and thirteenth centuries, acting as market centres for both domestic and overseas trade, administrative sites for royal and ecclesiastical officials, and centres of new technologies, fashions and ideas. Although the phrase 'community of the burgh' is not found in the records until the late thirteenth century (a development which paralleled the emergence of official recognition of the 'community of the realm'), the idea of the urban community was well established by this time. The Scottish burgh was a legal entity, defined by specific rights and privileges. The townspeople were bound together by laws and customs which, in theory at least, separated them from the countryside. Within this community, however, were various groups of people. The surviving records have given most prominence to a small group, the merchants involved in overseas trade, partly because they had most contact with royal government, partly because they saw themselves as the town leaders and the true representatives of the community. But equally important to the well-being of the community were those responsible for the crafts which fuelled the day-to-day functioning of the local economy. Just as the community of the realm could not survive without all the orders of society, neither could the community of the burgh.

The thirteenth century was a period of prosperity for Scotland, a prosperity fostered by royal encouragement of overseas trade and native industry and agriculture. As elsewhere in contemporary Europe, growth in the countryside fostered the growth of towns to provide goods and services.[1] One typical beneficiary was the burgh of Aberdeen, which during the century experienced a period of growth and expansion. During the reign of Alexander II, for example, the burgesses of Aberdeen were granted various privileges to give them a monopoly over the local cloth industry and the

---

1   S. Reynolds, *Kingdoms and Communities in Western Europe, 900-1300* (Oxford, 1984), p. 159; G.G. Astill, 'Archaeology and the smaller medieval town', *Urban History Yearbook*, 1985, p. 47.

trade in wool and hides.[1] As the centre of a large fertile hinterland, with excellent access to overseas and coastal trading routes, Aberdeen benefited from the favourable conditions in both its trade and industry. The story of its overseas trade has been discussed elsewhere.[2] The main focus of what follows here will be an area of the Aberdeen economy which is less often discussed for this early period, that is, the crafts which furnished a livelihood for many, if not indeed the majority, of the inhabitants of the burgh.

Craft activity generally involved the entire family, with children carrying out the simpler tasks while women often worked alongside their husbands or took charge of the selling of the finished product.[3] For immigrants to the burgh, labour in a craft workshop was one of the commonest ways to earn a living. Even those hired as domestic servants often lent a hand in the family workshop. In terms of the numbers of people involved, the crafts of the burgh probably far outstripped the importance of overseas trade as a source of livelihood for the inhabitants of the medieval town.

As has recently been pointed out, internal industry and local markets can play as crucial a role in a town's prosperity as overseas trade.[4] Although the records of such activities are not generally as numerous as those for long-distance commerce, this does not mean that they are of less importance than the better-documented export trade. Because the crafts in many towns did not receive much attention in the records until they had been formally incorporated, a process which generally did not occur until as late as the fifteenth century,[5] they have sometimes been downplayed in discussions of the early town economy. The picture will always remain incomplete, but archaeological evidence helps to fill in some of the gaps. In a sense, of course, the distinction between merchant and craftsman is somewhat artificial as craftsmen of the period and their families purchased their own raw materials and tools, employed journeymen and apprentices to help them, and sold their own products.[6] But while the retailing side of their operations was very important, the focus of this essay will be on the production part of their activities.

1   *Aberdeen Burgh Chrs.*, no. 3.
2   D. Ditchburn, 'Merchants, Pedlars and Pirates: A History of Scotland's Relations with Northern Germany' (Edinburgh University Ph.D. thesis, 1988); A.W.K. Stevenson, 'Trade between Scotland and the Low Countries in the Later Middle Ages' (Aberdeen University Ph.D. thesis, 1982), and 'Trade with the South, 1070-1513', in *The Scottish Medieval Town*, ed. M. Lynch *et al.* (Edinburgh, 1989). These works discuss Aberdeen in the context of Scotland's trade in general; the overseas trade of Aberdeen in particular will form part of a longer study I am undertaking of medieval Aberdeen.
3   R. Hilton, 'Towns in societies - medieval England', *Urban History Yearbook*, 1982, p. 10.
4   S. Reynolds, *An Introduction to the History of English Medieval Towns* (Oxford, 1982), pp. 19-20, 59, 64; E. Ennen, *The Medieval Town*, trans. N. Fryde (Amsterdam, 1979), pp. 1, 68; N. Goose, 'English pre-industrial urban communities', *Urban History Yearbook*, 1982, p. 25; Reynolds, *Kingdoms and Communities*, p. 200.
5   As, for example, at Chartres: ibid., pp. 71.
6   Reynolds, *English Medieval Towns*, pp. 62, 75; Ennen, *Medieval Town*, p. 132.

Although there are no population statistics for this period, Aberdeen's total population was probably about 1,000. Roughly a hundred of its thirteenth-century inhabitants are known to us by name, mainly through charters.[1] Most of these people were living in the town in the 1270s and '80s, although there are records of six inhabitants from the late twelfth and the early thirteenth centuries. Although the evidence of names is quite sparse — many Aberdonians appear in the records only once — it is possible to use it, along with other forms of evidence, to give a partial picture of the occupations of the townspeople and the economy of the town in the thirteenth century, especially during the reign of Alexander III. The picture which emerges is one of a town economy with a base varied enough to allow it to survive the vicissitudes of war and a declining market for its exports in the fourteenth century. The burgh was sufficiently prosperous in the 1300s not only to rebuild after a devastating fire, but also to extend the parish church and erect a new townhouse.

Although surnames were beginning to take on a more permanent form in the thirteenth century, surname evidence can still give an impression of place of origin and of occupation. The twelfth and thirteenth centuries were a period of growth for Aberdeen, with many immigrants starting a new life there, and it was not uncommon for such people to be identified by the place from which they came or by the way in which they earned their living. Occupational surnames and designations also help to give a picture of the economic activity of the thirteenth-century burgh, especially when combined with archaeological evidence from the same period, although by themselves they cannot tell us much about the economic standing of individuals, since the same surname could refer to both a wealthy entrepreneur and a poorer retailer.[2]

About 20 per cent of the people mentioned in the records of thirteenth-century Aberdeen have occupational surnames or designations. The names represent a variety of crafts: most of them are those that would be expected in a thirteenth-century burgh (or indeed most medieval settlements of any size), but others are of a somewhat more specialised nature, perhaps reflecting royal interest in the burgh and the activities involved with serving royal demands. After 1174, when control of three major southern burghs and castles was lost to the English king by William the Lion, new attention was paid to northern burghs such as Aberdeen, where royal councils and courts could be held. It is probably significant that Aberdeen's first surviving royal charter dates from about this period (1173 × 1184).[3]   Royal

1   The main sources are *Aberdeen Reg.*, i, pp. 35-7; ii, pp. 278-81; *RRS*, vi, no. 260; *Arbroath Lib.*, i, no. 140; Anderson, *Aberdeen Friars*, pp. 12-13; *CDS*, ii, no. 823; v, nos. 35, 70; *ER*, i, pp. 11-12; *Aberdeen-Banff Ills.*, iii, pp. 77-8.
2   A model approach was established by A.A.M. Duncan in 'Perth: The First Century of the Burgh', *Transactions of the Perthshire Society of Natural Science*, Special Issue (1974), which uses the evidence of about 30 charters and 100 names and occupations. See also his studies of the burghs in *Scotland: The Making of the Kingdom* (Edinburgh, 1975), chs. 18, 19.
3   Ibid., p. 468; *RRS*, ii, no. 153.

residence and the castles which were built in favourite places gave great impetus to a town's economy. The castle needed to be provisioned, the personnel of the court accommodated and fed. It is not clear when Aberdeen castle was built, although it was in existence by the 1260s;[1] but it seems likely that some royal residence was constructed in Aberdeen in the late twelfth or the early thirteenth century. Aberdeen also benefited from the establishment of the episcopal see of Aberdeen, a mile to the north of the town, in the twelfth century. The needs of a cathedral acted as a stimulus for both long-distance trade and local manufacturing in towns throughout Europe.[2]

Significantly, two of the wealthiest and most prominent burgesses were men with surnames suggesting their involvement in trades which would be useful to the crown: Martin Goldsmith ('Aurifaber') and Richard Mason ('Cementarius'). Possibly they owed their prosperity to royal favour. In twelfth- and thirteenth-century Perth, royal service brought rewards of urban property for a Flemish craftsman called Baldwin, a lorimer or maker of harness.[3] In Aberdeen, Richard Mason was engaged on building work on the king's castle at Aberdeen for which he was paid 20 merks in 1264;[4] unfortunately no record survives to show any connection between Martin Goldsmith and the crown, although he may have provided luxury goods for the royal court when it visited Aberdeen.[5] A late thirteenth-century balance arm found in excavation suggests that there was at least one worker in the burgh dealing with precious metal.[6] Goldsmith was a prosperous man, holding a variety of rents and lands throughout the burgh,[7] which implies both that his business was a profitable one and that burgesses did not necessarily limit themselves to one way of making a living. In fourteenth-century Edinburgh, a goldsmith was employed at the royal mint; perhaps Martin filled a similar function in Aberdeen, where a mint was active during the reign of Alexander III (1249-86). Local tradition speaks of a mint located at the Exchequer Row at the south-west corner of Castlegate.[8]

Some men, such as William 'Ballistarius', who was paid 20s. by the sheriff of Aberdeen for selling staffs ('baculi') and other such things, may have lived in the burgh only temporarily to carry out a royal project such as

---

1   Duncan, *Making of the Kingdom*, pp. 468-9; *ER*, i, pp. 11-12.
2   In Viborg, Denmark, for example, the success of the town has been linked to the building activities of the cathedral: T. Nyberg, 'Denmark', in *European Towns, Their Archaeology and Early History*, ed. M.W. Barley (London, 1977), p. 72. My thanks to Dr Leslie Macfarlane for discussing this point with reference to Aberdeen.
3   *RRS*, i, no. 171; Duncan, 'Perth', p. 37.
4   *ER*, i, pp. 11-12.
5   E.g., *Holyrood Lib.*, no. 71; *Bamff Chrs.*, no. 1; *Kelso Lib.*, ii, no. 396; *Moray Reg.*, no. 114.
6   *Excavations in the Medieval Burgh of Aberdeen, 1973-81*, ed. J.C. Murray (Edinburgh, 1982), p. 186.
7   *Aberdeen Reg.*, ii, pp. 278-9.
8   *ER*, ii, pp. cxiv, 160; I.H. Stewart, *The Scottish Coinage* (London, 1955), pp. 18-19; W. Kennedy, *Annals of Aberdeen* (London, 1818), i, p. 10. G.W.S. Barrow suggests a similar role for a Roxburgh goldsmith in 1296: *Kingship and Unity: Scotland 1000-1306* (London, 1981), p. 93.

the work on Aberdeen castle.[1] Medieval building-workers were often itinerant. Richard Mason, however, continued to live in Aberdeen until his death some time before 1294, and took an active role in burgh affairs.[2] Peter the Fowler ('Aucupex'), whose name suggests involvement in the hunting of birds, perhaps made his living by serving aristocratic patrons or the entourage of the king's court. He occupied a prominent enough place in burgh society to be accepted as one of two guarantors for a grant of land by a burgess's daughter in 1277;[3] the fact that he stood as guarantor for a local property implies his usual place of residence was the burgh.

Most Aberdonians probably made their living from rather less specialised pursuits, providing the goods which were needed in everyday life by their fellow townsfolk, by the inhabitants of the three religious houses (the Carmelites, the Trinity Friars, and the Dominicans[4]), and by people in the surrounding countryside. As in other Scottish burghs, the long burgage rigs provided space for both home and workshop for burgh craft families; products were often made in workshops in the rear of the property and sold in booths at the frontage, or in the market on market days.[5]

Several of the occupational surnames refer to crafts involved in food preparation. In late-medieval and early-modern English towns, about 30 per cent of the workforce was involved in the trades of food and drink, clothing, or building. Although many townsfolk could grow food in their gardens or on their plots of arable just outside the built-up area of the burgh, the processing was often left to others. Such processing formed an important part of the economy of small medieval towns.[6] Archaeological evidence from English market towns suggests that they were concerned more with processing country surplus than industrial production. While the nature of the surviving evidence for Aberdeen means that the relative importance of processing and industrial activity cannot be fully assessed, processing was certainly important to the burgh's economy.

In many countries, bakers were among the earliest to form guilds,[7] suggesting that they were one of the longest-established crafts. The Aberdonian William the Baker had land in Futy on which his widow paid an annual rent in the 1280s.[8] The baker's craft was common in early Scottish burghs. The Burgh Laws, a compilation of burgh customs dating mainly from the twelfth and thirteenth centuries, make frequent reference to bakers and attempt to regulate their activities closely.[9] One burgh law states that

1   *ER*, i, pp. 11-12.
2   *Aberdeen Reg.*, i, pp. 35-6.
3   *RRS*, vi, no. 260.
4   Anderson, *Aberdeen Friars, passim*.
5   E. Ewan, *Townlife in Fourteenth-Century Scotland* (Edinburgh, 1990), p. 25.
6   Goose, 'English pre-industrial communities', p. 24; C. Dyer, *Standards of Living in the Later Middle Ages: Social Change in England c.1200-1520* (Cambridge, 1989),pp. 196-7; Astill, 'Archaeology and the smaller medieval town', p. 48.
7   Reynolds, *English Medieval Towns*, p. 83.
8   *Aberdeen Reg.*, ii, p. 280.
9   *Ancient Burgh Laws*: 'Leges Burgorum', cc. 18, 59, 60.

only burgesses may have ovens; but if that was observed in Aberdeen (and it is a matter of contention how rigorously the laws were applied in the various burghs) later records suggest that bakers, despite their burgess status, came fairly low down in the burgh social hierarchy.[1] The baking trade continued to be one of the most closely-regulated by the burgh administration. That William rented his property from a prominent burgess and that he lived in Futy, on the outskirts of the town, suggest that he was not one of the more important inhabitants of the town — the suburbs of medieval European towns tended to be inhabited by the less well-off members of the population.[2] William's widow Mariota may have carried on his craft after his death, as she was responsible for the rent on the property. Women did act as bakers in late-fourteenth and fifteenth-century Aberdeen, although usually on a smaller scale than did male bakers.[3]

Some archaeological evidence suggesting baking activities has been uncovered in Aberdeen. An oven was found in the backlands of an Upperkirkgate property of the later thirteenth century. It was situated in the open air, well away from any buildings, presumably in order to minimise fire risks — a wise precaution, since it burned down a short while later. Small in size, it was probably mainly for domestic use: the lack of associated structures suggests that it belonged to the family living at the front of the site, probably a burgess family, as the frontage was the most prized area of a burgess property.[4] Larger ovens have been found in Perth,[5] and it seems likely that those Aberdonians who are identified as bakers would have been using larger structures of this kind in their craft. In 1399 there were at least seven bakers in the burgh.[6]

Flour for baking could be obtained at one of the local mills, either at the Justice Mills near the Denburn on the west side of the town, or at the Nether or Upper Mills in the town itself — one on the mill lade which ran from the Loch down to the east of St Nicholas' Church, the other on the Putachie Burn near the Trinity Friars. Henry the Miller was active in this area in the late twelfth or early thirteenth century.[7] There may also have been a windmill on the south side of the town during this time.[8]

Millers were as important to townsfolk as they were to country dwellers, for many inhabitants still engaged in rural pursuits to earn at least part of their livelihood. At Aberdeen, the burgh common lands stretched four miles to the west, and the burgh was surrounded by three areas of crofts: the

1   *Aberdeen Council Reg.*, i, p. 390.
2   C. Platt, *The English Medieval Town* (London, 1976), p. 38.
3   *Aberdeen Burgh Recs.*, pp. 92-3.
4   Murray, *Aberdeen Excavations*, pp. 53, 55, 81.
5   L.M. Blanchard, 'Kirk Close', in *Excavations in the Medieval Burgh of Perth, 1979-81*, ed. P. Holdsworth (Edinburgh, 1987), p. 39.
6   *Aberdeen Burgh Recs.*, p. 64.
7   G.M. Fraser, *Aberdeen Street Names: Their History, Meaning and Personal Associations* (Aberdeen, 1911), p. 3; E. Meldrum, *Aberdeen of Old* (2nd edn, Aberdeen, 1987), pp. 96, 196; *Arbroath Lib.*, i, no. 140.
8   Kennedy, *Annals of Aberdeen*, i, p. 410.

Denburn crofts to the west, the Gallowgate crofts to the north, and the Futy crofts to the east.[1] There the burgesses grew crops, for both themselves and their animals. The major crops were oats and barley, with the best land being reserved for wheat. Shortages of grain were not uncommon; it was often necessary to import it during the thirteenth century. The townsfolk were also limited in the amount of land that they could cultivate — the moorland beyond the croftlands was not suitable for growing crops.[2] The nature of animal bone finds in most medieval burghs suggests that any animals raised were used for meat rather than wool or hides.[3]

The rural character of the thirteenth-century burgh would have been clearly evident to any visitor. As late as 1661 there were still many fertile fields to the east of the town.[4] In the thirteenth century, the arable lands around the Denburn included an area to the east of the Denburn in the Green, thus blurring the line between burgh and countryside. Archaeological evidence of occupation along the western part of Upperkirkgate suggests similar quasi-agricultural land use in the suburbs.[5] Farming activities made an impression on the industrial life of the burgh. Many farmer-craftsmen would carry out their trades in such a way as to fit them around the responsibilities of the farming year, while others who might work full-time at a craft were dependent on farming-cycles for the supply of their raw materials and also additional labour provided by temporary immigrants from the countryside. Shoemakers, for example, did most of their work in the autumn, while fishers did so in the spring. Furthermore, although there might be a distinctive set of functions and range of occupations in the medieval town, town and country still shared many values, staying in touch through frequent contacts and visits, familial and commercial.[6] Country folk came each week to market to exchange their surplus produce for manufactured goods and imports, while others came to earn a living as servants or workers for those already established in the town.

Swine roamed the Aberdeen streets — as late as 1511 the council had to order that swine be kept off the streets for a fortnight in order to make the town presentable for a royal visit[7] — while cattle were taken each day by the common herd through the town to the burgh grazing lands, and livestock

1   The conjectural location of these crofts can be seen in P.J. Anderson's map at the end of *Aberdeen Burgh Chrs*.
2   Murray, *Aberdeen Excavations*, p. 241; Stevenson, 'Trade with the South', pp. 183-4; P. Marren, *A Natural History of Aberdeen* (Aberdeen, 1982), p. 20.
3   G.W.I. Hodgson, 'The Animal Remains from Medieval Sites within Three Burghs on the Eastern Scottish Seaboard', in *Site Environment and Economy*, ed. B. Proudfoot (British Archaeological Reports, International Series 173, 1983), pp. 9, 12.
4   James Gordon of Rothiemay, *Abredoniae Utriusque Descriptio: A Description of Both Towns of Aberdeen, 1661* (Spalding Club, 1842), p. 18.
5   Murray, *Aberdeen Excavations*, p. 112.
6   R.M. Spearman, 'Workshops, Materials and Debris: Evidence of Early Industry', in *Scottish Medieval Town*, p. 145. I.H. Adams says that the daily life of townsfolk would be far removed from the rhythms of the countryside (*The Making of Urban Scotland* [London, 1978], p. 29); but this might not be so pronounced in the thirteenth century: Dyer, *Standards of Living*, p. 223; Reynolds, *English Medieval Towns*, p. 89.
7   F. Wyness, *City by the Grey North Sea* (2nd edn, Aberdeen, 1972), p. 71.

bellowed from the backlands of the burgages where they were penned in each evening.[1] Vegetables were grown on the unbuilt-on areas of the burgh such as the backs of the burgage plots, or the extensive gardens surrounding the religious houses. Excavations in the Green, around the area of the Carmelite friary, have revealed large tracts of garden earth surrounding the conventual buildings in the thirteenth century and later; in 1571 the possessions of the former friary included garden lands, barns, a malt-house, and a kiln.[2]

It is not surprising that the earliest inhabitant of Aberdeen whose occupation is known was a miller. He probably lived at the mill, because his house is said to be near the stream near St Nicholas' Church, a description which would fit the mill lade running from the Loch to the town's Upper Mill in 1434.[3] The Justice Mills to the west of the Green were probably in operation by the late thirteenth century. In 1281 Henry 'Koc' (perhaps Cook), who is described as a baker, held some crofts in the area known as Justicefield to the west of the Denburn.[4] Perhaps we have here an early entrepreneur, renting lands to grow grain which can be milled near by and then baking bread from the flour produced, thus neatly combining agricultural and commercial pursuits. It has been suggested that early bakers might have been tied workers dependent for supplies on a victualler;[5] but this does not appear to have been the case with Henry Koc.

Interestingly enough, there are no references to brewers among thirteenth-century Aberdonians, although brewing was certainly carried out in Aberdeen — in 1306 Edward I ordered 26 quarters of malt from the royal store to be paid to the brewers of Aberdeen so that they could brew ale.[5] Barley was often parched to stop its germination for malting, and sometimes survives in excavations.[6] Possibly the lack of references to brewers is due to the low status of brewing as an occupation; most of the people mentioned in the charters are of high or middling status. But a more likely explanation is that brewing was often carried out as a part-time occupation, especially by married women who might sell the surplus after home consumption needs had been met. It has been suggested in studies of other European towns that the lack of surnames based on an occupation which is known from other evidence to have existed may indicate that the activity was generally part-time or temporary in nature, perhaps only one part of a diversified family economy. Brewers were thus unlikely to identify themselves as such in

1   *Aberdeen Burgh Recs.*, pp. 91-2; Ewan, *Townlife*, pp. 24-5.
2   Murray, *Aberdeen Excavations*, p. 248; *Three Scottish Carmelite Friaries: excavations at Aberdeen, Linlithgow and Perth, 1980-1986*, ed. J. Stones (Edinburgh, 1989), pp. 26, 35-51.
3   *Arbroath Lib.*, i, no. 140; J. Cripps, 'Establishing the topography of medieval Aberdeen: An assessment of the documentary sources', in *New Light on Medieval Aberdeen*, ed. J.S. Smith (Aberdeen, 1985), p. 22.
4   *Aberdeen Reg.*, ii, pp. 278-9.
5   Duncan, *Making of the Kingdom*, p. 500.
6   *CDS*, v, no. 475.
7   Spearman, 'Workshops, Materials and Debris', p. 142.

official charters.[1]

The Burgh Laws assume the presence of brewers (usually, although not always, women) within Scottish burghs, and when regular records begin to appear in Aberdeen in 1398, brewers are among the workers found most regularly before the courts. The Burgh Laws also assume that each burgess family would have brewing implements among its possessions. The part-time nature of the occupation is underlined in a law which states that all, except office-holders, might brew, as long as they paid 4d. and used different vats for brewing and for their other occupations.[2] An assize of bread and ale was common throughout England by the thirteenth century, and it seems likely that a similar assize was proclaimed regularly in Aberdeen long before 1398.[3] Women such as Mariota widow of William the Baker quite probably relied on brewing to make ends meet. Medieval towns usually had a high proportion of poor households, often headed by widows, which made do with casual earnings and the profits from selling foodstuffs or ale.[4]

Another source of food was fish. Aberdeen was known from an early period for its fish exports, especially salmon, and the burgh fishings, which were leased out by the town to individual burgesses, were one of its most valuable assets. In thirteenth-century Cologne and Flanders *L'Abberdaan* was the name used for a type of cod. In 1281 an English merchant was commissioned by Edward I to buy 100 barrels of salmon and 5,000 salted fish at Aberdeen. In 1290 400 fish of Aberdeen were among the provisions for a ship to bring the Maid of Norway to Scotland; fishing rights were among the privileges granted to the burgh in the early fourteenth century by Robert I.[5] Despite the undoubted importance of Aberdeen's fishing industry, however, there are no records of thirteenth-century townsfolk who made their living this way. Part of the explanation might be that fishing, like brewing, was a part-time occupation, or possibly the industry was largely in the hands of fishers who lived outside the burgh in Futy, which in later years was known as a fishers' community.[6] In the fifteenth century, despite the Aberdeen council's disapproval, fish was sold by local fisherman on the seashore at Footdee. There was also a fishing community at Torry to the south; in the seventeenth century fishing huts were set up at low tide on the low island known as the Inches.[7] The needs of Aberdeen itself and the

1   J.M. Bennett, 'The Village Ale-Wife: Women and Brewing in Fourteenth-century England', in *Women and Work in Pre-Industrial Europe*, ed. B.A. Hanawalt (Bloomington, 1986), pp. 21-2; J.A. Raftis, *A Small Town in Late Medieval England: Godmanchester, 1278-1400* (Toronto, 1982), p. 192.
2   *Ancient Burgh Laws*: 'Leges Burgorum', cc. 36, 59, 67, 94, 116; *Aberdeen Burgh Recs.*, pp. 87-9.
3   Bennett, 'Village Ale-Wife', pp. 20-1; *Aberdeen Burgh Recs.*, pp. 53-4.
4   Dyer, *Standards of Living*, p. 196.
5   Stevenson, 'Trade with the South', pp. 185-6; Ditchburn, 'Merchants, Pedlars and Pirates', p. 99; V.E. Clark, *The Port of Aberdeen* (Aberdeen, 1921), pp. 2-3; *RRS*, v, no. 247.
6   Wyness, *Grey North Sea*, p. 15.
7   H. Booton, 'Inland Trade: A Study of Aberdeen in the Later Middle Ages', in *Scottish Medieval Town*, pp. 150, 158; Gordon, *Abredoniae Utriusque Descriptio*, p. 18.

ecclesiastical community at Old Aberdeen, to the north, ensured that there was a good local market for fish; finds of fishbones are common on excavated sites.[1] When not fishing, the fishers acted as pilots for ships entering the Dee estuary, guiding them through the shifting sands of the harbour, either to anchorage at Torry, or, if they were small enough and it was high tide, to the Denburn anchorage at the Keyhead.[2]

Other occupations also made use of Aberdeen's streams and rivers. A spring rising in the Green (at the foot of Carnegie's Brae) ran along to the foot of Shiprow. The Putachie Burn, which may mean the 'cattlefield burn',[3] was perhaps an early grazing ground for cattle brought to the burgh. The Denburn itself ran down the west side of the Green and then turned east towards the Trinity Friars and past the foot of Shiprow, a street which wound its way down from the Castlegate along the east side of St Katherine's Hill. Running water was a valuable resource for several burgh crafts and may have led to a concentration of certain types of craft-workers in this area. Of the seven men of occupational surname or designation who are associated with specific properties, four lived or held property on Shiprow. Martin Goldsmith leased out his property, but three others — Nicholas Flesher ('Carnifex'), Simon Skinner ('Pelliparius') and John Cooper ('Cupar') — apparently lived on the street.[4]

Skinners and fleshers made great use of water in their trade. In later centuries there were complaints about such crafts defiling the water of the town;[5] such complaints were undoubtedly made earlier as well, but they do not survive in the records. Nicholas Flesher leased his land from the Trinity Friars, and, given the description of the property — it is described as to the west of the adjoining lands which suggests that it was on the part of Shiprow running east-west along the south part of St Katherine's Hill[6] — it seems possible that it lay beside the Putachie Burn. Here, before slaughtering them, he could graze the cattle he had bought at the market. Access to the grazing ground and a water source was perhaps Nicholas' main reason for leasing the land for, if the Trinity Friars followed the usual meatless diet of religious orders, he would not have found many customers there.[7] He probably did most of his selling in the market place; in the fifteenth century there was a special fleshhouse set aside for the fleshers to sell their wares.[8]

A burgh law required fleshers to go to their customers' houses to butcher

1   Murray, *Aberdeen Excavations*, pp. 229-30.
2   Kennedy, *Annals of Aberdeen*, ii, p. 289; J. Milne, *Aberdeen: Topographical, Antiquarian, and Historical Papers on the City of Aberdeen* (Aberdeen, 1911), p. 360; J.S. Smith, 'The physical site of historical Aberdeen', in *New Light on Medieval Aberdeen*, pp. 150, 158.
3   Milne, *Aberdeen Papers*, p. 34.
4   *Aberdeen Reg.*, ii, pp. 278-9.
5   *Ancient Burgh Laws*: 'Iter Camerarii', c. 23.
6   *Aberdeen Reg.*, ii, pp. 278-9.
7   Access to water, however, was not generally a problem in Aberdeen, because the hilly nature of the terrain meant that there were many lochs and marshy areas: Smith, 'Physical site of Aberdeen', p. 2.
8   *Aberdeen Council Reg.*, p. 11.

their animals at certain times of the year. This may have been due to a heavy period of demand, with burgesses with cattle wanting them to be slaughtered in the autumn so that they did not have to over-winter them;[1] in order to prevent congestion it was better for the flesher to go to individual properties than to have all the cattle come to him. Bones with butcher marks, found on a number of sites in Aberdeen show that there was a good demand for his services among the people of Aberdeen,[2] although it is not possible in most cases to tell whether or not the butchering process took place where the bones have been excavated. The king's household would also have needed fleshers — in 1263, 140 carcasses were bought for the king's larder.[3] Demand for them might have come from the countryside, too. In England, it was not uncommon for peasants to sell their animals to the butcher of a local town and then buy smaller quantities of meat when required.[4] The close connections between Aberdeen and its hinterland would have made such exchanges quite feasible.

Simon Skinner held land almost adjacent to Nicholas Skinner's property[5] — possibly the waste products of one craftsman served as the raw materials of another. A craft economy which made use of every part of a source of raw material was common to medieval towns, and Scottish burghs were no exception to this. Indeed, as has been pointed out elsewhere, the concentration in one place of crafts whose materials were drawn from similar sources was one of the strengths of the burgh economy.[6] In many European towns, there was also, from an early period, a concentration of craft groups in areas where their special needs could be met; for example, skinners, dyers and saddlers needed to be near a water source.[7] In thirteenth-century Aberdeen there may have been areas of craft concentration in the Green, along Shiprow, and along the Gallowgate near the Loch.

Moreover, Shiprow itself was not only a main route between the Keyhead where ships docked and the market place on Castlegate,[8] but was where the main road entered Aberdeen from the south; thus it would have been a good location for craft-workers anxious to offer their services to arriving visitors or departing merchants. This may have been why John Cooper, for instance, established himself there. Goods came by sea both from abroad and from other areas of Scotland. Exports by sea were

1   *Ancient Burgh Laws*: 'Leges Burgorum', c. 64. This was also common practice in the countryside: Barrow, *Kingship and Unity*, p. 3.
2   Murray, *Aberdeen Excavations*, p. 229; H. Murray, 'Excavation at 45-47 Gallowgate, Aberdeen', *PSAS*, cxiv (1984), additional microfiche A11.
3   *ER*, i, pp. 11-12.
4   Dyer, *Standards of Living*, p. 156.
5   *Aberdeen Reg.*, ii, pp. 278-9.
6   Spearman, 'Workshops, Materials and Debris', p. 136.
7   Ennen, *Medieval Town*, p. 133; Platt, *English Town*, pp. 45-6.
8   Hence the location of the Trinity Friars, who cared for sailors as part of their work — my thanks to Leslie Macfarlane for pointing this out.

commonly shipped in barrels;[1] in the 1480s an exporter ordered several good barrels from the town's coopers in which to pack his salmon for export.[2] John's location on Shiprow was a strategic one, although it has been suggested that many coopers may have preferred the outskirts of the burgh, near supplies of wood from the forest surrounding it. In the Aberdeen area these included oak, ash, hazel, elm, aspen, alder and willow.[3] The products of the cooper were also in common use by the local townsfolk. Barrels were used as containers, to line wells, and to catch rainwater; barrel staves, barrel tops and planks have been found from sites dating from the early thirteenth century onwards.[4]

John's near neighbour, Simon Skinner, was also engaged in an occupation which was important to the burgh economy. Leather goods fulfilled all sorts of functions for medieval people, from acting as armour and waterproof clothing to providing saddles, shoes, and even cooking pots. It is possible that local skinners could not keep up with demand and that skins were bought outside the burgh in the surrounding hinterland; from 1398, when its records begin, the burgh council almost annually prosecuted country-dwellers for the illegal selling of skins and hides.[5] The frequency of the prosecutions suggests that they had little effect. Skins came not just from cattle but also from the many wild animals living in the hinterland — bear, ox, boar, beaver and wolf.[6]

A property on Gallowgate in 1281 was rented by Colin Sellar.[7] He was probably a saddler, perhaps using the skins provided by Simon Skinner, after they had been treated by a tanner. Peter Armurer was living in the burgh by 1317.[8] His expertise may have been required by the men of Aberdeen after Edward I's incursions into Scotland began in 1296. It is not clear whether he worked in leather or metal or in both, for he is also referred to as Peter Wapinmaker, although that could refer to leather shields. In the thirteenth century, leather-workers were apparently plying their craft on Gallowgate towards the Loch, where they left scrap leather and offcuts in a boundary ditch. To the north of this site, an excavated midden reveals many fragments of cobbler's waste as well as some nearly complete shoes, suggesting that shoemakers or cobblers were at work near by.[9]

The site of the property, near the Loch which provided a source of

1   Booton, 'Inland Trade of Aberdeen', p. 148; Spearman, 'Workshops, Materials and Debris', p. 143.
2   Booton, 'Inland Trade of Aberdeen', p. 151.
3   Spearman, 'Workshops, Materials and Debris', p. 143; Marren, *Natural History of Aberdeen*, p. 13.
4   Murray, *Aberdeen Excavations*, p. 180; D. Evans, *Digging up the 'Coopie': Investigations in the Gallowgate and Lochlands* (Aberdeen, 1987), p. 11.
5   *Aberdeen Burgh Recs.*, p. 222; cf. Duncan, 'Perth', pp. 36-7.
6   Marren, *Natural History of Aberdeen*, p. 13.
7   *Aberdeen Reg.*, ii, pp. 278-9.
8   *Aberdeen Burgh Recs.*, pp. 13, 16-7.
9   Evans, *Digging up the 'Coopie'*, p. 11.

water, and the cattle hair found in the ditch-fill, suggest that there was a tannery in the area in the same period, although the probable tannery excavated in this area dates from the late fourteenth century. Similar finds of animal hair on a site further to the south in Gallowgate give further evidence of Aberdeen's tanning industry. Other evidence of leather-preparation near the Loch includes several leather scraps with pinholes, which suggest that they might have been hung or stretched, perhaps for cutting.[1]

In medieval towns, every part of the animal was used. From cattle, Aberdeen craft-workers produced and used neat's foot oil and gelatin from the hooves, meat, milk, blood, horn, bone, gut, leather, sinew and fat. Horn and bone were carved into a wide variety of objects, including combs, awls, picks and other tools, and even skates.[2] Other animals also produced useful by-products. Dog's dung was used to treat leather before tanning; the feathers of the goose could be used for the flights of arrows.[3]

One animal product that was of importance to merchants and craftworkers alike was wool. As has been pointed out elsewhere, twelfth- and thirteenth-century royal charters to the burghs suggest a crown anxious to foster a native cloth industry.[4] Cloth-making was an important industry in most contemporary European towns, although some historians have cautioned that the greater survival of its records may give an exaggerated idea of its importance.[5] One of the reasons for Scotland's growing prosperity in these years was the ever-increasing demand for Scottish wool from the looms of the cloth-weaving towns of Flanders. In 1180 the count of Flanders gave a grant of privileges to the monks of Melrose Abbey to facilitate their export of wool to his country.[6] Melrose and other monastic houses seem to have used their properties in the burghs as a centre for such exports. Arbroath Abbey, founded in 1178, may have used its property in Aberdeen for a similar purpose.[7]

A Scottish cloth industry developed as well, although it is impossible to determine how extensive it was as the records for such a study do not survive. The market for textiles was extensive, with demand not just for clothes but also for bedclothes, hangings, tablecloths, sacking for wrapping and many other products. In England, it has been estimated that the nobility spent as much as 10 per cent of their income on textiles.[8] The kings of Scots gave various privileges to the burgesses of royal burghs to allow them to maintain a monopoly over the production of cloth (or at least the better-

1   Murray, 'Excavation at 45-47 Gallowgate', microfiche F9, F13; Evans, *Digging up the 'Coopie'*, p. 11.
2   Murray, *Aberdeen Excavations*, p. 182.
3   Ewan, *Townlife*, p. 27.
4   Duncan, *Making of the Kingdom*, p. 510.
5   E.g., Reynolds, *English Medieval Towns*, pp. 61-2.
6   Duncan, *Making of the Kingdom*, p. 464; *CDS*, i, no. 880.
7   W. Stevenson, 'The Monastic Presence: Berwick in the Twelfth and Thirteenth Centuries', in *Scottish Medieval Town*, pp. 110-11; *Arbroath Lib.*, i, no. 140.
8   Dyer, *Standards of Living*, p. 78.

quality cloth). This may have served a two-fold purpose: it reduced dependence on imported fabrics, and, by giving control of woollen cloth production to the same group of people who were most involved in the export of wool, it helped to ensure a balance between the export of wool and its use in home production — although it seems that the emphasis increasingly shifted towards wool exports during the thirteenth century. By the century's end, new customs had been introduced on the export of wool.[1] When, in the fourteenth century, export records become available, it appears that the prosperity of the merchant burgesses was based primarily on the export of wool, rather than the manufacture of cloth; but the thirteenth century may have been a period of transition. There are several pieces of evidence from thirteenth-century Aberdeen which suggest the presence of a cloth industry, although it seems that this was mainly on a small, perhaps domestic scale.[2] It has been suggested that increasing demand for cloth in the thirteenth century made regulations against imported fine cloth and coarse rural cloth less effective.[3] As merchants turned to the increasingly profitable export of raw wool — by the later thirteenth century Aberdeen was recognised in St Omer as one of the four main Scottish centres of the wool export trade[4] — they may have been less concerned about competition in the cloth industry. Moreover, it was probably they who provided the vast bulk of the imported cloth, which they purchased with the proceeds of their wool sales. Most of the cloth found in medieval burghs was of poor-to-reasonable quality;[5] domestic cloth-making was probably carried out mainly on a small scale, and was perhaps of little interest to the merchant entrepreneurs.

Weavers and fullers are among the first townspeople to appear in the Scottish records, although they are mentioned because they are being specifically excluded from the merchant guild.[6] In many countries, weavers and fullers secured royal protection before the towns had established their own governments. This led both to resentment by other townsfolk and to attempts to suppress such independence, either by excluding weavers and fullers from the main body of townspeople, the merchant guild, or (as happened in Berwick in the mid-thirteenth century) by the banning of sectional guilds.[7] The social status of these crafts varied over time and from place to place. In the fifteenth century, for example, they were included in the merchant guilds of a number of towns.[8] William the Lion's Aberdeen charter suggests exclusion in the late twelfth century,[9] but this may no longer have been the case in the thirteenth century. Unfortunately, there are

1   *Aberdeen Burgh Recs.*, p. xcviii.
2   Murray, *Aberdeen Excavations*, p. 248.
3   Duncan, *Making of the Kingdom*, p. 510.
4   Stevenson, 'Trade with the South', p. 185.
5   Ewan, *Townlife*, p. 28.
6   *Ancient Burgh Laws*: 'Statuta Gilde', c. 1.
7   Reynolds, *English Medieval Towns*, p. 128.
8   Ewan, *Townlife*, p. 28.
9   *RRS*, ii, no. 153.

no names associated with these crafts surviving in the records for that period, although, as will be seen, archaeological evidence of looms shows that weaving was being carried on in the burgh.

One of the marks of better-quality cloth was that it was dyed. A native industry of coarse undyed cloth probably always existed in the burgh to provide clothing for the poorer inhabitants of the town, though it would have been of little concern to the cloth entrepreneurs. A burgh law referred to the provision of white or grey cloth (that is, undyed) for a poverty-stricken burgess.[1] Burgh charters suggest that the merchants' main aim was a monopoly over high-quality cloth, dyed and finished.[2] Two Aberdeen records from the thirteenth century refer to a property in the Green known as 'the Madder-yard'.[3] Madder, a plant which produced a red dye, was commonly used in medieval cloth production; it was usually imported to Scotland, as were most dyes, although some dye-plants such as weld and bog-myrtle could be found locally. Even weld, however, was relatively scarce.[4] It seems likely that this property was the site of a cloth- or wool-dyeing operation. Its situation in the Green meant it was near two water sources, the Denburn and the Putachie Burn, as well as near open space for drying the dyed cloth and also for hanging fulled cloth, and away from the noses of the wealthier burgesses in the residential areas up the hill to the east. The name the Green may come from Gaelic *grianan*, a drying-place.[5]

Dyeing was a rather noxious trade, and in many medieval towns dyers were concentrated together on the outskirts of the town. If, as has been suggested, the centre of the town shifted from the Green area to the Castlegate/Gallowgate area in the thirteenth century,[6] the location of such craft activity in this area would not be surprising, although the fact that the Madder yard was granted to the Carmelites in 1273, about the time they were establishing their house in Aberdeen, suggests that industrial activity had perhaps come to an end on this particular site.[7] Dyers were often among the elite of medieval towns.[8] In Scotland their craft required the import of materials from the Continent and thus involvement in overseas trade. The grant of the Madder yard to a friary suggests that the owner had become quite prosperous, and employed dyers rather than working at the trade with his own hands. Rentals in the Green were lower than other areas in the

1   *Ancient Burgh Laws*: 'Leges Burgorum', c. 42.
2   *Aberdeen Burgh Chrs.*, no. 3.
3   Anderson, *Aberdeen Friars*, p. 12; *RRS*, vi, no. 260.
4   Marren, *Natural History of Aberdeen*, p. 23. For other plants see Murray, *Aberdeen Excavations*, pp. 241-3.
5   The area was known as 'the road of the Green' in the thirteenth century, a name which probably refers to its being the way to the bleaching-place on the shores of the Denburn: G.M. Fraser, *Historical Aberdeen: The Green and Its Story* (Aberdeen, 1904), pp. 3-4, 6.
6   See the discussion in E.P. Torrie, 'The early urban site of New Aberdeen: a reappraisal of the evidence', *Northern Scotland*, xii (1992)
7   *RRS*, vi, no. 260.
8   Reynolds, *English Medieval Towns*, p. 80.

fourteenth century, which may suggest a concentration of less prosperous craftworkers there. Archaeological evidence, however, suggests that poorer houses were often interspersed with richer ones,[1] so it is dangerous to attempt to distinguish clear-cut occupational zones within the burgh at this early period. The Green continued to be used as a place of work by dyers and other crafts in the cloth trade, especially weavers. As late as 1758 they had to be ordered to stop scouring stockings in the Denburn.[2]

Archaeological evidence has suggested that craft-working activities could also be found in the more prestigious residential areas of the burgh such as Upperkirkgate. A spindle whorl suggests that yarn was being prepared here, although such activity would not be unusual as spinning was an expected part of the daily work of medieval townswomen of all social classes.[3] In the fourteenth century, a weaving sword and loom-weight found at the same site suggest the use of a vertical loom to weave cloth. The cloth was perhaps of fairly high quality as the wool used in one case had already been dyed red.[4] The use of the vertical loom supports the picture of a fairly unsophisticated small-scale cloth industry. In those parts of Europe where cloth production was a major industry, it had been largely replaced by the horizontal loom, but in contemporary Scotland this transition was just beginning to take place.[5]

There were also a number of crafts which provided a livelihood for thirteenth-century Aberdonians but which have left no documentary trace. One of these is pottery. The raised estuarine deposits of the Aberdeen area provided excellent clay. Local pottery is found in most sites, although it does not become the dominant type until the fourteenth century. A possible source for the town's potters was an area near the Virginia Street Steps. The clay from this site was removed in the thirteenth century and replaced by a midden. A rough pebbled track leading from the site may have been constructed to allow the removal of the clay, either for potters or for building purposes.[6] As in other Scottish burghs, no trace has yet been found of an urban pottery-working site, suggesting that this important activity may have been carried on outside the burgh limits, perhaps in the Clayhills area to the south-west.[7] In fourteenth-century levels, locally-made pottery becomes more predominant, indicating that the burgh suppliers were perhaps compensating for the loss of imports due to the disruption caused by the Wars of Independence.[8]

1   E. Ewan, 'The Age of Bon Accord: Aberdeen in the Fourteenth Century', in *New Light on Medieval Aberdeen*, p. 37.
2   Fraser, *Historical Aberdeen*, p. 7.
3   D. Herlihy, *Opera Muliebria. Women and Work in Medieval Europe* (New York, 1990), pp. 94-5.
4   *The Town Beneath the City*, ed. J.C. Murray (Aberdeen, 1978), pp. 18-19.
5   Murray, *Aberdeen Excavations*, p. 198.
6   Smith, 'Physical site of Aberdeen', p. 5; Murray, *Aberdeen Excavations*, p. 107.
7   J. Stones, *A Tale of Two Burghs: The Archaeology of Old And New Aberdeen* (Aberdeen, 1987), pp. 26-7.
8   Ewan, *Townlife*, p. 35.

Another industry, of which Martin Goldsmith is the only recorded representative, was metal-working. Clay moulds found in the Gallowgate suggest that there was a metal-casting workshop in this vicinity during the thirteenth century. The articles produced were probably mainly small ewers and plates, although a pin-mould suggests that jewellery was also fashioned in the workshop. The objects found were made of copper alloys. There was a metalworker to the north of this site in the fourteenth century. An annular brooch of the late thirteenth or early fourteenth century may have been made in this or another workshop.[1] There are also iron artefacts and iron slag which imply the work of smiths. The iron industry shows the close connection between town and countryside, for the nature of the iron waste implies that smelting took place outwith the towns. It was common for iron smelting to take place in wooded areas.[2] In 1306 William Fichet was paid 10s.3d. for 40 pieces of iron bought for the English king, although whether this was imported or locally worked is not stated.[3]

Burghs also participated in overseas trade and Aberdeen was no exception, serving as the major export and import centre for the whole north-east. Overseas merchants were among those who resided in Aberdeen, and their activities can sometimes be traced outside their home, especially if they had the misfortune to be shipwrecked on English shores or waylaid by pirates. Some had managed to secure safe-conducts from the rulers of the countries they intended to visit, giving them the right to sue for restitution in foreign courts if they suffered losses. Surviving evidence suggests that their exports consisted largely of raw materials, primarily wool, woolfells and hides.[4] It follows, therefore, that the products of the Aberdeen craftworkers were destined mainly for the domestic market.

It has been argued that the disruption caused by the Wars of Independence from 1296 dealt a severe blow to the Scottish economy. There seems to be no question that the country's export trade was badly affected. But the effect on the crafts of the country is less clear. Some historians have argued that they also went into decline.[5] Others have suggested that the impact may not have been as severe or as long-term as has sometimes been assumed.[6] In many burghs, including Aberdeen, the archaeological evidence suggests that in pottery, local products increasingly made up the shortfall in imports. It is difficult to determine for certain if a similar pattern took place in other crafts, but perhaps the picture presented here of the varied activities of the craftworkers of Aberdeen in the thirteenth century, might provide grounds

1   Murray, 'Excavation at 45-47 Gallowgate', pp. 306-8; Stones, *Two Burghs*, p. 27; Evans, *Digging up the 'Coopie'*, p. 14.
2   Murray, *Aberdeen Excavations*, p. 18; Spearman, 'Workshops, Materials and Debris', p. 144; Astill, 'Archaeology and the smaller medieval town', p. 50.
3   *CDS*, v, no. 472
4   Ditchburn, 'Merchants, Pedlars and Pirates', pp. 65, 32.
5   Stevenson, 'Trade with the South', p. 189.
6   A. Grant, *Independence and Nationhood: Scotland 1306-1469* (London, 1984), p. 76; Ditchburn, 'Merchants, Pedlars and Pirates', pp. 34-5.

for optimism that it did.

Overseas trade was only one element in a burgh's economy, and the fact that the burghs were able to achieve recognition of their political importance in a period of declining exports suggests that it was not always the most important one. The merchants' involvement in overseas trade was the main reason for the gaining of political rights — their willingness to accept new export taxes to pay for David II's ransom suggests that they had other sources of wealth as well. One of these may have been investment in urban real estate, after the pattern of Martin Goldsmith, but investment in local industry and regional trade may have been equally rewarding.[1] The strong basis established in the thirteenth century allowed the burghs to adapt to changed conditions in the fourteenth century.

Finally, one striking image which emerges from an examination of thirteenth-century Aberdeen is of all the interconnections within the burgh. No doubt the tensions and rivalries which appear when the town records begin also existed earlier, but the members of the community — the craft-workers, merchants, churchmen and royal officials — depended on each other for their livelihood, making it in their interests to preserve some sense of unity and community. Whatever the social hierarchy, thirteenth-century Aberdeen was in practice a highly integrated society. And in that respect, surely, it was a microcosm of the wider community of the realm of Scotland.[2]

1   For the increasing political recognition of burghs and burgess involvement in real estate, see Ewan, *Townlife*, pp. 35-6, 148-54, 107-9.

2   I am indebted to the Social Sciences and Humanities Research Council of Canada for financial assistance with the research for this essay, and to Gil Stelter and Pat Torrie for their helpful comments.

9

The Earls and Earldom of Buchan in the Thirteenth Century

ALAN YOUNG

> Now ga we to ye king agayne
> Yat off his wictory was rycht fayn,
> And gert his men bryn all Bowchane
> Fra end till end, and sparyt nane,
> And heryit tham on sic maner
> Yat eftir yat weile fyfty zer
> Men menyt ye herschip off Bouchane.[1]

The completeness of this 'harrying' of Buchan, after the defeat of John Comyn earl of Buchan by Robert Bruce at the battle of Inverurie (Barra Hill) in 1308, reflects Buchan's political significance at that time. Without the destruction of the Comyn power base in Buchan, Robert's kingship over the whole of Scotland could not be a reality. The event, as important for Bruce's reign as the 'Harrying of the North' in 1069 had been for William the Conqueror's kingship in England, has nevertheless distracted historians' attention from Buchan's wider significance in the development of politics and society in medieval Scotland, particularly during the period from c.1212 to 1308. This essay, therefore, explores the political anatomy of thirteenth-century Buchan by examining formative developments which have been overshadowed by the 'herschip', but which have significance in their own right and, in fact, help to explain and give added meaning to that 'herschip'.

In comparison with the 'herschip', relatively little emphasis has been placed on the marriage between William Comyn and Marjorie, the only child and heiress of Fergus earl of Buchan, in c.1212. The marriage meant that William Comyn became not only the first 'Norman' earl of Buchan,[2] but also the first 'Norman' earl in Scotland: it opened up a new phase in the development of Scottish political society. Accordingly, Buchan provides a good opportunity to consider the 'Normanisation' of a native earldom in the thirteenth century, and gives further evidence of the Celtic/feudal interplay highlighted elsewhere in this volume. The marriage also linked Buchan

---

1   *Barbour's Bruce*, pp. 221-2.
2   Bower refers to William Comyn as earl by 1212: *Chron. Bower* (Goodall), i, p. 532. The last reference to William witnessing a charter without the title was 9 October 1211 (*Dunfermline Reg.*, p. 32), and the first charter referring to him as earl of Buchan was 17 August 1214 (*Arbroath Lib.*, i, no. 69).

inextricably to the political career of the most powerful baronial family in thirteenth-century Scotland. The Comyns, with their widespread properties in Scotland and their extensive interests in England, played a key role in the development and definition of the Scottish kingdom in the thirteenth century.[1] Furthermore, study of Buchan in this period reveals how in its relationship with Scotland as a whole it had a politically strategic role, especially with respect to the consolidation of the kingdom's northern regions. Analysis of Buchan's government and examination of the earldom's extent also assists towards an understanding of how Comyn power operated in probably its most important base. Finally, the earldom of Buchan must be set firmly in the context not only of thirteenth-century Scottish politics but also of contemporary Anglo-Scottish relations.

The first requirement is to consider Buchan's place in the political geography of Scotland at the beginning of the thirteenth century. The need to strengthen royal authority and define the crown's territorial control is a major theme of twelfth- and thirteenth-century Scottish history. As Geoffrey Barrow has stressed, successive Scottish kings had attempted to consolidate power both in areas where royal authority was secure and in areas such as Argyll, Ross, Moray and Caithness where provincial particularism was especially strong.[2] Buchan can be seen as being in an area of relatively secure royal control whereas, further west, Moray was always regarded as a difficult region to control. Despite early planned feudal settlement under David I, Moray posed a regular and serious threat in the twelfth century. Donald MacWilliam claimed the throne from the 1160s until 1187, and, in fact, Highland rebellions continued until 1230. These rebellions were a danger not simply to the throne, but to all the royal efforts at defining and consolidating crown authority in the north.

Bower's *Scotichronicon* relates that when Guthred son of Donald MacWilliam landed in Ross in January 1211 to lead a rebellion through Ross and Moray, King William sent into Ross a great royal army of some 4,000 men, whose leaders included the earl of Atholl, the earl of Buchan, and representatives of the two families claiming the earldom of Mar, Malcolm son of Morgrund and Thomas Durward. It would appear that by 'earl of Buchan' Bower meant William Comyn, and that he was in charge of the operation which succeeded in suppressing the rebellion and capturing Guthred MacWilliam.

> Interea Gothredus MacWilliam proditione suorum captus et vinculatus, ad Willelmum Cumyn comitem Buchanie, justiciarium domini regis, adductus est usque Moraviam: erat enim tunc temporis ipse custos Moravie.[3]

1   A. Young, 'Noble Families and Political Factions in the Reign of Alexander III', in *Scotland in the Reign of Alexander III, 1249-1286*, ed. N.H. Reid (Edinburgh, 1990), pp. 14-16.
2   G.W.S. Barrow, *Kingship and Unity: Scotland 1000-1306* (London, 1981), pp. 49-50.
3   *Chron. Bower* (Goodall), i, pp. 531-2.

William Comyn's role as 'warden of Moray' is puzzling in view of Bower's apparently contradictory statement that Malcolm earl of Fife held that office in 1211.[1] Whether Bower confused the events of 1211-12 with those of the later rebellion in 1229-30,[2] or whether William Comyn was given the office on a short-term, temporary basis is not clear, though the latter seems more probable. What is evident is that the rebellion had some support from the nobility in the north, as King William came north to Moray to make a treaty of peace with the earl of Caithness whose daughter he took as hostage.[3] The events of 1211-12 caused the king to review and reinforce royal policy in the north and marked the beginning of a new phase in the history both of Buchan and of the Comyn family.

Despite the establishment of sheriffdoms in north-east Scotland,[4] the royal presence in the north and north-east was clearly in need of bolstering. As Barrow has remarked, a lack of royal demesne in the Buchan region as well as in other areas in northern Scotland meant that William the Lion was 'conspicuously absent' from this area.[5] Thus when in c.1212 William Comyn married as his second wife Marjorie, the only child and heiress of Fergus, last Celtic earl of Buchan, the significance was as much political as social and institutional. Even by 1286 only five earldoms were in the hands of families of Anglo-Continental origin.[6] It is a testimony to the deep-seated nature of the old Celtic earldoms that the Comyns — like, later, the de Umphravilles in 1243 and the Bruces in 1272 — gained the dignity only by marriage. But given the circumstances of 1211-12 it was more than a little convenient for the crown to introduce a powerful royal agent into the north-east, and actually establish him hereditarily in the earldom of Buchan.

Already by c.1212, the Comyns had made their mark in Scottish government. William Comyn held the important office of justiciar of Scotia from 1205,[7] and this can be seen as the first major sign of a deliberate royal policy to involve the family in the consolidation of royal authority in the north. The Comyn family had been in the vanguard of Anglo-Norman advancement in Scottish landholding and government since early in David I's reign.[8] William Comyn's father, Richard Comyn (d. c.1179), had built up sizeable estates in northern England and southern Scotland, and had also served as an important royal official, being justiciar of Lothian between

---

1   Ibid., i, p. 523; ii, p. 58.
2   G.W.S. Barrow, 'Badenoch and Strathspey, 1130-1312, 1: Secular and Political', *Northern Scotland*, viii (1988), pp. 6, 13 (n. 69).
3   *Chron. Fordun*, ii, p. 274.
4   The sheriffdoms of Aberdeen and Banff had come into existence in William the Lion's reign (*RRS*, ii, p. 36), if not earlier.
5   *RRS*, ii, p. 5.
6   G.W.S. Barrow, *The Anglo-Norman Era in Scottish History* (Oxford, 1980), pp. 157-8.
7   *Moray Reg.*, no. 18; *RRS*, ii, no. 465. Earl Duncan of Fife, the previous justiciar of Scotia, had died in 1204 (Anderson, *Early Sources*, ii, p. 362).
8   A. Young, *William Cumin: Border Politics and the Bishopric of Durham* (Borthwick Papers, no. 54, 1978).

1173 and 1178 and a prominent witness to royal charters.[1] William Comyn, Richard's oldest surviving son, inherited his father's estates and improved on this base by gaining the lands of Lenzie and Kirkintilloch, to the north of Glasgow.[2] Lenzie was a gift of William the Lion, and provides the first evidence of a royal grant to a member of the Comyn family for military service. His increasing status in royal administration in the 1190s can be seen by his appointment as sheriff of Forfar in *c*.1190, his role as a royal messenger to England in 1199-1200, and his frequent presence as a witness to royal charters.[3] Promotion to justiciar of Scotia in *c*.1205 was a natural extension of this royal service, and Comyn's elevation to the earldom of Buchan must be seen as further evidence of royal intention to use Buchan and the Comyn family to represent royal interests in the north. The justiciarship was 'the most significant bridge between the king's court and the localities',[4] and, when in *c*.1212 William Comyn as earl of Buchan became one of the most powerful landowners in northern Scotland, this fulfilled the king's need to have royal servants as substantial landowners in areas close to those of uncertain loyalty. Buchan's place at the forefront of Scottish political development in the thirteenth century had been firmly established.

Although the exact date of William Comyn's elevation to the earldom is uncertain,[5] it is possible to see his marriage to the heiress of Buchan as reward for his efforts against the MacWilliam rebels in Ross and Moray in 1211-12, while as earl of Buchan Comyn was very well placed to counter any other threats from those areas. When Moray again rose in revolt in *c*.1229-30, his success in 1211-12 prompted Alexander II to appoint him (again?) to the wardenship of Moray as 'a special emergency office', and furnish him with a great force of troops.[6] It was once more the rebellious MacWilliam family, this time represented by Gilleasbuig, who had disturbed the peace of Moray. The dependence of the crown on strong baronial support in the north is emphasised by the fact that it was after the king himself went against Gilleasbuig without success that he put the earl of Buchan in charge of Moray and held him responsible for Gilleasbuig's capture. This man and his two sons were killed and their heads brought to the king.[7]

Just as the elevation of a prominent royal servant to the earldom of Buchan was a long-term stabilising move for royal authority in the north after 1211-12, so the replacement of the temporary expedient of a warden of

1   A. Young, 'The Political Role of the Comyns in England and Scotland in the 13th Century' (Newcastle University Ph.D. thesis, 1974), pp. 40-64; *Newbattle Reg.*, p. 289, and *Melrose Lib.*, i, no. 13; *RRS*, ii, nos. 44, 125, and *Dunfermline Reg.*, p. 39.
2   *RRS*, ii, nos. 430, 557; SRO, Wigton Muniments, GD.101/1.
3   Young, 'Political Role of the Comyns', pp. 75-7; *Arbroath Lib.*, i, pp. 329-30; *CDS*, i, no. 292; *RRS*, ii, index at Cumin, William.
4   A.A.M. Duncan, *Scotland: The Making of the kingdom* (Edinburgh, 1975), p. 595.
5   See above, p. 174, n. 2.
6   *Chron. Bower* (Watt), v, pp. 142-3.
7   Ibid., v, pp. 142-5.

Moray by a hereditary lordship, that of Badenoch, represents a further extension of royal influence through the Comyn family after c.1230. By 1234 Walter Comyn, William's second son by his first marriage, was in possession of the lordship of Badenoch, having been granted it (almost certainly) by Alexander II.[1] Since Badenoch went together with Lochaber,[2] this meant that Comyn power stretched right across northern Scotland, from Buchan in the extreme east to Loch Linnhe on the west coast. Moreover, the Comyn earls of Buchan also held the castle and barony of Balvenie in Glenfiddich, which was significantly located only some twenty miles from Badenoch's eastern boundary.[3] In Buchan, Balvenie and Badenoch, therefore, the crown may be seen as having established its agents, the Comyns, in a strategic position which dominated the north of Scotland on its behalf.

The Comyns' power in the north, indeed, appears almost to have been viceregal. This is emphasised by their hereditary control of Buchan and Badenoch, and their long tenure of the justiciarship of Scotia — William, Alexander, and John, the three successive Comyn earls of Buchan, were justiciars of Scotia for no fewer than 66 of the 100 years between c.1205 and 1304.[4] The Comyns of Buchan and Badenoch were not slow to use their influence in the north against potential rivals, the Bissets in 1242-4 and, more especially, the Durwards in the 1240s and '50s.[5] This could be regarded as a natural consequence of the mutual self-interest and interdependence between key noble families and the monarchy as seen not only with the Comyns but with the Stewarts and the Murrays.[6] Yet thorough representation of royal interests in the north continued throughout the thirteenth century. An expansion of Comyn influence there can be seen in Alexander earl of Buchan's role as sheriff of Dingwall in 1264-6, a further extension of royal authority from the Buchan base, and an office which was kept in the Comyn family by means of another Alexander Comyn of Buchan (brother of Earl John Comyn), who was sheriff of Dingwall in c.1292.[7] This is a further sign of the tendency for hereditary offices to be created in areas more remote from royal centres. The Buchan role in the north was even more clearly shown in 1282 when Earl Alexander was sent on urgent royal business to the Northern Isles. Alexander III made it very apparent that the earl's activities in the north were indispensable to both the Scottish king and the kingdom.[8] And Buchan's key political and military position under

---

1   *Moray Reg.*, no. 76.
2   G.W.S. Barrow, 'The Highlands in the Lifetime of Robert the Bruce', in *Kingdom of the Scots* (London, 1973), p. 378.
3   Ibid., p. 377.
4   See the table in G.W.S. Barrow, 'The Justiciar', in ibid., pp. 137-8.
5   Matthew Paris, *Chronica Majora*, ed. H.R. Luard (Rolls Ser., 1872-83), iv, p. 200; Young, 'Noble Families and Political Factions', pp. 4-5.
6   Ibid., pp. 14-15; Barrow, *Kingship and Unity*, pp. 52-3, 149.
7   *ER*, i, pp. 19, 26; *Rot. Scot.*, i, p. 17a; the latter Alexander Comyn was also sheriff of Aberdeen in 1304-5: *CDS*, ii, no. 1617.
8   PRO, Special Collections, Ancient Correspondence, S.C.1.16/93; 20/158.

the Comyns continued until their destruction in 1308.

As has been said already, when William Comyn gained the earldom of Buchan through his marriage to Earl Fergus's heiress, this was the first 'Norman' acquisition of a Scottish earldom. The slowness of Anglo-Norman infiltration into Scottish earldoms in the thirteenth century may, in general, indicate a deep-rooted conservatism; and it has been noted how little affected Buchan, in particular, was by settlers from either southern Scotland or England.[1] Yet already by 1212 the native Highland magnates had shown themselves ready to adopt aspects of feudalism.[2] In the case of Buchan, that is reflected in a charter issued by Fergus, the last 'native' earl, which granted Fedderate and Ardendraught to John son of Uhtred, in exchange for Slains and Cruden. The charter is in strict feudal form: Fedderate and Ardendraught were granted 'as any earl or lord in the Scottish realm may infeft any vassal'; John and his heirs owed one archer's service plus the appropriate forinsec service, suit to Fergus's court at Ellon three times a year, and relief (when due) fixed at the large sum of £20 sterling; courts of life and limb were retained by the earl. The witness-list has a mixture of 'native' Scottish and 'Norman' names: Malcolm earl of Fife and his brother David, Thomas of Kinmalron, Alexander de Blare, Henry of Abernethy, William of Slains, Magnus son of Earl [Colbán of Buchan], Gilbride son of Lamund, Cospatric son of Maded and Malothem his brother, Norin son of Norman, Adam 'brother of the earl', and Robert de Montfort.[3] This indicates that in addition to new feudal terminology, Earl Fergus had a place for 'Normans' like Robert de Montfort in his following. And was the obviously local William of Slains, like Henry of Abernethy and indeed King William himself, essentially a native Scot with a French Christian name? Buchan appears already to have been changing, before the arrival of the Comyns.

The impetus for change possibly came from Fife. Earl Malcolm of Fife, his brother David, and Henry of Abernethy all witnessed Earl Fergus's charter, while Fergus himself appears to have had an interest in lands in Fife. This is shown by the fact that his daughter Marjorie, on her own behalf, confirmed grants of Kennoway church and nearby land, in Fife, to St Andrews Priory.[4] The grants were stated to have been made by Merleswain son of Colbán, presumably the Merleswain of Kennoway to whom William I gave Ardross in c.1173.[5] In Merleswain son of Colbán's

1   Duncan, *Making of the Kingdom*, p. 188.
2   Barrow, *Anglo-Norman Era*, pp. 137-41; for Strathearn, see C.J. Neville, 'The Earls of Strathearn from the Twelfth to the Mid-Fourteenth Century with an Edition of their Written Acts' (Aberdeen University Ph. D. thesis, 1983), ch. 3.
3   *Aberdeen-Banff Colls.* pp. 407-9; see Barrow, *Anglo-Norman Era*, pp. 139-44, for further discussion.
4   *St Andrews Lib.*, pp. 253-4. Her husband, Earl William Comyn, issued similar confirmations (ibid., pp. 251-2), but it is clear that the right to the church and lands was Marjorie's.
5   Cf. ibid., pp. 258-9; *RRS*, ii, no. 211.

own charter, the second witness (after the earl of Fife), is 'Colbán', presumably the donor's father; while in a subsequent confirmation by the donor's son, Merleswain son of Merleswain, the first two witnesses are Duncan earl of Fife and Earl Colbán.[1] The only Earl Colbán at that period was earl of Buchan; Merleswain of Kennoway would thus appear to have been his son[2] — though not by his only known wife, Eve, daughter of Earl Gartnait of Buchan (floruit c.1150). That marriage of Colbán's suggests he was probably not, himself, a member of the Buchan kin. Whether he was a forebear of Earl Fergus is unclear; but since Earl Fergus's daughter was confirming grants concerning Kennoway made by Merleswain son of Colbán, there certainly seems to have been a close relationship. The likelihood is that Earl Colbán had come to Buchan from Fife, and that through this Fife connection the earldom of Buchan was already linked to the royal establishment in the pre-Comyn era.[3]

Whatever the case, the trend towards change was, of course, intensified after William Comyn acquired the earldom. The greatly strengthened ties between Buchan and the crown have already been emphasised; internally, the Normanisation of the earldom can best be seen in the string of Comyn castles, which will be discussed below. Also, the relatively few charters of Earl William Comyn which survive are, naturally enough, in standard feudal forms.[4] But, while some 'Norman' incomers, in particular Richard de Mowat,[5] may be seen in their witness lists, a more interesting feature of these is continuity with the time of Earl Fergus. Such continuity was most obvious in the person of the Countess Marjorie, who witnessed three of her husband's charters,[6] and issued her own as well.[7] More significant are the appearances of sons of former earls of Buchan: Adam son of Earl Fergus (who, though presumably illegitimate, may have been head of the Buchan kindred, according to the Gaelic kin-based system[8]) also witnessed three Earl William charters;[9] Magnus son of Earl Colbán witnessed

---

1   St Andrews Lib., pp. 258-60.
2   As, of course, was 'Magnus son of Earl Colbán of Buchan', another of the witnesses to Earl Fergus's charter of Fedderate and Ardendraught.
3   The connection may have been established earlier: Gartnait earl of Buchan was married to Ete daughter of Gillemichael, who may well have been the earl of Fife of that name: Duncan, Making of the Kingdom, p. 165 (n. 56).
4   Only 15 charters (plus the gist of one other) issued by Earl William Comyn now survive; 14 of them have witness-lists.
5   Richard de Mowat witnessed seven of Earl William's charters: Arbroath Lib., i, no. 130; St Andrews Lib., pp. 250, 251*, 251-2*, 252; SRO, GD.101/2; SRO, Register House Transcripts, RH.1/2/32; he also witnessed two charters of Countess Marjorie: St Andrews Lib., pp. 252-3, 253* (* indicates that his brother Michael de Mowat witnessed as well). Also, for members of the Sinclair family witnessing, see SRO, RH.1/2/31, 32, and (for Countess Marjorie) St Andrews Lib., p. 254.
6   Aberdeen Reg., i, p. 14; Aberdeen-Banff Ills., ii, pp. 426-7, 427-8.
7   St Andrews Lib., pp. 252-4 (three charters), and Arbroath Lib., i, no. 132; cf. also ibid., no. 227. In these she is styled wife of Earl William, but in one of his she is also 'filia quondam Fergus comitis de Buchan' (Aberdeen-Banff Ills., ii, pp. 427-8).
8   See the essays elsewhere in this volume by John Bannerman on 'MacDuff of Fife' and by Hector MacQueen on 'The Kin of Kennedy, "Kenkynnol" and the Common Law'.
9   Aberdeen-Banff Ills., ii, pp. 426-7, 427-8; Glasgow Reg., i, no. 117.

two;[1] and Merleswain son of the earl witnessed once.[2] Then, of those associated with Earl Fergus in the Fedderate/Ardendraught charter, William of Slains and Robert de Montfort each appear in three of William Comyn's witness lists, while John son of Uhtred, the recipient of that charter, was in at least one.[3] In addition, Cospatric Macmadethyn, who, with his brother Malothem had witnessed Earl Fergus's charter, was himself the recipient of a charter from Earl William granting him Strichen and Kindroucht in feudal form:

> Tenendas sibi et heredibus suis de me et heredibus meis in feodo et hereditate cum molendino et omnibus aliis et iustis pertinentiis suis. Reddendo inde annuatim michi et heredibus meis duas petras cere ad Pentecostem et faciendo forinsecum servitium comitatus de Buchan quantum pertinet ad predictas terras de Stratheyn and Kyndrochet.[4]

Earl William's charters also included two *judices*: Farhard, *judex* of Buchan, and Kerald *judex*.[5] Clearly if, as seems likely, these witnesses represent a major part of William Comyn's affinity in Buchan, then his predecessor Earl Fergus would have found himself perfectly at home in it. And this continuity was maintained under the second Comyn earl, Alexander, whose charter witnesses included John of Kindroucht, possibly a son of Cospatric);[6] and in whose company the prominent Fife noble, Hugh of Abernethy, was regularly found.[7]

Thus it would appear that the Comyns combined the Celtic system of organisation based on personal relationships with the Norman feudal system based on tenure, each system supporting the other. Moreover, the Comyns themselves were strongly 'clannish'. In their surviving documents, the Countess Marjorie's kin — Adam, Magnus and Merleswain — are balanced by Earl William's family and relatives by marriage. William Comyn's children by his first marriage included Richard the eldest son, Walter lord of Badenoch, David (who married Isabel de Valognes, lady of East Kilbride),

---

1   *Aberdeen-Banff Ills.*, ii, pp. 426-7, 427-8. Magnus also witnessed the Fedderate/Ardendraught charter of Earl Fergus.
2   *Aberdeen-Banff Ills.*, ii, pp. 427-8. He may have been Merleswain son of Colbán, of Kennoway. But that is unlikely (a) because he appears in charters as late as 1236 (though Magnus son of earl Colbán can be found as late as 1219 × 33); (b) because, whereas Adam and Magnus come high up in the witness-lists, he is at the end, among the chaplains — was he perhaps a cleric?
3   William of Slains witnessed *Aberdeen Reg.*, i,, pp. 14-15; *Aberdeen-Banff Ills.*, ii, pp. 427-8; *Arbroath Lib.*, i, no. 130. Robert de Montfort witnessed *Glasgow Reg.*, i, p. 101; SRO, GD.101/2; SRO, RH.1/2/31. John son of Uhtred witnessed *Aberdeen-Banff Ills.*, ii, pp. 427-8.
4   *Aberdeen Reg.*, i, pp. 14-15.
5   Ibid.; *St Andrews Lib.*, p. 254.
6   *Aberdeen-Banff Ills.*, iii, p. 113; SRO, Forbes Muniments, GD.52/388. 'Native' witnesses for Earl Alexander include a number of clerks and officials such as Duncan *judex* (*Arbroath Lib.*, i, no. 247; *Lindores Cart.*, no. 124), Uchtred, seneschal (*Aberdeen-Banff Ills.*, iii, p. 113; SRO GD.52/388; *RMS*, ii, no. 1198), Morgrund, chaplain (SRO, GD.52/388).
7   *Chron. Bower* (Goodall), ii, p. 91; *CDS*, i, no. 2155; *Aberdeen Reg.*, i, pp. 29-30; ii, 272-3; SRO, Register House Charters, RH.6/58; Fraser, *Facsimiles*, no. 57; SRO, Dalhousie Muniments, GD.45/26/4.

William (a churchman), and Jean (who married William earl of Ross).
Those by his second marriage to Marjorie of Buchan were Alexander, the
heir to Buchan, William, Fergus, Idonea (who married Gilbert de la Hay),
Agnes (who married Philip de Fedarg, founder of the family of Meldrum of
Meldrum), and Elizabeth (who married William earl of Mar).[1] All appear in
the company of the Comyn earls of Buchan, some especially frequently. For
instance, Philip de Fedarg witnessed six of Earl William's charters[2] and was
subsequently to support Earl Alexander;[3] much the same can be said of Earl
William of Mar.[4] Family ties were strengthened further by feudal bonds, as
seen by Earl William's grant of lands in Peeblesshire to his sister on her
marriage to Adam son of Gilbert,[5] and also by his grant of Upper Coull in
Aberdeenshire to his daughter Idonea on her marriage to Gilbert de la Hay.[6]
In Earl Alexander's following his brothers, William and Fergus, and his son
and heir John figure strongly, and family involvement in Buchan was
reinforced by a feudal bond with another member of the family when Earl
Alexander granted to Jordan Comyn the land of Inverallochy 'pro eius
homagio et servitio' in 1277.[7] In broad terms, at least, the Comyns seem
probably to have been every bit as clannish as both their Gaelic predecessors
and their late-medieval successors.

What was the extent of the earldom of Buchan, and how was it governed
and organised under the Comyns? Unfortunately, the relatively sparse
survival of documents relating to Comyn-held Buchan makes it difficult to
give detailed answers. On the other hand, other thirteenth-century sources
provide some help, especially with respect to the boundaries of the earldom;
and it is also possible to supplement the thirteenth-century evidence by using
later material, dating from and after Robert I's dismemberment of the
earldom. What happened to Buchan after 1308 was as follows.[8] The last
Comyn earl, who had died childless by the end of 1308, left two nieces as
coheiresses. The younger of these, Margaret, was married to John Ross,
brother of Hugh earl of Ross (a brother-in-law of Robert I); she, too, died
childless, and her husband's nephew, William, son of Earl Hugh and
himself earl of Ross, eventually inherited her share of Buchan, which
consisted of what became the baronies of Kingedward and Philorth. These
estates went to Earl William's elder daughter, Euphemia, and her husband
Walter Leslie, but in 1375 they transferred Philorth to Euphemia's sister
Joan and her husband Alexander Fraser, with whose family it stayed; while
Kingedward was eventually inherited, with the earldom of Ross, by the

1   Young 'Political Role of the Comyns', pp. 107-112.
2   *St Andrews Lib.*, pp. 250, 251, 252; SRO, GD.101/2; SRO, RH.1/2/31.
3   *St Andrews Lib.*, p. 283; *Arbroath Lib.*, i, no. 254.
4   *Aberdeen Reg.*, ii, pp. 30-4, 276-7; *Inchcolm Chrs.*, nos. 25-7.
5   *Morton Reg.*, ii, pp. 4-5.
6   SRO, RH.1/2/31.
7   *RMS*, ii, no.1198.
8   See *Scots Peerage*, ii, pp. 258-61.

Lords of the Isles. As for the elder Comyn coheiress, Alice, she married the English lord, Henry Beaumont, and suffered forfeiture after 1314 (Beaumont was to be a prominent leader of the 'disinherited' in the 1330s). Alice's share of the earldom, therefore, is presumably represented by the Buchan lands granted out by Robert I, especially to Robert Keith the marischal and his brother Edward, Gilbert Hay the constable, and Archibald Douglas brother of 'the Good Sir James'.[1] By examining the later pattern of landownership in Buchan, therefore, an impression of the earldom's thirteenth-century extent can be gained. Also, as will be argued below, the fourteenth-century pattern of baronies probably reflects subdivisions within the earldom under the Comyns; and the nature of at least some of Robert I's grants is such as to suggest that only the comital demesnes were granted, which means that some exploration of Comyn demesnes can be made.

Buchan has been defined as the compact area 'which embraces the whole region between the sea and the [Rivers] Ythan and Deveron.'[2] While that is a good description of the province of Buchan, it does not quite tally with the evident bounds of the thirteenth-century earldom — as can be seen from Map (A), 'The Earldom of Buchan', in the Appendix, below. On the north coast, the earldom's lands stopped some seven miles short of the mouth of the Deveron; the area in between, the parish of Gamrie (until 1975 in Banffshire, not Aberdeenshire) contained the lordship of Troup (held in the thirteenth century by a family of that name), the lands of Lethnot (held in 1266 by Robert Corbet), and the royal thanage of Glendowachy.[3] Also, although the earldom did reach the Deveron about three miles from its mouth, and followed its banks southwards to Turriff and thence eastwards for several miles, the royal thanage of Conveth (now Inverkeithny) is then to be found on the Buchan (south) side of the river. Similarly, the River Ythan did not form an exact boundary to the south. On its upper reaches, a substantial part of the royal thanage of Formartine, including its burgh and castle of Fyvie, lay on the north and east banks, again the Buchan side. On the other hand, beyond Formartine was the parish of Bethelnie, now called Old Meldrum; since Earl William made a grant of the church of Bethelnie, it is reasonable to assume that this was a part of the earldom, albeit a detached part.[4] Moreover, for about the last eighteen miles of the Ythan's course to the sea, the earldom stretched over its southern banks; the area beyond the Ythan included Kelly (now Haddo), a major base of the Comyn earls, and also their urban foundation at Newburgh near the coast. The southern boundary, indeed, was probably formed by the royal thanage of Belhelvie, not by the Ythan. But the most striking anomaly in Buchan's borders was probably the lordship (later barony) of Inverugie, a compact coastal territory to the north of the River Ugie (near modern Peterhead),

1   G.W.S. Barrow, *Robert Bruce and the Community of the Realm of Scotland* (3rd edn, Edinburgh, 1988), pp. 271-4.
2   J.F. Tocher, *The Book of Buchan* (Aberdeen, 1910), p. 137.
3   *Aberdeen-Banff Colls.*, pp. 488-92; *Aberdeen-Banff Ills.*, ii, pp. 363-8.
4   *Arbroath Lib.*, i, no. 130; cf. *Aberdeen-Banff Colls.*, p. p. 311-2.

right in the middle of Buchan's eastern coastline. In the early thirteenth century Inverugie seems to have been possessed by Ralph le Nain,[1] and thereafter it went to the family of Cheyne;[2] in neither case does it appear to have been held of the earls of Buchan. The Cheynes were, admittedly, regular companions and political supporters of the Comyn earls,[3] but there is no evidence of any formal tenurial relationship with them. What is more, Inverugie came to be a detached part of the sheriffdom of Banff,[4] which appears to indicate that it was separate from Aberdeenshire and Buchan. If we put Inverugie with Troup, Glendowachy, Conveth and Formartine, we can see that the thirteenth-century earldom of Buchan did not quite correspond to what look like its 'natural frontiers'.

In that context it is worth returning to the grant made by Earl Fergus to John son of Uhtred, of Fedderate and Ardendraught in exchange for Slains and Cruden.[5] Fedderate and Ardendraught were to be held of Earl Fergus — but had that originally been the case with Slains and Cruden, which together were a block of territory running north along the coast from the mouth of the Ythan, or was the earl extending his earldom and in effect buying what had been an independent holding (plus imposing feudal lordship at the same time)? It should be noted that the earldom of Fife was extended considerably in the twelfth century; originally even Falkland was not a part of it.[6] Similarly, Dingwall, in the fifteenth century one of the main bases of the earldom of Ross, was originally separate, and although granted to the earls of Ross in 1321, was still not incorporated into the earldom in 1382.[7] Perhaps, therefore, the Slains/Fedderate exchange shows the earl managing to take over what may have been one of several enclaves on the fringe of his earldom. If so, the later Comyn earls followed suit. By the end of Alexander III's reign, Earl Alexander had gained the thanage of Conveth (ferming it for 80 merks a year).[8] Then, during the reign of John Balliol, Earl John surrendered his hereditary claims on the lordship of Galloway in exchange for 'terra theinagii de Fermartyn et de Dereleye';[9] there is a 'Derley' near Fyvie,[10] but it is not otherwise found as a thanage, and Fyvie was clearly the

1   *RRS*, ii, no. 513, p. 463: a confirmation dated 1213, which would suggest that Ralph le Nain was in Inverugie before William Comyn became earl of Buchan. Cf. Barrow, *Anglo-Norman Era*, pp. 188-9, for the family of le Nain.
2   *Aberdeen-Banff Ills.*, ii, pp. 72-5; *RRS*, vi, no. 262.
3   *Aberdeen-Banff Ills.*, iv, p. 602; *Aberdeen Reg.*, i, p. 34; ii, p. 277; NLS, MS Advocates, 24.1.10, vol. ii, p. 24; *CDS*, iv, no. 2155.
4   *Aberdeen-Banff Colls.*, pp. 422-3.
5   Ibid., pp. 407-9. Fedderate lay inland, east of Deer; Ardendraught was near Cruden Bay (i.e. John son of Uhtred kept part of his land there); Slains and Cruden together form a block of land running northwards for about ten miles from the mouth of the Ythan, and inland over Cruden Moss.
6   See Alexander Grant's essay, 'Thanes and Thanages, from the Eleventh to the Fourteenth Centuries', above, p. 46.
7   *ER*, i, pp. 19, 26; *RRS*, v, no. 196; *RMS*, i, no. 741.
8   *Memoranda de Parliamento, 1305*, ed. F.W. Maitland (Rolls Ser., 1895), p. 191; cf. Grant, 'Thanes and Thanages', Appendix, no. 17.
9   Maitland, *Memoranda de Parliamento, 1305*, p. 191; also *CDS*, ii, no. 1541.
10   *Aberdeen-Banff Colls.*, p. 504.

chief place in Formartine. It seems more likely that 'Dereleye' is a clerical mistake for Belhelvie;[1] that, in fact, John Comyn was not only rounding off his earldom but extending it considerably to the south by acquiring both Formartine and Belhelvie. The forfeiture of the earldom only a few years later, however, makes the point impossible to prove.

In the early modern period the *caput* of the earldom of Buchan was considered to be at Ellon. In October 1476 James Stewart, newly created earl of Buchan, was given sasine of the earldom 'super montem de Ellane'; thereafter the mound was known as the 'Earlshill' (or, later 'Earlshillock'!).[2] The accuracy of the later tradition would seem to be confirmed by the charter of Earl Fergus to John son of Uhtred, according to which John owed annual suits to the earl's three chief courts at Ellon;[3] in this particular case, continuity with the Celtic past spanned the twelfth and sixteenth centuries. Remarkably, however, Ellon does not feature with that significance in any of the Comyn earls' documents — though the fact that hardly any of these have place-dates reduces the value of this point. On the other hand, we do, at least, know that two of Earl Alexander's charters were issued at Kelly (now Haddo), in 1261 and 1272.[4] The later charter was witnessed by King Alexander III himself, together with the earl of Mar and the Comyns' great political rival, Alan Durward. That Kelly was where the king and court were entertained surely indicates that it was an important — perhaps the most important — centre of the Comyn earldom.

If Ellon and Kelly were the secular centres of the earldom, the religious centre was obviously the abbey of Deer, in the very heart of Buchan. This was originally a Celtic monastery, said to have been founded by St Columba, which was given lands and privileges by various mormaers of Buchan;[5] but it was refounded as a Cistercian abbey by Earl William Comyn, probably in 1219.[6] Few documents relating to the medieval abbey survive (we actually have more evidence of grants to the Celtic monastery); the thirteenth-century documents were probably lost during the 'herschip of Buchan', for in 1315 Robert I, 'in recompense for war damage', gave the abbey a blanket confirmation of all the grants it had had from Earls William (and his wife Marjorie), Alexander and John Comyn, but without giving a list of them.[7] Clearly, however, Deer's main estates were in the old parishes of Deer and Peterugie, and ran in a swathe through central Buchan.[8] Teinds from these parishes, and from Kingedward in the north of the earldom and

1   As argued in Grant,'Thanes and Thanages', above, p. 69.
2   *Aberdeen-Banff Ills.*, iii, pp. 5-6.
3   *Aberdeen-Banff Colls.*, p. 408.
4   *Aberdeen Reg.*, ii, pp. 276-7; i, pp. 30-4.
5   Jackson, *Gaelic Notes*, pp. 33-6. The grantors run down to Earl Colbán in the later twelfth century.
6   I.B. Cowan and D.E. Easson, *Medieval Religious Houses: Scotland* (2nd edn, London, 1976), p. 74.
7   *RRS*, v, no. 48.
8   See Map (A), 'The Earldom of Buchan', in the Appendix, below.

Foveran in the south-east, together with the lands of Fechil (on the Ythan near Newburgh) and Barry in Strathisla (across the Deveron, beyond Aberchirder) were also part of its endowment.[1] Fechil and Barry were given by Earl William Comyn,[2] while the churches of Kingedward and (probably) Foveran were bestowed at the end of the century by Earl John Comyn.[3] But, otherwise, what the abbey received from Earl Alexander, and how much of its lands were really Comyn grants rather than simply taken over from the old Celtic monastery, is uncertain.[4]

In addition to Deer, the great abbey of Arbroath was the recipient of William Comyn's patronage, in the form of the church of Bethelnie, now Old Meldrum.[5] Arbroath Abbey was also given half a merk a year by Earl Alexander in 1286, but that was as compensation for a piece of land in the abbey's estate of Tarves (adjacent to Kelly but outside the earldom), which had been incorporated into the earl's park of Kelly.[6] Alexander made a similar, if more valuable, deal with Lindores Abbey, giving it 10 merks a year from his *tenementum* of Kelly, in exchange for the (unidentified) lands of 'Kyncardinebegg'.[7] And one of the Comyn earls gave two stones of wax a year (the annual render from Strichen and Kindroucht) to the chapel at Rattray, on the north-east coast.[8] Then, there were Earl Alexander's foundations of two almshouses. The first, in 1261, was in the extreme south-east of the earldom, at Newburgh, and was for one chaplain and six poor people living in the burgh; it was supported by half an acre near the burgh gate, four chalders 14 bolls of oatmeal a year from the earl's mill at Forvie (just across the Ythan estuary) delivered by the bailie of Slains, and 18 shillings a year for clothing paid by the burgh's provost from its fermes.[9] The second, which had its foundation witnessed by Alexander III in 1272, was on the earldom's western edge, at Turriff, and was on a rather larger scale: it was to maintain a master, six chaplains, and thirteen poor, chosen by the earl 'ex pauperis husbandis de Buchane', and was endowed with the nearby land of 'Cnockikuby', plus two chalders of meal and three chalders of malt a year, to be provided by the constable of Kingedward castle.[10] Here, the thirteen 'poor husbandmen of Buchan' perhaps give a sense of the earl's feeling for the community of his earldom — or was the foundation intended to impress the king and court? There is a hint of less concern for

1   Mid-16th century rentals (*Aberdeen-Banff Ills.*, iv, pp. 19-30) probably give a fair impression of Deer's possessions in the 13th century.
2   *Aberdeen-Banff Ills.*, ii, pp. 426-8.
3   *RMS*, i, app. II, nos. 28, 29.
4   Certainly some of the early grants noted in the Book of Deer can be equated with possessions of the later abbey: Jackson, *Gaelic Notes*, pp. 37ff.
5   *Arbroath Lib.*, i, no. 130.
6   Ibid., no. 319. But cf. *RRS*, v, no. 49, for disputes between Earl Alexander and Arbroath Abbey.
7   *Lindores Cart.*, no. 124.
8   *Aberdeen Reg.*, i, p. 14.
9   Ibid., ii, pp. 276-7.
10   Ibid., i, pp. 30-4.

the lower orders in Earl Alexander's other recorded dealing with the Church. At Ellon, the kirkton was the property of the bishops of St Andrews, outside the feudal earldom; it was valued at £20 a year, but on it lived a number of peasants known as the 'scolocs of Ellon', who had to supply four properly clad clerks for Ellon parish church and pay the bishop 16 shillings annually. In 1265 Earl Alexander leased these 'Scologlands' for three lifetimes at an annual rent of two merks; that gave the bishop a rather better return from the land, effectively filled in a small enclave in the lands of the earldom — and, we might guess, gave Alexander some under-used land which could be exploited more effectively.[1]

The 'scolocs of Ellon' have brought us back to the ancient centre of the earldom, where Earl Fergus (and probably his Comyn successors) held their courts. In these, judgement would have been made by the *judex*; in the first half of the thirteenth century Farhard was *judex* of Buchan,[2] and he was probably succeeded by Duncan *judex* later in the century.[3] Another central official was the seneschal: a man called Uchtred was seneschal of Buchan in the 1270s.[4] More locally, the earls employed bailies and constables; the bailie of Slains and the constable of Kingedward castle have been encountered in the foundation charters of the Newburgh and Turriff almshouses. But these, unfortunately, are the only subordinate officers of the earldom found in the surviving records.

Slains and Kingedward, however, both had castles, and from the more solid evidence of castle-building we can learn something about the earldom's organisation.[5] There seems to have been a thirteenth-century castle at Ellon;[6] but, somewhat surprisingly, there is no record or physical evidence of one at Kelly, probably the other main base of the earldom[7] — though it would be remarkable if one had not been built there. A second known inland castle of the Comyns was Kingedward, in the north-west of the earldom, whose constable is mentioned in 1273; this was certainly an important stronghold, with a prominent site on a bold precipitous rock and protected by the Kingedward burn on the west.[8] Then, being a compact coastal earldom, Buchan was naturally defended by an impressive grouping of well-sited castles along its coastline. Moving clockwise from north-west to south-east,

1   *Aberdeen-Banff Colls.*, pp. 310-12; cf., for 'scolocs', Duncan, *Making of the Kingdom*, pp. 329-31.
2   *Aberdeen Reg.*, i, p. 15 (1212 × 33); *Arbroath Lib.*, i, no. 227 (1251).
3   Ibid., no. 247 (1265); *Lindores Cart.*, no. 124 (1273 × 89).
4   *Aberdeen-Banff Ills.*, iii, p. 113; SRO GD.52/388; *RMS*, ii, no. 1198.
5   I am very grateful to the architectural section of the National Monuments Record of Scotland (RCAHMS) for their assistance with bibliographies and database entries for castle sites in Buchan.
6   G. Stell, 'Provisional List of Mottes in Scotland', in *Essays on the Nobility of Medieval Scotland*, ed. K.J. Stringer (Edinburgh, 1985), p. 14; J.D. Galbraith, 'Castles and Strongpoints 1296-1306', in *An Historical Atlas of Scotland, c.400-c.1600*, ed. P. McNeill and R. Nicholson (St Andrews, 1975), Map 54; J. Godsman, *A History of the Burgh and Parish of Ellon, Aberdeenshire* (Aberdeen, 1958), pp. 49-57.
7   'The House of Kelly', *Transactions of the Buchan Field Club*, ii (1891-2), p. 179.
8   J. Ferguson, 'On the Old Castles of Buchan', ibid., x (1925), p. 71.

the first is the imposing Dundarg, built in the thirteenth century on a rock of red sandstone looking northwards over the outer reaches of the Moray Firth.[1] Next, on the corner of Buchan, so to speak, there is Cairnbulg (originally named Philorth). Standing on a fairly prominent mound which was possibly a motte, later cut down to form a platform for the stone building, Cairnbulg has an impressive tower house which probably dates to the middle of the century.[2] Further down the coast at Rattray, commanding what is thought to have been the port of Buchan, there was another Comyn castle.[3] South of Rattray there was the Cheynes' castle of Inverugie;[4] then, finally, we come to the castle of Slains, later one of the main bases of the Hays' earldom of Errol, but originally, once again, a thirteenth-century Comyn stronghold.[5]

This pattern of castles makes obvious strategic sense. But, if we look at what happened to Buchan after the fall of the Comyns, we may also see an administrative pattern emerge. The main grants of forfeited Comyn land in Buchan made by Robert I were the barony of Slains, to Gilbert Hay the constable; the barony of Aden, to Robert Keith the marischal; the barony of Kelly, to the marischal's brother Edward Keith; and the barony of Crimond, or Rattray, to Archibald Douglas.[6] Then, as has been seen earlier, the unforfeited (Ross) share of the earldom eventually became the baronies of Kingedward and Philorth.[7] And there was one other barony in fourteenth-century Buchan, the barony of Aberdour; in 1378 this was held by Archibald Douglas 'lord of Galloway and Bothwell', a style which he generally employed when dealing with the Murray lands which he had acquired through his wife.[8] But how Aberdour came to the Murrays is a mystery.[9] What is clear, on the other hand, is the relationship of most of these baronies with the Comyn castles. Apart from Ellon (which, being the *caput* of the suppressed earldom, was probably kept in crown hands after

1   W.D. Simpson, *Dundarg Castle: A History of the Site and record of the Excavations in 1950 and 1951* (Aberdeen University Studies, no. 131); W. Beveridge, 'Notes on Excavations at Dundargue Castle, Aberdeenshire', *PSAS*, xlviii (1913-14), pp. 184-92.
2   W.D. Simpson, 'Cairnbulg Castle, Aberdeenshire', *PSAS*, lxxxiii (1948-9), pp. 32-44.
3   H.K. Murray and J.C. Murray, 'Old Rattrey, burgh and castle', *Discovery and Excavation, Scotland* (1986) p. 10. Between Cairnbulg and Rattray there is Inverallochy castle, but although the land of Inverallochy was granted by Earl Alexander to Jordan Comyn in 1277, there is no evidence that a castle was built there in this period: D. Macgibbon and T. Ross, *The Castellated and Domestic Architecture of Scotland from the 12th to the 18th centuries* (Edinburgh, 1887-92), ii, pp. 331-3.
4   Galbraith, 'Castles and Strongpoints 1296-1306.
5   Tocher, *Book of Buchan*, p. 188; W.D. Simpson, 'Slains Castle' *Trans. Buchan Field Club*, xvi (1940), pp. 39-40.
6   *RMS*, i, no. 602; app. I, no. 47; *RRS*, v, no. 347; *RMS*, i, app. I, no. 66.
7   Above, pp. 182-3.
8   *Aberdeen-Banff Ills.*, iv, pp. 113-14; also ii, pp. 375-8.
9   There seem to be two possibilities: (a) that it was granted by Robert I to one of the Murrays, most probably Andrew Murray of Bothwell (as Barrow has remarked, it is surprising that no grant of patronage by King Robert to Andrew Murray is known: *Robert Bruce*, p. 282); or (b) that it was granted to a Murray in the 13th century by one of the Comyn earls, and came to be held directly of the crown after 1314 (that is perhaps the more likely explanation).

1314), all the castles listed above were the head points of the baronies: Kingedward, Rattray and Slains are self-evident, while Dundarg was the main stronghold in Aberdour barony, just as Cairnbulg was in Philorth. Of the new baronies, only Kelly and Aden did not have castles; but Kelly's likely significance as a Comyn base has already been stressed, while Aden, near Deer, was created out of what had formerly been the earls' forest of Aden.

From this, therefore, we can envisage the likely organisation of the earldom of Buchan under the Comyns. As said already, the legal centre was presumably at Ellon, while what can perhaps be thought of as the domestic centre — complete with enclosed park for hunting — was at Kelly. The religious centre was at Deer, and nearby was the larger hunting reserve of the forest of Aden. Newburgh, at the southern extremity, nearest to Aberdeen, was probably the main burgh. And, around the fringes of the earldom, there were five local subdivisions, Kingedward, Dundarg (Aberdour), Cairnbulg (Philorth), Rattray, and Slains: each one based on a major castle, and each run, presumably, by constables and/or bailies.

The structure of the later baronies also helps to elucidate the likely pattern of landholding within the Comyn earldom; an attempt at this is given in Map (A) and the accompanying list of probable demesnes of the earldom (in the Appendix). We may start with the baronies created out of the forfeited half of Buchan. Two of these were compact blocks of land: Slains, which comprised 100 librates, and was roughly equivalent to the adjoining parishes of Slains and Cruden;[1] and Aden, which was created out of the earls' forest of Aden. The other two, in contrast, appear to have been made up of isolated and dispersed settlements: fourteen in the case of Kelly, eight in the case of Rattray. The components of Rattray all lie fairly close to the castle, within a four-mile radius, and most of Kelly's are within a six-mile radius of Kelly, except that the land at Newburgh [Ky9][2] is ten miles away, and Gonarhall [Ky6] and Saithly [Ky10] are both about sixteen miles distant (and, themselves, seven miles apart). Then there was the barony of Aberdour; no creation charter of it exists, but the Douglases granted, or sold, eighty merklands in it to Alexander Fraser of Philorth, which consisted of thirteen pieces of land:[3] most were within a five-mile radius, but Little Drumwhindle [A6] and Quilcox [A11] are some eighteen miles to the south, near Kelly. Finally, there was the Ross share of the earldom. Initially, this was all counted as being within the barony of Kingedward; but in 1375, as already stated, Alexander Fraser, husband of the younger of the two daughters of

---

1   Slains castle is at the north end of Cruden Bay, in the middle of Cruden parish; Ardendraught, a separate tenancy, is within Slains parish.
2   *RRS*, v, no. 347, and index, for grid references for the place-names. The text is somewhat ambiguous, but, e.g. *Aberdeen-Banff Ills.*, iv, p. 560, confirms that 'ac terciam partem' applies to all the places after Newburgh. Letters and numbers in square brackets serve to locate lands in these baronies on Map (A) in the Appendix.
3   *Aberdeen-Banff Ills.*, ii, p. 376-7; iv, pp. 113-4.

William earl of Ross, was given what became the barony of Philorth as part of his wife's share of the Ross inheritance, and as compensation for giving up any claims on that earldom.[1] The Buchan lands which Fraser gained seem to have been the majority of the demesnes of Kingedward, amounting to some twenty-one territories: they are almost all in the north (except, again, the anomalous Drumwhindle [P13]), but while several of them lie in a cluster near Philorth, Cairneywhing [P11] and Bracklawmore [P8] are about twelve miles away in the midst of north Buchan, and ten others are grouped within a three-mile radius of Kingedward, nineteen miles from Philorth on the other side of the earldom. As for Kingedward itself, in 1478 John Lord of the Isles (formerly earl of Ross) granted away all the lands of his barony of Kingedward, namely the demesne of Kingedward commonly called the Castletown [K2], and the lands of Easter Tyrie [K8], Kinharrachie [K6], and Faithlie [K3].[2] Castletown apart, these are far from Kingedward: Faithlie is now Fraserburgh, on the north-east coast, Easter Tyrie is near there too, and Kinharrachie is close to Ellon in the south — the earls of Ross, as lords of Kingedward, had obviously kept a piece of land beside what had been the *caput* of the original earldom. In addition, a few other lands attached to Kingedward in the fourteenth century can be identified;[3] Byth [K1] is about six miles away, and the others are no more than a mile or so from the castle.

Although it is impossible to prove the point, these lands were probably demesne territories of the Comyn earls. When half the earldom was forfeited, it looks as if Kelly and Rattray were put together out of a number of the earls' farmtouns; just as David I had done almost two centuries earlier in Moray, Robert I probably used demesnes of the forfeited earldom to endow his supporters.[4] It is likely that that applies to Aden (the earls' forest), and also quite possibly to Slains, which presumably corresponded closely to the territories of Slains and Cruden acquired by Earl Fergus from John son of Uhtred. As for Aberdour, no matter when it was created the lands which went to the Frasers look like a collection of farmtouns similar to those in Kelly and Rattray; much the same can be said of the barony of Philorth, the Frasers' share of Kingedward. With Kingedward, on the other hand, some feudal tenancies are to be found, which suggests that the earls of Ross retained the feudal superiorities within their share of Buchan;[5] but the lands noted above appear to be demesnes.

In that case, we can begin to see how Comyn Buchan might have looked.[6] Most, though not all, of the lands considered above were clustered around the castles (or in Kelly's case, perhaps manor) on which their later baronies were based. It seems reasonable to assume, therefore, that these

1 Ibid., iv, pp. 87-8.
2 *Acts of the Lords of the Isles*, no. 113.
3 From *Aberdeen-Banff Ills.*, iii, pp. 521, 525, 531-3.
4 Barrow, 'Badenoch and Strathspey, 1', pp. 2-3.
5 Including, for example, Kindroucht, which Earl William Comyn had granted to Cospatric Macmadethyn: *Aberdeen Reg.*, i, pp. 14-15; *Acts of the Lords of the Isles*, no. 72.
6 The components of the new baronies are listed and mapped in the Appendix.

were the lands attached to and run from those castles, by the Comyns' bailies and constables, during the thirteenth century. Whether or not that applies to the more distant components of the new baronies, however, is impossible to say, although it would appear likely in the cases of Kingedward and Philorth that the eastern lands originally went with Kingedward castle and the western ones with Cairnbulg. Another interesting question that may be raised is whether this pattern of, mostly, clustered demesne farmtouns goes back to before the Comyn earls? Does the map actually reveal a system of multiple estates in Buchan deriving from the distant past? And while no clear answer can be given to that, the overall geographical pattern of these lands is worth commenting on. They are almost all either in the north or in the south of the earldom; very few are in the middle. If these lands were the earls' demesnes, then they fall into two obvious halves, separated by the possessions of Deer Abbey, the earls' forest at Aden, a number of feudal tenancies, and on the coast the enclave of Inverugie. Buchan, so far as the earls' own lands are concerned, was made up of two distinct halves, separated from each other by other landlords' holdings.

It is now necessary to consider those other landlords. The Church lands belonging to Deer Abbey and to other ecclesiastical bodies have already been discussed. For estates held feudally by lay landlords, the most straightforward are those where the earls' original grants survive. The first, obviously, were the lands of Fedderate and Ardendraught, granted by Earl Fergus to John son of Uhtred.[1] John's successors (probably his descendants) as lords of Fedderate took the name from their estate; William of Fedderate, son of Magnus of Fedderate, appears at the end of the thirteenth century, while another William (or perhaps the same man) was fighting against the English in 1336 and seems to have left only a daughter in the 1360s.[2] By the end of the thirteenth century, however, it appears that Ardendraught had been separated from Fedderate, for Robert I gave it to John de Bonneville.[3] Secondly, there were Kindroucht and Strichen, which Earl William Comyn granted to Cospatric Macmadethyn.[4] Later in the thirteenth century there was a John of Kindroucht, possibly his son, but what happened thereafter is unclear; by the 1460s Kindroucht and Adziel were held feudally of the barony of Kingedward by a Thomas Cumming, which perhaps suggests continuity with the Comyn period (Adziel is in Strichen parish, so probably this is the same estate).[5] When John of Kindroucht appears, it was as a witness to a charter by Earl Alexander Comyn granting the estate of Fiddes ('Fothes') to Fergus, son of John of Fiddes;[6] this was a more durable

1   *Aberdeen-Banff Colls.*, pp. 407-8.
2   *Moray Reg.*, nos. 130-1; *Aberdeen-Banff Ills.*, iv, pp. 612-13; *RMS*, i, app. II, no. 1283.
3   *RRS*, v, nos. 174, 204.
4   *Aberdeen Reg.*, i, pp. 14-15.
5   *Aberdeen-Banff Ills.*, iii, p. 113; SRO, GD.52/388; *Acts of the Lords of the Isles*, no. 72.
6   *Aberdeen-Banff Ills.*, iii, pp. 112-13; SRO GD.52/388.

family, for another John, son of Alan of Fiddes, was active in the 1390s.[1]
Lastly, Earl Alexander granted Inverallochy to his kinsman Jordan Comyn
in 1277, but this feudal tenancy probably did not last beyond 1308; in 1375,
Inverallochy was included in what became the barony of Philorth.[2]

Those are the only feudal holdings in thirteenth-century Buchan for
which charters exist, but, no doubt, other estates were held of the earls as
well. One was probably Crichie: in 1246 William Pratt, who witnessed Earl
William's charter to Cospatric Macmadethyn, granted four shillings a year
for a candle at Pentecost to Deer Abbey; presumably he was the lord of
Crichie.[3] Then there was Meldrum, in the detached portion of the earldom
which contained the parish of Bethelnie, now Old Meldrum. In 1263 Philip
of Meldrum and his wife Agnes Comyn were in dispute with Arbroath
Abbey over the patronage of the church of Bethelnie, which had been given
to the abbey by Earl William Comyn; it is therefore extremely likely that
Philip of Meldrum had received that estate, probably the whole parish of
Bethelnie, from one of the earls.[4] And later, in 1342, one of his
descendants, William Meldrum, was lord of Waterton near Ellon;[5] that, too,
may well have been a Meldrum holding from the earls of Buchan.
Furthermore, if the Meldrums received feudal holdings from the earls of
Buchan, then surely the Mowats, who were so prominent as witnesses to
Comyn charters, also did so. One possibility is that they held Balquhinochy,
(now Hatton castle). Robert I granted Patrick Mowat 'the lands of Loscragy
and Culpedauchis',[6] and Lescraigie is less than a mile from 'Balquholly, the
seat of Mowat of Balquholly, who is reckoned chief of this name' (c.1727).[7]
Perhaps the purpose of Robert I's charter was to confirm the Mowat holding
there; or to transfer it to another member of the family; or to add Lescraigie
and 'Culpedauchis' to what the Mowats held already. Some of Robert I's
other grants of land in Buchan may also have been of lands held feudally of
the Comyns, perhaps to their pre-1314 owners, but there is insufficient
evidence to be sure.[8]

Because of the difficulty of establishing just what was the extent of
feudal tenancies in thirteenth-century Buchan, and their relationship to the
fourteenth-century grants by Robert I and in later reigns, no overall analysis
of this aspect of the earldom's history can be attempted. It is interesting,
however, that on the map the feudal holdings turn out mostly to be either in
the middle of the earldom's territory, or on the fringes; Buchan's territorial
structure seems to have consisted of broad strips. And one other point

1   *Aberdeen-Banff Ills.*, iii, pp. 93, 112.
2   *RMS*, ii, no. 1198; *Aberdeen-Banff Ills.*, iv, p. 88.
3   *Aberdeen Reg.*, i, pp. 13-14; *Aberdeen-Banff Ills.*, iv, p. 3.
4   *Arbroath Lib.*, i, nos. 130, 192. As Philip de Fedarg (which was also in the earldom:
St Andrews Lib., p. 361), he was given land in Tarves, south-east of Kelly, by Arbroath
Abbey: *Arbroath Lib.*, i, no. 257.
5   *Aberdeen-Banff Ills.*, iii, p. 43.
6   *RMS*, i, app. II, no. 36.
7   *Aberdeen-Banff Colls.*, p. 464. In 1727 it was sold 'to Duff of Hatton'.
8   *RMS*, i, app. I, no. 3; app. II, nos. 26, 69.

emerges from the previous two paragraphs. Of seven landowning families which have been discussed, four (Fedderate, Fiddes, Meldrum and Mowat) continued to be landowners in Buchan after the collapse of the Comyn earldom. The fate of the other three is unclear, but the Kindrouchts were replaced by a family of Comyns, which had presumably been in Buchan since the thirteenth century. Thus although the male line of the Comyn earls of Buchan was removed by Robert I, continuity with the Comyn era is clearly evident among the earldom's lairds or gentry.

So far the examination of the lands of the earls of Buchan has been limited to the earldom itself. It must be remembered, however, that the Comyn earls also possessed considerable amounts of land elsewhere. Starting in the north-east, they had an interest in Strathisla (Banffshire), about ten miles west of Kingedward, as is evidenced by Earl William's grant to Deer Abbey of Barry 'in the barony of Strathisla'.[1] Then, further to the west, William can be seen acquiring a small piece of land — half a carrucate, a toft and a croft — in the toun of 'Dunbernyn' (probably Dunbennan), just beyond Huntly in Strathbogie, from a John son of Geoffrey, who received land at Gartshore in Kirkintilloch in exchange.[2] More significantly, Earl Alexander Comyn exchanged territory in Tranent (East Lothian) for Mortlach and its important castle of Balvenie, in highland Banffshire; a charter of 1285 shows that this transaction had taken place a generation earlier.[3] As has been remarked already, Balvenie was a valuable strategic site, making a bridge between Buchan and the lordship of Badenoch, which belonged to what after Earl William's death was the senior branch of the Comyn family. And Alexander also made an important gain on the earldom's western border when he took over the thanage of Conveth at an annual ferme of 80 merks.[4] Then, to the south of Buchan, Earl William had property in or near the Durward barony of Coull, in Mar; when his daughter Idonea married Gilbert de la Hay he granted them the land of of Upper Coull.[5] His wife Marjorie, meanwhile, gave the monks of Arbroath Abbey permission to lease the pond on her land of Fordoun, in the Mearns; that indicates that she and presumably her father Earl Fergus had part of the thanage of Kincardine, which was probably the same as the parish of Fordoun.[6] And the Kincardine property was increased quite significantly in 1264 by Earl Alexander, when he leased nine pieces of land in the neighbourhood from Arbroath Abbey.[7] Also in the Mearns, one of the Comyn earls acquired superiority over, and a 10-merk annual rent from, the thanage of Uras; this was said in 1331 to have been forfeited by Earl John Comyn.[8] In the summaries of the Great Seal Register, it is

1   *Aberdeen-Banff Ills.*, ii, p. 427.
2   SRO, RH.1/2/32.
3   SRO, RH.6/59.
4   *CDS*, ii, no. 1541; Grant, 'Thanes and Thanages', above, Appendix, no. 17.
5   SRO, RH 1/2/31.
6   *Arbroath Lib.*, i, no. 132.
7   *Arbroath Lib.*, i, nos. 247, 311; *RRS*, v, no. 49.
8   *RRS*, vi, no. 5; Grant, 'Thanes and Thanages', above, Appendix, no. 28.

recorded that Earl John had 10 merks and (probably) the lands of Durris, but Uras may well have been meant.[1] At the end of the century, however, Earl John was keeper of the forests of Durris and Cowie.[2] And he made probably the biggest Comyn acquisition in the north-east (apart from the earldom), when, as we have seen, King John Balliol granted him the thanages of Formartine and (very probably) Belhelvie, which together were worth over 250 merks a year.[3]

Further south, Earl Alexander, and possibly his predecessors, had a manor house on the banks of the River Tay, probably near Scone.[4] South of the Tay, the Comyn earl of Buchan estates can be divided into three sections. First, there were lands in Fife which had probably come to the Celtic earls of Buchan through Earl Colbán. Countess Marjorie's confirmation to St Andrews Priory of grants in Kennoway, made by Merleswain son of Colbán, have already been mentioned;[5] Marjorie and William also gave the priory half a merk a year from the fermes of 'Inverine'.[6] And they almost certainly had the lordship, or barony, of Fithkil (now Leslie) as well. In c.1263 Scholastica, daughter of Merleswain of Ardross, was in dispute with Earl Alexander over the patronage of the church of Fithkill;[7] since later Fithkill barony belonged to the earls of Ross, heirs to half the Comyn of Buchan possessions,[8] it should presumably be included among these, and be linked with Kennoway as Fife property of the last Celtic earls of Buchan.[9] Secondly, there were the original Comyn estates in southern Scotland, which William Comyn inherited from his father Richard and to which King William I added the substantial lordship of Lenzie and Kirkintilloch (Dunbartonshire).[10] On William Comyn's death, however, these lands went to his eldest son by his first marriage, and so to the family of Comyns of Badenoch. But this loss to the Comyn earls of Buchan was more than made up for by the third set of estates, gained through the marriage of Earl Alexander to Elizabeth, one of the three daughters and coheiresses of the great Anglo-Scottish magnate Roger de Quincy. When de Quincy died in 1264, the earl of Buchan acquired wide estates in Fife, Galloway, Dumfriesshire, and Lothian;[11] in addition,

1   RMS, i, app. II, no. 876. In one index the land is given as 'Barris [Werress] alias Derres]' (sic); in the other simply as 'Durris'. Uras could well be meant; but Robert I is said to have given it to Robert Wallace, whereas RRS, vi, no. 5, states that the king gave Uras to Thomas Charteris.
2   Rot. Scot., i, p. 10.
3   See above, pp. 184-5; and Aberdeen Reg., i, p. 55, for the values.
4   Near 'Drumyog': Scone Lib., no. 146.
5   Above, pp. 179-80. See also Dryburgh Lib., pp. 11-12.
6   St Andrews Lib., pp. 250, 252.
7   Inchcolm Chrs., nos. 25-7.
8   RMS, i, no. 742; HMC, 4th Report, p. 494, nos. 9, 10.
9   The notion that the Celtic mormaers and earls devoted all their time and energies to their own provinces seems to be refuted by the evidence for these estates.
10   Above, p. 177.
11   ER, pp. 22-23, 31; Stevenson, Documents, i, pp. 329-30; RMS, i, app. II, nos. 308, 319, 361, 497; SRO, RH.6/59/6; SRO, GD.175/24. For de Quincy, see G.G. Simpson, 'An Anglo-Scottish Baron of the Thirteenth Century: The Acts of Roger de Quincy, Earl of Winchester and Constable of Scotland' (Edinburgh University Ph.D. thesis, 1965).

the office of constable, which also came to Earl Alexander as a result of the de Quincy inheritance, brought with it lands in Perth, Clackmannan, Inverness, and Cowie near Stonehaven.[1] The combination of de Quincy and Buchan territories made Earl Alexander one of the greatest magnates in Scotland.

Furthermore, there were the de Quincy estates in England. The three-way division of these was a complex, litigious and protracted affair, but in the end Earl Alexander gained widespread properties, in northern England (in Cumberland, Yorkshire, Lincolnshire and Derbyshire), in the central Midlands (especially in Leicestershire, Warwickshire, Northamptonshire and Huntingdonshire), to the south-east (in Cambridgeshire, Bedfordshire, Buckinghamshire, Hertfordshire and Essex), and in the south-west (stretching through Oxfordshire, Berkshire, Gloucestershire, and Wiltshire as far as the south coast).[2]

This inheritance made Earl Alexander into a cross-Border baron almost on a par with Roger de Quincy or the Huntingdons, Balliols and Bruces. It has been shown that these magnates involved others in their Anglo-Scottish interests: the *familia*, retainers and administrators of both Roger de Quincy and Earl David of Huntingdon transcended the national frontiers in their lords' service.[3] To a degree, this can be seen in the case of Earl Alexander Comyn. With, for example, the long drawn-out transactions over the division of the de Quincy inheritance, he could not be travelling continually from Buchan to England, and generally appointed attorneys to see to his English affairs: some of these were English, but many northern Scots, such as Ralph and William de Lascelles (from Fife), Thomas and Gilbert of Kinross, Robert of Leslie, Nicholas of Slains, Maurice de Murray and John of Buchan, were working for him in England in 1266, 1268, 1271, 1281 and 1282.[4] But as for Earl Alexander himself, he appears, perhaps because of the distance involved, to have been reluctant to travel to England. He did appear in person at Shepshed in 1282 to receive the homage of William of Bridport.[5] But he was slow to come to the English court to do homage for his wife's inheritance; as a result her share was still in the king's hands at the end of 1274.[6] Later, he preferred to involve his sons in the family's English interests. The *post mortem* inquisition on his death reveals that by then he held no lands in chief of the English crown, for more than seven years previously he had enfeoffed John, his son and heir, in the manor of

1   Ibid., pp. 69-70.
2   Young, 'Political Role of the Comyns', pp. 261-4.
3   G.G. Simpson, 'The *Familia* of Roger de Quincy, Earl of Winchester and Constable of Scotland', in Stringer, *Nobility of Medieval Scotland*; K.J. Stringer, *Earl David of Huntingdon, 1152-1219: A Study in Anglo-Scottish History* (Edinburgh, 1985), chs. 8, 9.
4   *CDS*, i, no. 2513; ii, no. 187; *Calendar of the Close Rolls, 1272-9*, pp. 136, 429; *Calendar of the Patent Rolls, 1266-72*, p. 300.
5   *Calendar of Inquisitions Post Mortem*, iv, no. 138.
6   *Cal. Close Rolls, 1272-9*, p. 138. He had been granted safe-conducts to come to England in person in 1265, 1266, and 1267, but sent attorneys (*Cal. Patent Rolls, 1258-66*, p. 460; *1266-72*, pp. 17, 117).

Whitwick (worth £100 a year) and all his other lands in Leicestershire and Warwickshire.[1] Also, he gave his second son Roger the manor of East Farndon in Northamptonshire;[2] and wrote to Edward I asking that, if Roger's right to that property were questioned, then the king's agents should come to Elizabeth in Scotland to obtain her recognisance and testimony in the matter, and that Elizabeth should not have to go to Westminster.[3] That probably gives a good insight into Earl Alexander's attitude towards his English holdings.

As that suggests, Earl Alexander's interests remained essentially Scottish. And within Scotland — despite being given offices in the south-west, in particular that of sheriff of Wigtown in the 1260s,[4] which complemented his new landed power there — it appears that he was chiefly concerned with the north-east. The majority of his surviving charters, for instance, dealt with land in that region,[5] while their witness-lists have a clear north-of-Scotland bias. Thus the Buchan families of Mowat of (probably) Balquholly, Meldrum (Fedarg) of Meldrum, and Cheyne of Inverugie were prominent in these, as well as being firm political supporters of both Earl William and Earl Alexander.[6] Other north-easterners, such as William of Slains, John of Kindroucht, Andrew of Garioch, and Walter and Maurice de Murray, can also be found in the earl of Buchan's following.[7] Similarly, the ecclesiastical witnesses to Earl Alexander's charters were almost all connected in some way with the north-east: they include Bishop Richard de Pottun of Aberdeen, Roger of Derby, canon (1259) and precentor (1264-5) of Aberdeen, Roger 'Paternoster', later chancellor of Aberdeen (1321), Roger of Scartheburg, 'official' of Aberdeen, and Robert of Leslie, rector of Slains.[8]

At the same time, however, Earl Alexander's witness-lists included a number of lairds from Fife, such as William of Abernethy, David of Lochore, Richard of Bickerton, Michael of Arnot, Duncan of Crombie and John of Kinnear.[9] Especially prominent were members of the Lascelles family, who held 'Balmonethe' in Fife of the earls of Buchan, and appear not only in Earl Alexander's entourage but also as attorneys to Earl John

1   Cal. Inquis. Post Mortem, ii, no. 753.
2   Placita de Quo Warranto (Record Comm., 1818), p. 559.
3   PRO, Special Collections, Ancient Petitions, S.C.8/197, no. 9816.
4   ER, i, pp. 22, 30.
5   Aberdeen Reg., i, pp. 30-4; ii, pp. 276-7; Lindores Cart., no. 124; Arbroath Lib., i, no. 319; RMS, ii, no. 1198; SRO, GD.52/388.
6   SRO, RH.1/2/32; St Andrews Lib., pp. 250-3; Lindores Cart., no. 124; CDS, i, no. 2153; SRO, GD.175/24.
7   Aberdeen Reg., i, p. 34; Arbroath Lib., i, p. 266; Inchcolm Chrs., nos. 25-7; SRO, GD.52/388; SRO RH.6/52.
8   Aberdeen Reg., i, pp. 33-4; i, p. 277; Lindores Cart., pp. 152, 158; Arbroath Lib., i, no. 310; Fasti Ecclesiae Scoticanae Medii Aevi ad annum 1638, ed. D.E.R. Watt (2nd Draft, St Andrews, 1969), p. 9.
9   British Library, MS Additional 33,245, fo. 160; Arbroath Lib., i, nos. 310-11; Inchcolm Chrs., no. 26.

Comyn in 1292 and as Comyn supporters in 1296.[1] Conversely, in 1265 Thomas of Meldrum, presumably from Buchan and a Comyn kinsman through marriage, was involved with Earl Alexander in the dispute over the patronage of Fithkil church.[2] And Roger Comyn possessed land in Leuchars in Fife (a de Quincy estate) by grant of his father, Earl Alexander.[3]

Judging by the evidence of his charters, therefore, Earl Alexander Comyn is to be located essentially in north-east Scotland, but with a significant sphere of influence in Fife. These regions seem to have been much more important to him than either England or, indeed, southern Scotland, despite his de Quincy inheritance. But that, relatively speaking, may not have been quite so spectacular as appears at first sight. The value of Roger de Quincy's Scottish lands has been estimated at around £400 *per annum*, and the third share which went to another of the coheirs was valued at £128 in 1296;[4] thus Earl Alexander's income from his wife's Scottish lands can hardly have been much more than £150 a year. That is probably no more than his successor, Earl John Comyn, would have had from the thanages of Formartine and Belhelvie;[5] and hence, presumably, is considerably less than both earls must have had from the earldom of Buchan. The likely economic balance of Earl Alexander's interests matches the evidence of his charters: Buchan, supplemented by the earldom lands in Fife, easily constituted the most significant part of his possessions.

The discussion of Comyn possessions outside Buchan has thus brought us back to the importance of the earldom and the north-east. Moreover, far from being diverted southwards by the de Quincy inheritance, Earls Alexander and John surrendered parts of that in order to build up their north-east possessions. Tranent had been a de Quincy estate in East Lothian; Earl Alexander, as we have seen, exchanged his lands there for Mortlach in Banffshire, on the edge of Moray.[6] And when Earl John in effect extended the borders of Buchan well to the south through John Balliol's grant of Formartine and (probably) Belhelvie, it was in return for surrendering the claim which he had, through the wife of Roger de Quincy, to a share of the lordship of Galloway.[7] In addition, Earl Alexander had gained control of the thanage of Conveth.[8] At the end of the thirteenth century, the Comyn earls' interests were being concentrated even more on Buchan.

The two exchanges raise one final point. In the grant of Formartine,

---

1   SRO, GD.52/388; *Aberdeen Reg.*, ii, pp. 276-7; *Arbroath Lib.*, i, no. pp. 247; *Inchcolm Chrs.*, nos. 25-6; *St Andrews Lib.*, pp. 282-3; *CDS*, ii, nos. 635, 736, 1870.
2   *Inchcolm Chrs.*, pp. 25, 26, 141.
3   NLS, MS Advocates 16.2.29, fo. 159.
4   Simpson, 'Roger de Quincy', pp. 214-6; Stringer, *Earl David of Huntingdon*, p. 192 (it was the la Zouche share).
5   See above, p. 194.
6   SRO, RH.6/59.
7   *CDS*, ii, no. 1541.
8   Ibid.

John Balliol kept the burgh and castle of Fyvie for the crown; but with Mortlach, Earl Alexander acquired the strong castle of Balvenie. The network of at least six Comyn castles in Buchan (Ellon, Slains, Rattray, Cairnbulg, Dundarg and Kingedward)[1] gave the family military as well as territorial might, to which the acquisition of Balvenie added considerably. Furthermore, the Comyn of Buchan castles should not be seen in isolation; the Comyns were a close-knit kin which acted together in political crises, especially during the first stages of the Wars of Independence. Thus in political and military contexts the earls of Buchan's own castles should be considered in conjunction with those held by other members of the family. As Map (B), 'The Castles of the Comyns' (in the Appendix) shows, these collectively gave the family a formidable grip on the north of Scotland. If the north-east was controlled by those of the earldom of Buchan, then the province of Moray was equally dominated by those of the Comyns of Badenoch: Lochindorb in the heart of Moray, Castle Roy in mid-Speyside; Ruthven guarding the Upper Spey and the Pass of Drumochter; and Blair Atholl at Drumochter's southern end.[2] Between Buchan and Badenoch, as already said, lay the newly acquired Balvenie castle; while much further to the west, the entrance to the Great Glen was commanded by the Comyns of Badenoch's castle of Inverlochy in Lochaber.[3] Furthermore, this military complex was strengthened significantly in the 1290s and early 1300s through appointments as sheriffs and keepers of royal castles. By c.1290 Earl John Comyn had the castle of Banff, and by 1303 his brother Alexander Comyn had Aberdeen, both as sheriffs.[4] Also, Alexander Comyn had taken over the castles of Tarradale in the Black Isle and Urquhart on Loch Ness by 1303, and was trying to gain Aboyne as well.[5] At the beginning of the fourteenth century, the Comyns seemed to have a stranglehold on the major castles, and with them the lines of communication, right across Scotland from Buchan to the Moray Firth and along the length of the Great Glen; only Inverness seems to have been out of their grasp.

In the event, of course, the Comyns' lands and castles did not save them from having to submit to the English; Edward I was too formidable an opponent. Then from 1306, with Robert Bruce's killing of John Comyn of Badenoch and seizure of the crown, the fortunes of the Comyn earls of Buchan and indeed Buchan itself were inextricably linked with those of the English. Earl John Comyn and his kin became the king of England's loyal adherents. Two years later, when Robert Bruce attacked them, the Comyns of Buchan and their supporters were acting as representatives of Edward II.[6] Buchan had been a royal base from c.1212, but now, in 1308, it was one for

1   See above, pp. 187-8.
2   Barrow, 'Badenoch and Strathspey, 1', pp. 8-9; Barrow, 'Highlands in the Lifetime of Robert the Bruce', pp. 377-8.
3   Ibid., p. 378.
4   ER, i, p. 49; CDS, ii, no. 1633.
5   Ibid.
6   CDS, ii, no. 43; Barrow, Robert Bruce, p. 172.

the English king. That meant that the Comyns of Buchan had to be thoroughly defeated if Robert Bruce was to have any chance of consolidating his kingship in Scotland. And after Bruce's military genius had nullified and defeated the Comyn might, the 'herschip' of Buchan was inevitable. Buchan's significance in the thirteenth-century kingdom was emphasised by Bruce's infeftments for military service of key supporters and government officers, including Sir Robert Keith the marischal and Sir Gilbert Hay the constable, in this strategically significant region. But while these grants irrevocably broke up the Comyn earldom, it should be remembered that below the Keiths, Hays, Douglases, and Rosses, there were lesser landowners like the Meldrums, Mowats, Fedderates, Fiddeses, Cheynes, and even some Comyns, who although erstwhile followers of the Comyns, kept their lands in Buchan. The Comyn lordship in Buchan may have been destroyed; but in the local landowners who survived to accept the Bruce regime we may perhaps see the community of the province or earldom taking the side of the community of the realm.[1]

1   I must express my warmest gratitude to the editors for helping me to improve and structure the first draft of this essay, and in particular to Dr Grant for lending me his essay on 'Thanes and Thanages', for giving me the benefit of his knowledge of the 14th-century material, and for spending many hours working with me on the territorial analyses.

Appendix

Lands of the earls of Buchan

This appendix gives details of the Comyn earls' lands, in three sections: first a list of the probable demesnes of the earldom (with National Grid references), derived from the lands which were component parts of the new fourteenth-century baronies; second, Map (A), which shows these lands and also feudal and ecclesiastical tenancies; and third, Map (B), which shows the major Comyn possessions and castles elsewhere in Scotland.

I.   Probable Demesnes of the Earldom of Buchan

	ABERDOUR/Dundarg	NJ8964		KELLY	NJ8634
A1	Ardlaw	NJ9363	Ky1	Asleid	NJ8441
A2	Auchlin	NJ9163	Ky2	Auchmaliddie	NJ8844
A3	Auchmacleddie	NJ9257	Ky3	Auchnagatt	NJ9341
A4	Bodychell mill	NJ9562	Ky4	Barrack	NK8941
A5	Coburty	NJ9264	Ky5	Cairngall	NK0447
A6	Drumwhindle (Little)	NJ9236	Ky6	Gonarhall	NJ8658
A7	Glasslaw	NJ8559	Ky7	Knaven	NJ8943
A8	Memsie	NJ9761	Ky8	Logierieve	NJ9127
A9	Pitsligo	NJ9367	Ky9	Newburgh	NJ9925
A10	Pittulie	NJ9567	Ky10	Saithlie	NJ9756
A11	Quilquox	NJ9038	Ky11	Tillydesk	NJ9536
A12	Rathen	NK0060			
	KINGEDWARD	NJ7256		PHILORTH	NK0063
K1	Byth	NJ8156	P1	Ardglassie	NK0161
K2	Castleton	NJ7156	P2	Ardmachron	NJ9861
K3	Faithlie	NJ9967	P3	Ashogle	NJ7092
K4	Fishry	NJ7658	P4	Auchentumb	NJ9258
K5	Fortrie	NJ7359	P5	Auchmacleddie	NJ9257
K6	Kinharrachie	NJ9231	P6	Balchers	NJ7158
K7	Scattertie	NJ6957	P7	Blackton	NJ7257
K8	Tyrie (Easter)	NJ9363	P8	Bracklawmore	NJ8458
			P9	Brakans	NJ7553
			P10	Cairnbulg	NK0163
			P11	Cairneywhing	NJ8757
	RATTRAY	NK0858	P12	Delgatie	NJ7550
R1	Cairnglass	NK0462	P13	Drumwhindle	NJ9236
R2	Crimond	NK0557	P14	Fintry	NJ7554
R3	'Crimondbell' [Bilbo?]	NK0656	P15	Inverallochy	NK0462
R4	Crimongorth	NK0455	P16	Kinbog	NJ9962
R5	'Rothnathie'[1]	NK0360	P17	Kinglasser	NJ9963
R6	Tullikera	NK0259	P18	Luncarty	NJ7153
	'Kindolos'	unidentified	P19	Plaidy	NJ7254

SOURCES.   Aberdour: *Aberdeen-Banff Ills.*, ii, p. 377.   Kelly: *RRS*, v, no. 347.
Kingedward: *Aberdeen-Banff Ills.*, iii, pp. 521-35; *Acts of the Lords of the Isles*, no. 113.
Philorth: *Aberdeen-Banff Ills.*, iv, p. 86-8.   Rattray, *RMS*, i, app. I, no. 66.

1   'Rothnathie' is not on the modern map, but is on Gordon of Straloch's (see *Aberdeen-Banff Colls.*, at end of preface).

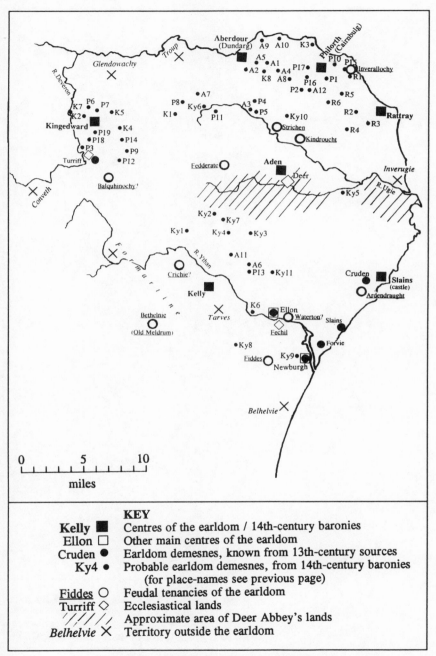

KEY

Kelly ■	Centres of the earldom / 14th-century baronies
Ellon □	Other main centres of the earldom
Cruden ●	Earldom demesnes, known from 13th-century sources
Ky4 •	Probable earldom demesnes, from 14th-century baronies (for place-names see previous page)
Fiddes ○	Feudal tenancies of the earldom
Turriff ◇	Ecclesiastical lands
/////	Approximate area of Deer Abbey's lands
Belhelvie ✕	Territory outside the earldom

II. **Map (A): The Earldom of Buchan**

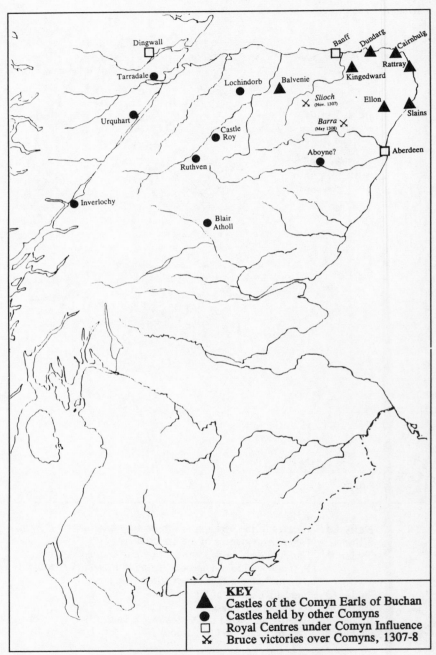

KEY

▲ Castles of the Comyn Earls of Buchan
● Castles held by other Comyns
□ Royal Centres under Comyn Influence
✗ Bruce victories over Comyns, 1307-8

III.   **Map (B): The Castles of the Comyns**

10

Crown and Community under Robert I

NORMAN H. REID

Over fifty of the *acta* of Robert I make reference to Alexander III, or to the time when he ruled Scotland.[1] All but a handful of them refer to him as King Robert's 'predecessor who last died' ('predecessor noster ultimus defunctus'). Other kings who gain occasional mentions — Edgar, Alexander I, David I, Malcolm IV, William I, Alexander II — are also frequently described as Robert's 'predecessors',[2] and indeed the phrase 'our predecessors the kings of Scotland' ('predecessores nostri reges Scocie') is not unusual.[3] Clearly, the words are carefully used: the earlier kings of Scots in general, and Alexander III as the last of them, were those whom Robert I followed, and for whom he carried on the traditions of Scottish monarchy. The stylistic references to kings long dead thus stresses continuity of kingship. This was policy: it was a court style, employed by at least eight different scribes.[4]

A not insignificant number of *acta* also refer to the earlier kings as Robert I's 'ancestors' ('antecessores nostri reges Scocie'),[5] although this expression is never found with reference to Alexander III, for such a claim would at best have been tendentious. The purpose of the expression is again continuity, but this time dynastic continuity. The stress here is on Robert I's personal right to the throne.

Robert I was not, of course, unique in associating himself with his predecessors. References in the *acta* of Malcolm IV to his 'antecessores' and 'predecessores' do not always relate to specific people,[6] but the occurrence of the terms in relation to named kings is frequent enough to make it clear who these 'predecessors' and 'ancestors' were deemed to be.[7] Interestingly, the term 'predecessor' is sometimes used in relation to Queen Margaret, Aethelred (brother of Edgar), and Alexander I's queen, Sybil.[8] Conversely, the term 'ancestor' is sometimes used to refer exclusively to previous kings.[9]

---

1   See, for example, *RRS*, v, nos. 9, 15, 19, 28, 47, 50-4, 60, 64, 71, 73, 87-8, 91-3.
2   E.g., *RRS*, v, nos. 241, 336, 340.
3   This, or similar phrases, can be found in, e.g., *RRS*, v, nos. 28, 52-4, 191, 389.
4   Scribes identified by A.A.M. Duncan in *RRS*, v, as 'A', 'B', 'C', 'D', 'K', 'L', and two others, unique.
5   E.g., *RRS*, v, nos. 71, 124, 266.
6   E.g., *RRS*, i, nos. 112, 114, 122-4.
7   E.g., *RRS*, i, nos. 115, 118.
8   E.g., *RRS*, i, no. 118.
9   E.g., *RRS*, i, no. 150.

Similar usages are to be found in the *acta* of William I.[1] It appears, then, that in the context of the royal *acta* at least, the two terms are virtually interchangeable, and should not be taken too literally. This formula for reference to previous kings as ancestors and/or predecessors continues throughout the reigns of Alexander II,[2] Alexander III,[3] and John.[4] It would thus seem that by using this formula Bruce was not only stressing his dynastic links with the past and his inheritance of a traditional role, but also displaying a further continuity in the kingship by adopting the stylistic norms of his predecessors. There is, however, a difference which betrays Bruce's weaker dynastic position. The older kings make considerable use of specific terms of relationship — 'pater', 'avus', 'atavus', 'ava', for example.[5] Such terms, for obvious reasons, are missing from Bruce's *acta*. The dynastic *caesura* of the late thirteenth century meant that Bruce's usage necessarily linked him with the entire dynasty, rather than, in a more personal way, with specific members of it.

A striking factor in Robert I's reference to his predecessors is the complete absence of any reference to King John, his immediate predecessor. In one or two acts, in which John's forfeited estates were regranted, he is referred to simply as John Balliol.[6] Nowhere in the *acta* is there any hint that there had been a king in the two decades following the death of Alexander III. Yet John had been fully inaugurated; he had as good a claim to call the old kings his ancestors as had Robert I; and in his capacity as Guardian of the realm from 1298 to 1300, Bruce had even issued acts 'nomine preclari principis domini Johannis dei gratia regis Scocie illustris'.[7] In the early days of Robert I's reign, John (who survived in exile until 1313) may have been seen as an embarrassment, even as a threat, both constitutional and dynastic: although stripped of his kingly insignia by Edward I of England, he had never been formally repudiated by the Scottish community, who for some years viewed his abdication as invalid because of English coercion.[8] After Balliol's death, however, it seems unlikely that the Balliol episode can have been regarded as a threat to Robert I's kingship. Nonetheless, presumably for political reasons, there seems to have been a pretence that the Balliol kingship had never existed.

It is intriguing, for instance, that the Declaration of Arbroath[9] did not adopt the same approach as had been used in the declarations from the St

1   E.g., *RRS*, ii, nos. 8, 28, 30.
2   E.g., *Arbroath Lib.*, i, no. 100; *Melrose Lib.*, i, no. 174; *Scone Lib.*, no. 59.
3   E.g., *Melrose Lib.*, i, nos. 309-10.
4   E.g., *Glasgow Reg.*, i, no. 249; Stevenson, *Documents*, i, pp. 421-2 (note especially the references to Alexander III as 'praedecessor noster, ultimus rex Scociae').
5   E.g., *RRS*, i, nos. 109, 118, 148; *RRS*, ii, nos. 8, 19.
6   *RRS*, v, nos. 59, 67, 95.
7   *APS*, i, p. 454.
8   See, e.g., Baldred Bisset's *Processus*: *Chron. Pluscarden*, i, p. 217.
9   *APS*, i, pp. 474-5; translated in G. Donaldson, *Scottish Historical Documents* (Edinburgh, 1970), pp. 55-8.

Andrews parliament of 1309,[1] in which the Balliol episode had been explained away as an aberration. According to those documents, a king, wrongfully chosen, had been ineffectual, and had abandoned the kingdom to the encroaching enemy, thus exposing the land to ruin, and forfeiting the right to rule. To explain the change, we must examine the circumstances in which the documents were produced. The 1309 documents (better, more authentically, known through the reissue of the clerical declaration in 1310) were intended purely to express support for Robert I. Barely five years had passed since the Scottish governing community had strongly upheld Balliol's right to rule. It was thus necessary to explain the change of direction, and to that end Balliol was turned into the villain of the piece. Wrongful election and negligence by the previous king justified support for this 'new man'. This is the earliest written reference to the idea that there had been general reluctance to accept the Balliol claim to the throne in 1290-2, and it is an idea, like many other propagandist coups, which has adhered well in the popular memory.

By 1320, however, Bruce had less need to justify his position. Instead, the Declaration of Arbroath is a statement of his *de facto* leadership of the realm, and of the Scots' acceptance of his dynastic, personal and constitutional fitness to rule. It is also, more importantly, an attempt to lay the blame for almost a quarter of a century's warfare at the door of the English monarchy. In 1309/10 it had suited the Scots' purpose to blame John Balliol for the turmoil; in 1320, as in 1301 (Baldred Bisset's *Processus*[2]), the purpose in hand was to blame the English, so that papal pressure could be brought to bear on them in the hopes that an end to the war and recognition of Bruce's status would be hastened. In 1320, therefore, there was no need for Balliol to be a central character. Nonetheless, the Declaration of Arbroath's sketchy account of the events of 1290-1306 seems deliberately to avoid the Balliol issue. Similarly (and much more strikingly, given the length and detail of the document), the Scottish case put forward at the negotiations with the English at Bamburgh in 1321[3] also avoids Balliol, who is mentioned only in the context of English claims, which are said to be unfounded. This retort to the English version of recent history is a lengthier account than that contained in the Declaration of Arbroath, but is equally fudged. Veiled reference is made to Balliol only in the context of 'the nullity of the judgements given by this king [Edward I], as if judgement had not been made by him'.[4] Balliol's deposition is not mentioned. The whole Balliol episode is thus reduced to the level of malicious English

1   *APS*, i, p. 460; translated in Stones, *Anglo-Scottish Relations*, no. 36. For discussion of the documents, see N.H. Reid, 'The Political Role of the Monarchy in Scotland, 1249-1329' (Edinburgh University Ph.D. thesis, 1984), p. 315 (n. 151), and the sources cited.
2   *Chron. Pluscarden*, i, pp. 205-18. (See also the *Instructions* on which the *Processus* was based: ibid., pp. 178-201.)
3   P.A. Linehan, 'A Fourteenth-Century History of Anglo-Scottish Relations in a Spanish Manuscript', *Bulletin of the Institute of Historical Research*, xlviii (1975).
4   'nullitatem sententie dicti regis tanquam a non suo judice late' (ibid., p. 121).

fiction. At a time when the 1309/10 version would presumably have had a more authentic ring to it, it is curious that the Scots apparently lacked the confidence to assert their rights to deselect a king (rightly or wrongly elevated in the first place) who had failed in his task of leadership, and support in his stead a man of equal dynastic legitimacy, who had kingly attributes, and who had moreover proved his ability by force of arms. Such a right was expressed with regard even to Robert I in the Declaration of Arbroath itself, and so why was John Balliol subject to such taboo in the secure days of the 1320s? Certainly, there was no real need to justify Bruce's position, but, equally, there can have been no danger in doing so.

Once again, it is interesting that in this Bruce displays continuity with his royal predecessors. One scans in vain the published *acta* of Malcolm IV and William I for the name of Donald Ban, although his reign was subsequent to that of Malcolm III, who does find mention as ancestor and predecessor.[1] Donald Ban was, of course, regarded as a usurper, as is made clear by the *Processus* of Baldred Bisset, which refers to his ousting of Duncan II and Edgar.[2] It would seem, then, that the ruling line did not publicly recognise the dynastic aberrations which occasionally interrupted its hold on the kingdom. As the principles of primogeniture gradually became more firmly ensconced in the *mores* of the ruling dynasty from the mid-twelfth century onwards, its desire to prove its antiquity and 'right' became stronger. This need was, of course, exacerbated by the political struggles of the twelfth and thirteenth centuries. As English claims to ecclesiastical and political supremacy were expressed, so the Scots' own expression of their 'ancient liberty' was modelled in the clay of origin myths.[3] The transposition of the Pictish king Angus's conversion back in time by three centuries to *c*.400 (and the later notion that the Scots had been converted by the bones of Saint Andrew himself) gave the Scottish Church an antiquity which made it impossible for the supposedly younger see of York to claim an ancient superiority. Similarly, as political pressure increased from the south (and, during the eleventh and twelfth centuries, as internal conflicts over the succession flared), the myths of *Scota* and of the Scythian origins of the Scots were formalised and embellished to give an unimpeachable authority to this line of kings, unbroken by usurpation, it was claimed, since time immemorial.[4] The canonisation of Queen Margaret and the translation of her relics in 1250 (along with the creation of a mythology around it) have been described as 'a celebration of an ancient royal cult, assiduously fostered by the royal line since the time of Margaret and Malcolm's sons'.[5] Robert I was

1   E.g., *RRS*, i, no. 118; ii, no. 30.
2   *Chron. Pluscarden*, i, p. 214.
3   See W.F. Skene, 'Development of the Scottish Fable', in *Chron. Picts-Scots*, pp. clxix-cxcv, for a full discussion of the Scottish origin myths. See also E.J. Cowan, 'Myth and Identity in Early Medieval Scotland', *SHR*, lxiii (1984).
4   The use of origin myths to this end is well exemplified in both Bisset's *Processus* (*Chron. Pluscarden*, i, p. 214) and the Declaration of Arbroath (*APS*, i, p. 474).
5   M. Ash, 'The Church in the Reign of Alexander III', in *Scotland in the Reign of Alexander III, 1249-1286*, ed. N.H. Reid (Edinburgh, 1990), p. 31.

not the king to disparage such a cult, and neither was the early fourteenth
century the time to do so.

The 'self-conscious antiquarianism' of Bruce's *acta* and the refusal to
acknowledge Balliol are not, however, merely attempts to affirm his own
dynastic legitimacy and administrative consistency; the repeated insistence
on Alexander III as his immediate predecessor has a further significance of
its own. Since Alexander's death in 1286, the land had been subject to
warfare, invasion, usurpation of native rule, many changes of government,
and the dislocation of all areas of administration, with consequent lack of
security for all sections of the community. It is thus understandable that
Alexander III's reign seems to have been revered. Inevitably, Bruce's
primary task was to restore Scotland to the state which it had achieved
during Alexander's personal rule, which, with its relative freedom from
external or internal strife, and its reputation for effective justice and at least
relative prosperity and ecclesiastical stability, could be seen as a model for a
medieval king. The anonymous versifier responsible for the fragment of
poetry preserved in Wyntoun's chronicle, who must have written it before
1314, described the expected fruits of the good ruler. Even in later days it
seems that John of Fordun, for example, used Alexander III as a blue-print
for ideal kingship.[1] Alexander, then, may well have been regarded in the
early fourteenth century as representing the zenith of his dynasty; it should
therefore come as no surprise that in administration, as is exemplified in the
stylistic norms of his *acta*, Bruce was not a great innovator, but one whose
aim was to restore. His task was to regain a lost normality, not to
revolutionise existing practice.[2]

As shall be seen, however, in one important aspect it was impossible for
Bruce to reassert the kingship of his predecessors. The success of the
community in quickly establishing an interim government in 1286, in
contrast with the situation during Alexander III's minority, has been
frequently commented upon as a manifestation of the developing
constitutional awareness of the Scots. It may indeed be seen as one of the
most important achievements of the 'Age of the Alexanders' that the
community adopted an increasingly prominent stance in the rule of the land,
which culminated in the remarkably concerted effort to work through the
potential difficulties of 1286, and the crisis of 1290.

Although it could hardly be claimed that the community was united in its
efforts, the years between the deposition of John Balliol and the inauguration
of Robert I did indeed display a 'dour and dogged persistence in the
common enterprise'.[3] The repeated 'experiments in guardianship'[4] exhibit

1   N.H. Reid, 'Alexander III: The Historiography of a Myth', in ibid., pp. 186-91.
2   This general assertion is not intended to imply that there was no innovation in the
administration of Robert I, and should be taken in conjunction with the detailed
interpretation of the evidence of chancery working, and of administrative style and
procedure in the Introduction to *RRS*, v, and in G.W.S. Barrow, *Robert Bruce and the
Community of the Realm of Scotland* (3rd edn, Edinburgh, 1988), esp. ch. 14.
3   Ibid., p. 109.
4   Ibid., p. 90.

a rising development in constitutional theory — a reliance on the concept of the 'royal dignity' as the all-enveloping source of kingly authority and governmental right. It was an authority which the community would under normal circumstances bestow upon a king of their choice, but which on occasions they would be forced to wield themselves.[1] The turmoil of the years of Alexander III's minority, and perhaps particularly the level of external interference in Scottish affairs which they witnessed, created an environment ripe for political maturation,[2] encouraged by the largely peaceful, stable and prosperous period which followed. That process resulted in fresh constitutional possibilities which, in the years following 1286, held the country together, but ultimately failed in averting repeated foreign intervention, due to the 'nightmare scenario' of dynastic collapse. The continued pressure on national identity exerted by the English from 1291 had the effect of encouraging a further strengthening of the community's perception of its own role in the realm's government.[3] A consequence of the essential self-reliance of the community was that the king himself became, in a sense, marginalised; to an increasing extent he became the servant of the kingdom, the one appointed to guide and safeguard it, through the wielding of the royal dignity which had been bestowed upon him through election and inauguration by the community.

The kingship which Bruce assumed in 1306 was, therefore, far removed from that of his model, Alexander III. Bruce had indeed to fight to establish his authority in both military and administrative terms, in order to reduce the internal and external political opposition to his rule; but, more importantly, he had to gain sufficient standing in the realm to earn the approbation of its community. In 1309, at his first parliament, the declarations of nobles and clergy apparently gave him that approbation, but in reality that parliament was barely a representative gathering.

The Declaration of Arbroath is a much more significant document than the 1309/10 declaration, for, running in the name of thirty-nine leading nobles (and carrying the seals of yet more), it does indeed seem to be representative of at least the lay community. It carries, for our purpose, two main themes. The first, clearly, is the support of the community for the king. That a number of the names associated with the document were soon to be associated with a treasonable plot against the king does not negate that conclusion.[4] The document was most certainly an attempt to display the unity of the realm; that there was sufficient unanimity to ensure that even the names of those who were perhaps doubtful in their support could be included, strengthens rather than weakens our perception of Bruce's support

1   For a discussion of the community's wielding of the 'royal dignity', see N.H. Reid, 'The Kingless Kingdom: the Scottish Guardianships of 1286-1306', SHR, lxi (1982).
2   See Grant Simpson's essay, 'Kingship in Miniature: A Seal of Minority of Alexander III', above.
3   For further discussion, see Reid, 'Political Role of the Monarchy', esp. ch. 6 and Conclusion.
4   See below, pp. 215-16.

from the community of the realm. Here, in a much more plausible context than in 1309, we have the wholehearted approbation of the king by the community. The second theme is the power which the community had assumed, and which, for perhaps the first time, it explicitly expressed. The community had demonstrated its ability to assume government between 1286 and 1290; it had at least limited the independent actions of King John in 1295; it had again assumed control between 1297 and 1304; and its approval of the intended royal succession had been sought by the king in 1284, 1315 and 1318.[1] But here for the first time it was stated that the king was but the servant of the community itself; in effect, that the realm, the royal dignity, was embodied within the community, and that the king was a dispensable part thereof, elected in order to provide a symbolic unity and leadership for the realm.

That this developed theory is found in such a document, instigated by the king himself, implies either that the king approved, and even encouraged such constitutional limitation of his power, or that he had little option but to accept it. It may of course have suited Bruce to have it appear that the community would not allow him to make peace without recognition by the English of his kingship and his realm's independence. (A similar approach had been adopted in 1317, when on the advice of his council the king refused to negotiate with papal envoys, or to receive the letters they carried, since these did not address him correctly.)[2] Against the background of the events of 1284-1318, however, there is a ring of truth to the powerful claim of the community to have authority over the king.

For all the antiquarianism of Bruce's reign and administration, then, he did not and could not re-establish the kingship of Alexander III. He lived in a different age, and it is therefore instructive to examine Bruce's kingship after the expulsion of the English from Scotland, when he was at the height of his personal power, when he had re-established the firm administration of Alexander III's time, and when, also, the community of the realm was in a stronger position than ever before. What was the nature of the new relationship between king and community, and what was its significance for Scottish kingship?

By 1320 Bruce claimed to have been on the Scottish throne for fourteen years. With respect to the first few of those years, the claim was tentative: exile followed by errant warfare, largely against the king's 'own' people, was not an auspicious start to a reign. By 1310, however, the civil war was largely over, and, with most of his foes either subjugated or won over, Robert was able to complete the establishment of his own administration, and concentrate on removing the occupying forces in Scotland and on pursuing the war which he hoped would bring about English recognition of his kingship and, consequently, normalisation of foreign relations and trade.

1   See below, pp. 213-14.
2   *Foedera*, ii, pp. 340-1, and see below, p. 211.

(Despite the strength of Bruce's position after 1314, English disruption of
Scottish trade remained a constant impediment to the Scots, who nonetheless
persistently ran the gauntlet of the English blockade on both the east and the
west coasts.[1]) The weakness of the English kings, particularly after 1314,
left Bruce in a strong position in Scotland. Success in battle, and regular
income from tribute taken in northern England,[2] helped to strengthen both
his personal status and his ailing treasury. The apparent inability of Edward
II to oppose the Scots' ravaging of the northern counties left him in an even
weaker position, and it must have seemed that Scottish policies were largely
effective. Clearly, this was a period when Bruce's position in Scotland was
consolidated, and it is therefore worth examining his conduct of the realm's
affairs from 1314 to 1320.

Soon after Bannockburn, Bruce attempted to open peace negotiations
with the English,[3] and although ambassadors were appointed on both sides
(Edward II having been encouraged to participate by Philip IV of France),
and both kings declared their sole ambition to be the achievement of peace,
by the end of the year there had apparently been no progress.[4] Certainly, the
English could be forgiven for seeing little sincerity in the Scottish overtures,
since this was a period of active aggression in the north of England. The
military pressure was doubtless intended to force the English to the
negotiating table, a pattern of activity which was to continue for many years
to come. Further negotiations in the spring of 1316 led to an exchange of
prisoners and a further truce which was, however, short-lived.[5] These
negotiations took place during the Scottish invasion of Ireland, at a time
when an assault on Berwick and a further invasion of England were planned,
and against a background of constant endeavour by the English to raise
forces and supplies for the defence of the north.[6]

A papal attempt to bring the two kings to terms in 1317-18 is of some
significance for later events. In January 1317 John XXII, citing as his main
concern the war's disruption of crusading plans, declared a two-year Anglo-
Scottish truce, under pain of excommunication.[7] Neither side seems to have
paid much attention. By April, preparations were being made to send two
cardinals to England, empowered to negotiate for peace, and to proceed
against Robert Bruce and his supporters, excommunicates 'who govern
Scotland at present'.[8] Pressure on the Scots was unabating: in the same

1   For continued Scottish trade, and English attempts to stop it, see, e.g., *Calendar of the
Close Rolls, 1313-18* (London, 1893), pp. 218-19; ibid., *1318-23* (London, 1895), p. 132.
2   For an interesting analysis of the Scottish raids on the north of England, see J.
Scammell, 'Robert I and the North of England', *English Historical Review*, lxxiii (1958).
3   *RRS*, v, no. 40; for the English response, see *Rot. Scot.*, i, pp. 131-4.
4   *Historical Papers and Letters from the Northern Registers*, ed. J. Raine (Rolls Ser.,
1873), nos. 150, 153.
5   *Rot. Scot.*, i, pp. 153-5; *Foedera*, ii, pp. 289, 295-6; *CDS*, iii, no. 490.
6   For the assault on Berwick, see *CDS*, iii, no. 486; for the English defence of the
north, see, e.g., *Foedera*, ii, pp. 280-2, 286.
7   Ibid., p. 308.
8   Ibid., pp. 317-18.

month a tithe of the churches in England, Ireland, Wales and Scotland was granted to Edward II for crusading purposes, and bulls were issued excommunicating all those invading the kingdom of England. Robert and Edward Bruce were accused of fomenting war; Robert, it was said, had usurped the rule of the kingdom ('gubernacula regni ab ipso suscepta'); and mendicant friars of Ireland were criticised for encouraging rebellion against English authority. Those who continued the fight against the Scots received papal thanks.[1]

The papal envoys arrived in June 1317 and attempted to speak with Bruce, but were attacked on their way north, in an ambush which may well have involved collusion between the Scots and English dissident lords.[2] The messengers sent by the cardinals to Bruce received polite but firm refusals to negotiate, unless he was approached as a king.[3] They left, empty handed, to return to the Curia in August 1318.[4] In the meantime the pope had refused to accept Robert I's nomination for the see of Glasgow following the death of Bishop Wishart in 1316; and a further threat of excommunication was issued against the invaders of England, followed by a sentence of excommunication and further vilification of Bruce after the Scots' capture of Berwick.[5] Towards the end of 1318 the Scots seem to have made an attempt to boost their flagging fortunes at the Curia; and it is possible that in certain quarters (if not by the pope himself) they were received quite sympathetically. On 9 January 1319, whilst engaged in serious efforts to raise men against renewed Scots aggression on the border, Edward II wrote to the pope advising him to ignore the case put by a Scottish mission to the Curia, and only a few days later, having intercepted letters from certain people at the Curia, he protested about their encouragement of the Scots' rebellion.[6] By March, however, it seems that at least some of the Scots at the Curia had been silenced.[7] Late in 1319, with further negotiations in progress which led to a two-year truce,[8] Edward II continued to use the antagonistic papacy as a lever on the Scots. He refreshed the pope's memory regarding the crimes of Bishop Lamberton, and encouraged the pope to continue exerting pressure on the Scots.[9] Lamberton and the bishops of Aberdeen, Dunkeld and Moray were summoned to the pope in November 1319,[10] a summons which they ignored. January 1320 saw fresh sentences of excommunication for Bruce and his followers, and the bishops, along with the king, were again excommunicated in June 1320.[11] By then, of course,

1   Ibid., pp. 320-1, 325, 327.
2   See J.R. Maddicott, *Thomas of Lancaster, 1307-1322* (Oxford, 1970), pp. 204-5, and the sources cited.
3   *Foedera*, ii, pp. 340-1.
4   Ibid., p. 372.
5   Ibid., pp. 337, 353, 362-4.
6   Ibid., p. 384.
7   Ibid., p. 390.
8   Ibid., pp. 388, 391, 404, 409-12.
9   Ibid., p. 406.
10   *Calendar of ... Papal Letters, 1305-42* (London, 1895), p. 191.
11   *Foedera*, ii, pp. 412-13; *Cal. Papal Letters, 1305-42*, p. 199.

the Scottish diplomatic campaign had reached its zenith, and the Declaration of Arbroath, along with other letters from the king and clergy, were well on their way to the Curia. In response, the papal tone was soon to become less menacing,[1] and if the Scottish king still had a long wait before papal recognition of his status was granted, at least the pressure was relieved to some extent. Indeed, it may have been due to some papal (and French) pressure on Edward II that peace negotiations took place at Bamburgh in 1320-1.[2] On this occasion the English explicitly complained that the Scots, while stating they wished for peace, showed little inclination towards it.[3] Clearly, the latter's military position was secure enough for them to feel able not to compromise. On the central issue of recognition of Bruce as the king of an independent kingdom, there can have been nothing but obstinacy on either side.

Eventual Scottish success in the diplomatic arena, and on the home military front, was not, however, matched beyond the Irish Sea. John of Argyll, with a fleet based in Dublin, had recaptured the Isle of Man (taken by the Scots in 1313) for Edward II around the turn of the year 1314/15.[4] The loss of Man probably occurred too late for it to have been a motivating factor in the Scots invasion of Ireland in May 1315, but the policy which led the Scots to desire the security of Man and control of Ireland must have been one and the same. That the force which took Man was based in Ireland reinforced the perception that dominance of the Irish Sea was vital, both for the restoration of safe Scottish trade and for the successful pursuit of the war with England. Pressure on the north of England was in some respects a distant threat, but the prospect of this second front, and the consequent likelihood of attack from Ireland through Wales[5] to the English heartland must have been of great concern to Edward II, who, despite continued activity by the Scots, made strenuous efforts to retain mastery of the Irish Sea.[6]

The expedition to Ireland was, of course, a failure, ending after three years with the death of Bruce's brother at the battle of Faughart in October 1318.[7] The Irish escapade had certainly posed a serious threat to English supremacy in the west, and had doubtless for a time limited Ireland's usefulness to the English as a source of supplies. The Scots, however, had paid a high price: the squandering of men and resources and, more importantly, the death of the heir to the throne.

To gauge the effects of these developments in foreign affairs, it is worth returning to the internal affairs of Scotland in the period from Bannockburn

1   See G. Donaldson, 'The Pope's Reply to the Scottish Barons in 1320', *SHR*, xxix (1950), pp. 119-20; see also *Foedera*, ii, p. 432.
2   Ibid., pp. 441-2, 450.
3   Stones, *Anglo-Scottish Relations*, no. 38 (h).
4   *CDS*, iii, no. 420.
5   *Cal. Close Rolls, 1313-18*, pp. 186, 253; *Rot. Scot.*, i, p. 159; see also Barrow, *Robert Bruce*, p. 380 (n. 17).
6   E.g., *Rot. Scot.*, i, p. 139.
7   *Chron. Pluscarden*, i, p. 250.

to 1320. The supremacy of Bruce over his enemies within Scotland is demonstrated by the act passed at the parliament of November 1314, which allowed for the forfeiture of the estates and other possessions or rights within Scotland of those who had died outwith the king's peace, or who had not yet submitted to him.[1] The resulting redistribution of lands must have helped to cement the king's relationship with the 'whole community of the realm', both clerical and lay, which had 'agreed' on the action. It also had the effect of further reducing the rump of resident opposition. David earl of Atholl, for one, not only lost his lands and the office of constable (regranted to Gilbert Hay at the same parliament), but was also forced into exile.[2]

The succession, as always in a medieval realm, was a matter of grave concern. The settlement of the succession on Edward Bruce in April 1315, in the event of the king's death without direct male heirs, was a measure of the need for Robert to bow to the 'popular' opinion of the community.[3] Previously, the entail states, the king's daughter Marjorie had been recognised as heir; but (as in 1284, when it seems that the community was less than enthusiastic about accepting the succession of Margaret, Maid of Norway) there can be little doubt that, particularly in the context of continuing warfare, there was a strong preference for a male heir. Thus for no reason, other than that implied by approval of Edward's virility and expertise in war ('vir strenuus et in actibus bellicis ... expertus'), Marjorie, with her supposed concurrence, was replaced as heir apparent. This was, in name at least, an act not of the king but of the community itself, carried out with the assent of all the major parties for the defence and security of the kingdom. In contrast, the entail of 1284 had been an undertaking by named parties to uphold the succession of Margaret, quite clearly at the insistence of the king himself.[4] The wording of the 1315 document is thus another indication of the development of the community's perceived role in government. The desire that Robert should be succeeded by a 'vir strenuus' is what one would expect of the fourteenth-century community. The timing of the change, however, is significant. The king's first parliament had been held in 1309, and if his hold on the realm had not then been sufficiently secure for the succession to have been an important issue, surely it had become so before 1315. Why was the matter not dealt with until 1315? Was it merely that Bannockburn had made certain Robert's continued hold on the throne, and produced a recognised breathing space during which such constitutional niceties could receive attention? Such an explanation seems unlikely. Bannockburn itself finds little place in the contemporary Scottish sources, and can scarcely have been regarded with the overwhelming significance which hindsight has tended to bestow upon it.[5] It can, in fact,

1   *RRS*, v, no. 41.
2   *RRS*, v, p. 4 and no. 42.
3   *RRS*, v, no. 58.
4   *APS*, i, p. 424.
5   For a brief discussion of the significance of Bannockburn, see Reid, 'Political Role of the Monarchy', pp. 354-6.

be no coincidence that the entail was made only weeks before Edward Bruce left for Ireland, and probably immediately before the marriage of Marjorie to Walter Stewart.[1] If, as the king's letter to his 'friends' in Ireland[2] and subsequent events would indicate, the royal policy was indeed to unite Scotland and Ireland under one crown, then the succession of the same king to both realms was of some importance. Edward Bruce was clearly intended to take the crown of all Ireland; and the marriage of Marjorie to Walter Stewart, with the possibility that their heirs would succeed to the Scottish throne, would have cut across dynastic union, had the Irish venture succeeded. By establishing Edward Bruce as the heir apparent, the Scots not only gained a strong, battle-hardened heir, but also moved towards dynastic unity, or at least to closer dynastic links, with Ireland. It is also probable that in the early fourteenth century the principles of primogeniture were not yet so firmly established that, in the conservative minds of the community (and perhaps also the king), the succession of the brother and his heirs before the daughter and hers was regarded as abnormal. That a formal written agreement was made may be another indication of Bruce's use of Alexander III as a model. More importantly, the entail was an attempt to pre-ordain events, in order to establish right and thus minimise the potential for civil strife in the event of the king's premature death. It is thus tempting to read into it an acknowledgement that the community was not, in fact, united behind the king.

The Irish venture, of course, did not succeed and, with Edward Bruce's death in 1318, the Scottish succession had to be settled again. A parliament met for the purpose on 3 December 1318, when another entail was made.[3] Failing heirs male to the king, the succession was to revert to the heirs of Marjorie (she herself having died in 1317, leaving one infant son, Robert). An attempt was made to circumvent any repetition of the tribulations of Alexander III's minority by naming the earl of Moray, Thomas Randolph, (failing whom, James Douglas) as Guardian should the young Robert succeed as a minor. There must have been fervent hopes that he would not do so, for although Randolph and Douglas were high in the king's favour and prominent members of the community, there must still have been those who would either oppose the succession of a Bruce/Stewart minor in favour of an adult of an alternative family, or at least jostle for a share in the fruits of power. It is significant that this entail ran under the name of the king and community, and bore the seals of an eminent cross-section of the senior clergy and nobility, but not, apparently, that of the king.

By the time the Declaration of Arbroath had been dispatched to the Curia, then, the situation in Scotland was far from being the same as it had

1   *RRS*, v, no. 391.
2   *RRS*, v, no. 564; conveniently translated in Barrow, *Robert Bruce*, p. 314; see also ibid., p. 379 (n. 9).
3   The text of the 1318 entail is printed, with a discussion, in *RRS*, v, no. 301. See also *RRS*, v, pp. 62-3.

been in 1314. Certainly, there had been military success, and the king did appear to have Scotland well under his control. But despite this apparent success, the relationship between king and community must nonetheless have been fraught with tension. Seemingly endless warfare, with its consequent drain on both economic and human resources, and its disruption not only of regular commercial and agricultural activity, but also of lucrative cross-Border landholding; the diversion of resources to the futile Irish expedition; the shadow of continuing ecclesiastical censure; the lack of a sure succession and the apparently increasing likelihood of another minority (the king was forty-seven in 1320); perhaps also the mistrust and jealousies inherent in a community dependent on royal patronage in the aftermath of war — all these factors must have imposed strain, and perhaps a sense of foreboding. By 1320 King Robert had indeed led the country through the trials and perils described so vividly in the Declaration, and for almost a decade and a half had single-mindedly pursued a policy which, it was claimed, they all espoused. That policy, they declared, was paramount, more important than the man. It was, however, pursued at a cost which had been, and continued to be, borne by them all.

Robert I's sense of realism in his dealings with the Scottish community is evident in his recognition of the part it must play in, for instance, policy regarding the royal succession. As stressed in the Declaration of Arbroath, that power was real enough; the settlement of the great affairs of state was of necessity more of a communal task than it seems to have been in ages past. When the tensions within the community manifested themselves in conspiracy and revolt, however, the personal position of power which the king had assumed became important, and it was in his interest to assert it.

The various independent accounts of the conspiracy involving some prominent members of the community in 1320 do not provide a clear picture of what happened. Taken together, they indicate that a plot to dispose of the king and enthrone in his place Sir William de Soules (son of the Nicholas de Soules who had registered a claim to the throne in 1290) had been foiled by the disclosure of the conspiracy by Murdoch earl of Menteith.[1] There are, however, various difficulties with this story, most obviously the far-fetched notion that the Scottish community would have accepted de Soules as its king. It is not impossible that if there was an English involvement in this plot (which seems likely, given that Murdoch is supposed to have returned to Scotland from English exile specifically to betray the plot to Robert I[2]), de Soules may have been offered some sort of lieutenantship, held of the English crown (as had been held by various English magnates after 1297), in return for ousting Bruce. De Soules came from a family which had supported the Comyns in the minority of Alexander III, had provided two auditors for Robert Bruce 'the Competitor' in the Great Cause, and had been

---

1   *Scalacronica* (Maxwell), p. 59; *Chron. Fordun*, i, pp. 348-9; *Barbour's Bruce*, iii, pp. 208-12.
2   See Reid, 'Political Role of the Monarchy', p. 417 (n. 38).

a bastion of the pro-Balliol Guardianships of the early years of the fourteenth century. Perhaps, then, the English link is less persuasive than a Balliol one. Was de Soules's position to have been analogous with that of his more famous great-uncle — Guardian of Scotland in the name of Edward Balliol? There seems to be no evidence on which to base a conclusion. We can, however, speculate on the reasons for the conspiracy, and examine the king's reaction to it.

Dissatisfaction with the prolonged warfare, and the other tensions associated with the post-Bannockburn period outlined above, may well have given enough incitement for an attempt to replace Bruce with another who, it must be assumed, would adopt a more conciliatory line with the English, and who might avert the looming spectre of a minority. When Edward Balliol made his bid for the throne in 1332, his support came mainly from the lords who had lost Scottish lands through disinheritance by Bruce. It may not be unreasonable to assume that in 1320 there would also be those who might hope for the recovery of lost English lands.

The sword of justice was wielded firmly by Robert I.[1] Three of the conspirators were executed (and also Sir David Brechin, who was not, it seems, a conspirator, but had failed to inform the king of the plot), two were imprisoned for life, one had died before trial, and one escaped trial by fleeing to England. What is perhaps more interesting, however, is that five others were acquitted. This does not, therefore, bear the hallmark of a witch-hunt; there was no panic-reaction on the part of the king, but rather a swift and efficient use of the judicial process in a manner which undoubtedly asserted his personal authority, but which, equally importantly, underlined the judicial authority of the community in parliament. A small section of the community had apparently tried to assume responsibility for a radical change of direction in the country's government — a change which the community did believe it had the right to make, but only with the consent and support of the community as a whole. The actions of this group were therefore perceived to have been aimed not only at the king's person, but also at the royal dignity — the concept of sovereign authority wielded and delegated by the community.

It is tempting to see the year 1320 as a turning point in the career of Robert I. The fulsome support proclaimed by the community in the Declaration of Arbroath and the personal triumph of the king over the conspirators seem to indicate that at last he had completed his mastery of the community; now, perhaps for the first time, he had the kingdom fully and irresistibly under his control. To make too clear a distinction between the years before 1320 and those which followed, however, would be a mistake, for the evolution of the relationship between king and community which is so clearly evident in 1320 had taken place gradually in the years between c.1249 and c.1315, and no fundamental change in the relationship can be

---

1   The narrative sources complement rather than contradict each other with regard to the punishments meted out to the conspirators. There is no formal record of the proceedings.

discerned in the 1320s.

For example, the main corpus of legislation extant from the reign — from the parliament of 1318 — contains few surprises in the context of later events. As well as the standard protection of the rights and liberties of the Church, and the declaration that justice was for rich and poor alike, one finds protection of royal rights in pleas pertaining to the crown, the treatment as treason of rumour or conspiracy 'through which discord may spring up between the lord king and his people', and the requirement that nobles in dispute should bring their disagreements to court rather than create civil disturbance. (It is perhaps significant that this last act makes specific reference to the strife which followed Alexander III's death.) Other acts regarding the execution of justice had, of course, the dual function of increasing stability through a strengthened judicial arm of government, and of boosting royal revenue. There can be no doubt that much of the legislation of 1318 was intended to bolster the position of the king. Other acts, however, were plainly concerned with the well-being of the community: those concerning the inheritance (and disinheritance) of land were intended to resolve some of the confusion which resulted from decades of strife; those concerning the punishment of offenders from within the king's host, and payment for the supplies taken for its sustenance, were measures aimed at protecting the people of the land from some of the scourges of warfare; and those concerning the prevention of the abuse of royal office in judicial matters seem to be similarly in the interests of the 'common weal'.[1]

The legislation, then, is generally in the interests of the community, but undoubtedly enhances the position of the king: if by promoting effective trade and a firm and efficient judicial system, for instance, the king was able to benefit the realm, then those acting in prejudice of the king became common foes of the kingdom as a whole, and the community at large would welcome their subjugation in the common interest. The recognition and achievement of this mutual dependence was crucial to the successful establishment of a king's authority: without the community's support, which was won by good government, there could be no effective kingship. If that had been true under Alexander III, it was all the more true several decades later. Fordun claimed that under Alexander III malefactors would 'put a rope round their necks, ready for hanging, were that his will and pleasure, and bow themselves under his rule'; the king's leadership was based not on the sword, but on firm and prosperous rule, on the 'fear and love' with which he was regarded and on the 'security in steadfastness of peace and quiet' which his reign achieved.[2] Robert I could not claim a peaceful reign, but he nevertheless sought to re-establish the governing balance which created the conditions for at least internal stability. That the de Soules conspiracy did not command more support, and that its vigorous suppression

1   *RRS*, v, no. 139.
2   *Chron. Fordun*, i, pp. 304-5.

seems to have caused little stir, perhaps signify his success.

The parliament held at Cambuskenneth in July 1326 provides further evidence of the cooperation between king and community. Just before the parliament met the king ratified a treaty with Charles IV of France.[1] The treaty, negotiated by envoys on behalf of the kings, their successors and the whole communities of both realms, was made in the interests of peace and stability, and in the common interests of both realms against their long-standing adversary, the king of England. In providing for mutual aggression and defence against England, the treaty represented a resolute adherence to the policy espoused by Bruce throughout his reign, and gave no ground to those who may still have favoured a more pliable approach to the question of English recognition of Bruce's kingship.

The acts of the 1326 parliament shed additional light on the balance of power between king and community. On 15 July two measures were agreed by the parliament: an annual tax of a tenth of revenues from land was granted to the king for the sustentation of the court (and presumably its policies) in return for the cessation of various other exactions; and the succession was once again settled, in a reaffirmation of the entail of 1318.[2] By this time the king had a two-year-old son, who, under the terms of the 1318 entail, would succeed his father. It seems likely that in 1326 the re-issue of the 1318 document was intended to confirm the positions of both David as heir-apparent (perhaps a contentious issue) and Robert Stewart as regent. The firm insistence on the succession of Robert I's infant son is a sign of the personal power and leadership of the king. Here was a ruler by conquest now establishing his dynasty. That it was done in parliament, again with the explicit approval of the community, demonstrates not only the king's ability to carry the community with him, but also his realistic assessment of his own position: for the king's personal ambition to be realised, the power and consent of the community was necessary.

The indenture relating to taxation, which was transcribed for the 1328 parliament 'for future memory' ('in testimonium et memoriam futurorum'),[3] was a manifestation of the personal authority of the king. In agreeing to the tax only for the king's lifetime, the parliament explicitly recognised the special deserts of Robert's personality, as opposed to the needs of 'the crown'. The wording of the document, stressing as it does the arguments put forward by the king, and the discussion of their merits, before recording that the grant was made 'unanimously, thankfully and cheerfully', further testifies to the personal respect commanded by Robert I. His agreement to abide by the conditions that no grants were to be made out of the tax and that the money must pass through the Exchequer, along with the cessation of prises and carriages without payment, indicates his own equal respect for the community. The stated reason for the royal penury was that the transfer of

1   *RRS*, v, no. 299.
2   *RRS*, v, nos. 300-1, 335.
3   *RRS*, v, no. 335; translated in Donaldson, *Scottish Documents*, pp. 59-60.

lands and revenue on account of the war had reduced the crown's income. The winning or rewarding of support through patronage must have been a primary element in this, and the community were here extracting an undertaking that this grant, specifically for the king's own use, would not be similarly used. The revenues of the kingdom were not to be diverted into baronial hands. The involvement of the burgesses in this agreement adds further weight to the point. Burgesses may well have been attending parliament for several decades by this time, but are rarely specifically referred to.[1] Their close involvement here is indicative of the fact that this grant was far from pure rhetoric: it was a financial deal, in the interests of both king and community, requiring the authority and support of all of those who would be party to it. In a sense it represents a modernisation of the system for the support of the crown: in lieu of the sometimes unfairly burdensome prises and carriages, a single blanket levy was to be made, of a previously agreed amount.

The final achievement of Bruce, the high-point of his reign, came on 17 March 1328, with the conclusion of the treaty of Edinburgh.[2] This brought an end to the war and, finally, recognition by the English king of Bruce's royal status. Such a satisfactory conclusion had not been inevitable. The years after 1320 saw several attempts at negotiating a settlement. As mentioned above, the negotiations at Bamburgh early in 1321 seem to have been taken seriously, but apparently foundered on Scottish intransigence. A planned second round of talks in September of that year do not appear to have taken place. Late in 1321 the Scots had discussions with the rebellious earl of Lancaster, presumably aimed at providing him with military assistance in return for concessions in the event of his success against Edward II.[3] At the end of the year, the existing truce having expired, the Scots again invaded: on this occasion Edward II gave a commission to Andrew Harcla to treat for peace or truce, and once again prepared a counter-attack.[4] Throughout 1322 there was military action on the Border, and in June an English force reached Edinburgh, only to retreat in disorder and suffer rout at Byland.[5] Late in the year Harcla, perhaps representing the frustration of many northerners with the futility of Edwardian policy towards the Scots, carried out negotiations resulting in a personal treaty, remarkably similar in its terms to the one made five years later.[6] For this act of treason, Harcla lost his life. Further negotiations followed in 1323, leading to the thirteen-year truce made at Bishopthorpe in May between 'us [Edward II] and the people of England, Wales, Gascony and Ireland on one side, and the nuncios of Scotland and the land and people of Scotland on the

1 See Barrow, *Robert Bruce*, p. 300.
2 *RRS*, v, nos. 342-3.
3 *Foedera*, ii, pp. 463, 472, 474-6, 478-9.
4 Ibid., pp. 473-5.
5 For a description of the events of the summer of 1322, see Barrow, *Robert Bruce*, pp. 243-4.
6 The treaty is printed in Stones, *Anglo-Scottish Relations*, no. 39.

other side'.[1] The initiative for these talks came from Bruce, and in that context the English form of words implies a recognition of unity among the Scots, and also of Bruce's power to represent them (if not his title of 'king').

The truce seems to have held fairly well, and was clearly seen as a staging-post towards a final peace, for which negotiations continued. Eminent embassies met for discussions at the end of 1324,[2] and there was, in accordance with the terms of the truce, an English attempt to lessen the papal pressure on the Scots,[3] as well as further Scottish diplomacy in that quarter, which had resulted, as early as January 1324, in the pope at last agreeing to accord Bruce the title 'king'.[4] Deadlocked again, however, the talks were abandoned early in 1325, with the Scots refusing to submit to papal arbitration.[5] In autumn 1325, and again in the spring of 1326 the Scots broke the truce with action in the northern counties,[6] which must be viewed in the context of the treaty of Corbeil, signed in April 1326. Edward II, again at war with France and increasingly beleaguered at home, tried to maintain the truce. In the end, he appears to have planned to accede to Bruce's demands in return for help against Isabella and Mortimer — but the plans did not come to fruition before his deposition early in 1327.[7]

Edward III, 'counselled' by Isabella and Mortimer, also ratified and attempted to uphold the Bishopthorpe truce, and soon prepared for further negotiations.[8] The Scots, however, used the well-worn tactic of attempting to force the English into a settlement through military pressure. The English responded in kind, the talks collapsed, and there immediately followed a Scottish invasion and embarrassment for Edward III in Weardale.[9] While this campaign was in progress, Bruce himself was exerting pressure once again in Ireland, where he forced the English seneschal of Ulster to seek a truce.[10] On his return, Bruce personally led a further attack on Norham.[11] The combined effect exerted enough pressure on the weak English government to force it to the negotiating table again, resulting in the English climbdown and the treaty of Edinburgh.

The treaty is a reflection of the political situation in both countries in 1328. The English accepted the abandonment of their claims to dominion

---

1   *Foedera*, ii, p. 521.
2   Ibid., pp. 570, 577-8.
3   *CDS*, iii, no. 855.
4   *Foedera*, ii, p. 541. See also Reid, 'Political Role of the Monarchy', p. 375 and n. 107.
5   *Foedera*, ii, p. 595.
6   Ibid., pp. 609-10, 626-7; *CDS*, iii, nos. 882-3.
7   *Chron. Lanercost*, p. 259.
8   *Foedera*, ii, pp. 650, 689, 695-6; PRO, Scotch Rolls, C. 71/11, m. 10. See also Reid, 'Political Role of the Monarchy', pp. 377-8 and nn. 118-20.
9   *Foedera*, ii, p. 712. A detailed description of the campaign is given in R. Nicholson, 'The Last Campaign of Robert Bruce', *English Historical Review*, lxxvii (1962); but see also Reid, 'Political Role of the Monarchy', pp. 378-9, and particularly n. 128.
10   *CDS*, iii, no. 922.
11   *Rot. Scot.*, ii, pp. 221-2; *Historical Papers and Letters from the Northern Registers*, no. 221.

over Scotland because they could no longer sustain a war in the north. The Scots, on the other hand, were prepared to accept less favourable terms than those of Bishopthorpe, and to pay a large sum of money. For this final settlement of the one basic contention, they were prepared to pay, perhaps galvanised at last into action by the serious illness of their king.[1] Finally, the king and community together had restored the kingdom to the stability, and in some measure to the prosperity, of Alexander III's reign, and had re-established international recognition of Scottish sovereignty.

The success of Bruce's kingship lay precisely in the balance of personal power with regal authority which he was able to maintain. In recognising, and accommodating, the increased self-reliance of the community, and its heightened sense of political involvement, Bruce was able to command its support and respect as the one to whom the community delegated its sovereign authority. He further gained personal power and respect through conquest and warlike ability, resolute and consistent pursuit of foreign policy, and firm, effective administration of the fiscal and judicial affairs of the land. His success in reversing the catastrophe of the latter years of the thirteenth century, and apparently returning to Scotland to at least a reflection of what were seen as the halcyon days of Alexander III, gave added charisma not only to his personal authority, but also to the monarchy, and thus to the standing of the community itself.

It should not be thought that this was the achievement of Robert I alone. Throughout the previous two centuries the increasing centralisation of authority, with the spreading influence of Anglo-Norman custom, the development of a European trading economy, and the geographical unification of the kingdom, had increased the status of the monarch as the focal point of the community. The community itself had also been strengthened, particularly during the reign of Alexander III and the periods of Guardianship which followed it. The difficulties of 1286-1304, when the Scots had repeatedly been thrown back on their own resources, seem to have created a self-reliance within the community which made impossible general acceptance of the English claims to superiority. The strength of the community was, admittedly, eclipsed in 1304-5, but by means of the successful uprising led by Robert Bruce between 1306 and 1309, that eclipse was reversed. At the same time, a strong personal monarchy was established, asserting dynastic continuity, which was important for justifying the war both at home and abroad. The monarchy needed more than powerful bases of personality and dynasty, however; it also required a theoretical basis. In the context, therefore, of the personal power accorded the king, and of his theoretical right to rule, the community played a vital role. As the king needed the community, so the community needed the king, and thus a powerful inter-reliance developed as both parties gained strength and

---

1   For discussion of Robert I's illness, see Barrow, *Robert Bruce*, pp. 322-3, and the sources cited.

security one from the other — the community as the enabler of the king, the king as the upholder of the community. By the 1320s this inter-reliance was at its strongest, and it finds its clearest expression in the Declaration of Arbroath. A useful comparison can perhaps be made here with John Balliol. Balliol, in his inability to resist the English king's interference, had failed to uphold the community of the Scots. The community in turn did not continue to allow Balliol the freedom and power to rule, and although his name was used as the figurehead of the community's own governmental efforts for many years, the final abandonment of him in favour of another who could fulfil the role was probably inevitable.

The relationship of crown and community in the 1320s thus represents a step forward in the constitutional history of Scotland. In referring to his ancestors and predecessors on the throne of Scotland, Robert I called upon a dynastic legitimacy and continuity with the past. But his achievement was not to restore to Scotland the rule of those older kings. They were in the past, and the nation had moved on to find new strengths. In subsequent decades and centuries, however, the contract nurtured between king and community during the first quarter of the fourteenth century appears to have been diluted; it is for that reason that the 1320s should be regarded as a high-point in the history of the Scottish monarchy.

11

Scotland without a King, 1329-1341

BRUCE WEBSTER

The 1330s were a traumatic time for Scotland. The death of Robert I left a successor of five who could not take up his father's role for many years. Whether he ever could remained to be seen, but it was hardly clear that he would get the chance. Robert's leading supporters were no longer young (both Douglas and Randolph had been active in the 1290s, nearly thirty years before) and the prospects for Scotland worsened drastically in October 1330 when the regime of Isabella and Mortimer was overthrown by the young Edward III, who was unlikely to feel committed to the former government's settlement of 1328. Even more ominously, this brought into favour a number of the 'disinherited', dispossessed by Robert I of lands they claimed in Scotland. This potentially formidable group, with a prospective claimant to the throne in Edward Balliol, son of the deposed King John, had naturally been opposed to Isabella and Mortimer and to the treaty of 1328 (some were in exile in France after the revolt of 1329), and the October revolution not only put on the throne a young and vigorous king who might wish to win his spurs in Scotland, but also opened up for the 'disinherited' the chance to push their claims with at least English acquiescence.[1] What followed is well known: two disastrous Scottish defeats, the exile of the young king to Château Gaillard for seven years, and the establishment of Balliol on the throne, nominally in 1332 and somewhat more effectively in 1333, as head of an English-backed regime, which ceded a large part of southern Scotland to the direct rule of Edward III and tried to hold the rest with recurrent English support. The consequence was a series of large-scale English invasions in 1334, 1335 and 1336, with further efforts in 1337 and 1338; the establishment of English garrisons in several prominent castles; and a guerilla war of resistance fought by scattered and not particularly well-coordinated bands of fighters, in a series of struggles which were able to deny effective rule to either side, but whose principal result was widespread devastation in the Scottish countryside, a devastation repeatedly mentioned by all the chroniclers. In the end the English effort faltered. Balliol lost whatever bases he possessed, and the successor of Robert I was able to return in June 1341 to his shattered kingdom — and celebrate by launching retaliatory attacks on northern England, attacks which landed him in 1346

---

1   R.G. Nicholson, *Edward III and the Scots* (London, 1965), pp. 61-5.

once more in exile as a prisoner in England, and opened Scotland yet again to English attacks and internal disturbance, though on a far lesser scale than between 1332 and 1341. The object of the present essay is to look again at these traumatic years, to discover as far as possible what was actually happening in Scotland, and to try to assess the chances and the achievements of each side. What did the English occupation and its puppet regime amount to? What was the state of, and the problems facing, the Scottish resistance? How far was it able to maintain or establish an effective government while the king was a minor in exile? And behind this lies the need to establish these years in their proper place in the Wars of Independence, which were so important in developing a Scottish national identity and in creating the myths which were to reinforce that identity for centuries.

These questions are not made any easier by the very unbalanced sources on which we have to rely. English records are abundant for the English interventions in Scotland, and have been very thoroughly used.[1] Evidence from the Scottish side is much less satisfactory. There are great gaps in the series of Exchequer Rolls, perhaps not surprisingly, though something can be learned from those that do exist. The *Acts of David II* contain only ten documents issued in Scotland between the battle of Dupplin and David's return in May 1341;[2] and there are no contemporary Scottish chronicles for this period: all we have date from the late fourteenth and the fifteenth centuries and are heavily affected by nationalist mythology in their approach, though they contain a great deal of useful detail embedded in what is sometimes very confusing chronology. Our most useful near-contemporary accounts come from two English chronicles, that of Lanercost Priory and the remarkable *Scalacronica* of Sir Thomas Gray. The chief task in trying to understand this period is to penetrate behind the multifold biases of all this material, to try to see what the situation really was.

Those who remained loyal to the Bruce dynasty had very grave problems to overcome. The Scottish nobility suffered losses at Dupplin and Halidon which were to be paralleled only in the ultimate disaster of Flodden. On or around those two disastrous days, no fewer than eight earls, Atholl, Carrick, Lennox, Mar (the Guardian at Dupplin), Menteith, Moray (the son of the first Randolph earl), Ross and Sutherland, were killed, as were — among others — Archibald Douglas, brother of Sir James and Guardian at Halidon, Robert Bruce of Liddesdale (illegitimate son of Robert I), Nigel Bruce, three Frasers, three Stewarts and three Setons.[3] It was a shattering series of losses. After the death of Thomas Randolph in July 1332 the Scots, already deprived of Sir James Douglas in 1330, had had problems in any event in

---

1   E.g., in D. Dalrymple, *Annals of Scotland* (3rd edn, Edinburgh, 1819), ii; Nicholson, *Edward III*; R. Nicholson, *Scotland: The Later Middle Ages* (Edinburgh, 1974).
2   *RRS*, vi, nos. 11-19, 22.
3   Nicholson, *Edward III*, pp. 89, 137; see also *Chron. Wyntoun* (Laing), ii, pp. 388, 402; *Chron. Bower* (Goodall), ii, pp. 305, 311; *Chron. Lanercost*, p. 274.

finding suitable leaders, particularly of a rank to act as Guardians during the inevitably long minority. The council that met under the threat of invasion by the disinherited was wracked by 'gret and lang dyssentyown'[1] before it agreed to make Donald earl of Mar Guardian. Although Mar was the young king's uncle, it was a remarkable choice because he had been largely brought up in England at the court of Edward II, with whom he became friendly; he had refused a chance to return home in 1314, and was even alleged to have encouraged Balliol to pursue his claim! He had returned to Scotland only after Edward's deposition, when Robert I generously restored him to the earldom.[2] True, the Lanercost Chronicle says that, once elected, he attached himself to the side of David II;[3] but he does not seem to have had any military talents to compensate for his dubious history, and he was in any case killed at Dupplin after what can have been only a few days in office. We have no evidence about how any of his successors emerged, or about the principles that guided their choice. The heir to the throne after David II, Robert the Stewart, was presumably regarded at this time as too young. The only possible earls would seem to have been Patrick earl of March and Hugh earl of Ross; but although Earl Patrick is described as Guardian of southern Scotland (that is, south of the Forth) in two accounts of the Dupplin campaign[4] and got his troops to Auchterarder, he failed to join up with Mar and so contributed to the disaster. Hugh earl of Ross is only mentioned to record his death at Halidon. The next Guardian, Robert I's companion Andrew Murray 'le Riche', was in fact the one who seems to have had some capacity for leadership. Married in 1326 to Robert's sister Christiana, he was eventually to pull together at least some of the threads of Scottish resistance before his death in 1338; but on this occasion he was given little chance, for he was captured in an attack on Balliol at Roxburgh in the autumn of 1332[5] and remained a prisoner till the autumn of 1334. He was replaced by Archibald Douglas, brother of Sir James, who was in turn killed at Halidon. The succession then becomes unclear. The other obvious candidates were Robert the Stewart, now aged seventeen, the son of Walter the Stewart and Marjorie Bruce and heir to the throne after David II, and John Randolph, younger son of Thomas Randolph the first earl. There is no evidence, however, of any Guardian in 1334; the next references are in three charters of 1335, one issued in April by the Stewart as Lieutenant, and two in May by the Stewart and John Randolph jointly.[6] Randolph, however, was captured in the summer of 1335; and even before that there were problems. A Bruce parliament at Dairsie, early in 1335, was bedevilled by

1   *Chron. Wyntoun* (Laing), ii, p. 384; see also *Chron. Fordun*, i, p. 354.
2   For Mar's earlier career and involvement with the 'disinherited', see G.W.S. Barrow, *Robert Bruce and the Community of the Realm of Scotland* (London, 1965), pp. 385-6; Nicholson, *Edward III*, pp. 63-4, 81-2.
3   *Chron. Lanercost*, p. 267.
4   *Chron. Wyntoun* (Laing), ii, pp. 385-6, 389-90; *Chron. Pluscarden*, i, p. 265.
5   *Chron. Lanercost*, p. 270; *Chron. Wyntoun* (Laing), ii, pp. 396-7.
6   *RRS*, vi, nos. 11-13.

the complex consequences of a quarrel in the previous year among the 'disinherited' which left David earl of Atholl vulnerable to attack and so brought him temporarily and under compulsion on to the Scottish side! He was vehemently hostile to John Randolph, and contrived to make the Stewart join him at this parliament in a violent quarrel with Randolph, a quarrel which is reflected in a cryptic reference in the Exchequer Rolls.[1] Randolph was captured by the English in the summer of 1335, but the consequences of the split remained. When Atholl made his peace with Edward III that August he seems to have persuaded the Stewart at least to sound out the possibilities of his coming over also. Indeed, according to one chronicle the Stewart actually did so, though we have no other evidence of how far things went.[2] It seems at least clear that the Stewart ceased to be Guardian around this point.

So far the tale of organised Scottish resistance was a sorry one. No Guardian seems to have survived long, and the Stewart was not in fact the only noble to have played a doubtful part at some point. Duncan earl of Fife went over to Balliol's side immediately after Dupplin and actually took part in his coronation, though he was soon recaptured and apparently imprisoned by the Scots.[3] He, too, may have been in touch with Edward III in the autumn of 1335,[4] but seems consistently to have opposed the English from the summer of 1336, taking part then in attacks on Cupar castle,[5] and, in the following year, on Falkland, St Andrews and Leuchars.[6] Patrick earl of March also made his peace with Edward III after Halidon and remained a visibly active supporter of Balliol till the autumn of 1334.[7] These uncertainties and hesitations among the Scottish leaders scarcely betoken an effective national resistance. Not until Andrew Murray had resumed the Guardianship late in 1335 and won the striking victory at Culblean, is there much sign of any consistent leadership. And when Murray died in 1338, he was succeeded by the Stewart, some years older by now, but not, it would appear, much more effective as a leader.

Not surprisingly in the years between 1333 and 1338 there is little indication that the formal processes of Scottish government could continue. Only a handful of acts issued on behalf of the king survive: the three by the Stewart and Randolph in the earlier part of 1335 and one by Murray in December of the same year.[8] Thereafter no act is found before September 1338. The first surviving sequence of Exchequer records ends in March 1333; the next audit on record is one held at Aberdeen at the turn of 1337/8,

1   *ER*, i, p. 435; see also *Chron. Fordun*, i, p. 358.
2   *Scalacronica*, p. 166.
3   *Chron. Lanercost*, pp. 269-72.
4   *Rot. Scot.*, i, p. 380.
5   *Chron. Lanercost*, p. 285.
6   *Chron. Fordun*, i, p. 362; *Chron. Bower* (Goodall), ii, p. 324; *Chron. Pluscarden*, i, p. 283.
7   Nicholson, *Edward III*, pp. 127-8; *Chron. Lanercost*, pp. 277-8; *Chron. Bower* (Goodall), ii, p. 317.
8   *RRS*, vi, no. 14.

with most accounts going back to Martinmas 1336.[1] In only one case, that of the chamberlain, does the account claim to cover the period from Martinmas 1334; and it asserts that there was no return to be made for the 'year in the time of the king of England', that is, for the period from Halidon to the autumn of 1334. The receipts are very sketchy — payments from the sheriffs of Aberdeen and Kincardine, and various sums collected by an under-chamberlain, with whom indeed the chamberlain seems to have had a serious dispute over his accounts which provoked a detailed reply.[2] The records betoken some attempt to get accounting going again from 1336, but it was confined to the north-east from Banff to Kincardine and the records, such as they are, reveal more about the gaps and failings in administration — understandable in the circumstances — than about the workings of any system. It is clear from references that in the difficult year 1335 the Stewart and Randolph seized as Guardians much of such revenues as could be collected into their own hands, without its appearing in the accounts.[3] They could probably do little else.

There was, nevertheless, a consistent line of Scottish resistance in this period, for at no point were Edward Balliol and Edward III allowed to dominate without opposition. Of what did this resistance consist?

The immediate consequence of Halidon was to set up an English-backed 'Quisling' regime under Balliol which ceded a substantial part of southern Scotland to Edward III. (Perhaps Vichy would be a closer parallel!) To this, as we have seen, the natural leaders could oppose only a faltering resistance, though with difficulty a skeleton of a government on behalf of David II was maintained. It is true that very few of the Scottish leaders rallied genuinely to Balliol, though many of those who survived Dupplin and Halidon contemplated a deal at some time, whether seriously or not. Every source stresses that in the winter of 1333 the only castles in the hands of the 'Scots' were Dumbarton, held by Malcolm Fleming (where David himself and his wife were then sheltering); Loch Leven, held by Alan Vipont; Kildrummy, held by Christiana Bruce, Andrew Murray's wife; Urquhart, held by Thomas Lauder; and the pele of Lochdoon, held by John Thomson.[4] At this point Andrew Murray and William Douglas of Lothian were prisoners; the Stewart seems to have been in hiding;[5] and the earl of March was actively cooperating with Edward III. The very situation determined that there could be no united and organised defence against Balliol and the English. The only option was isolated and uncoordinated guerilla warfare, a resistance movement with all the difficulties and conflicts with which we have become familiar in the present century. As is shown by the experiences of France

---

1   *ER*, i, pp. 440-54.
2   *ER*, i, pp. 435-9.
3   *ER*, i, p. 435.
4   *Chron. Wyntoun* (Laing), ii, p. 404; *Chron. Bower* (Goodall), ii, p. 311.
5   Ibid., p. 313.

and Yugoslavia during the Second World War, in such circumstances political uncertainties and personal feuds are almost inevitable. It was that kind of resistance movement which developed in Scotland. The Stewart, escaping from his own estates, took refuge in Dumbarton, where he was joined by various supporters from Renfrewshire, Ayrshire and Argyll; then, what seems to have been a spontaneous rising in Bute and Ayrshire gave him the chance to put himself at its head.[1] By late 1334, William Douglas of Lothian, who had been captured at Lochmaben early in 1333, was released and made sporadic attacks on the Galwegian supporters of Balliol, and on English garrisons within his home territory of Lothian.[2] Also in 1334, Patrick earl of March reverted to the Scottish side, providing it with another centre of resistance at Dunbar (there were English garrisons at Edinburgh, Berwick and Roxburgh). Farther north, Andrew Murray, following his victory at Culblean in November 1335, moved into Fife and the Mearns.[3] And in Lothian another leader, Alexander Ramsay, emerged and eventually based himself at Hawthornden near Roslin, where many others joined him, some to have distinguished careers later on in David II's reign.[4] All this amounts indeed to a substantial resistance movement; but it consisted of isolated and often uncoordinated guerilla groups, joining together for occasional dramatic efforts, like the relief of Kildrummy in 1335 and the battle of Culblean that followed, but occupied mainly in scattered guerilla fighting and the inevitable widespread ravaging which reduced all the areas involved, if we may believe the chronicles, to a wilderness. John of Fordun (himself probably from the Mearns) describes the consequences of Andrew Murray's campaign of 1336, in which he recovered the castles of Dunnottar, Kinnef and Laurieston, as follows: 'and thus by the continual depredations of both sides the whole land of Gowrie, Angus and Mearns was reduced to almost irredeemable devastation and extreme poverty.'[5] His otherwise glowing obituary of Murray returns to the same theme: 'but every region which he traversed in his campaigns he reduced to such desolation and barrenness that more people died of famine and hunger than those who were destroyed by the sword in the fighting.'[6] Similar comments were made of other campaigns and leaders; destruction is an inescapable consequence of such warfare.

The picture, however, is of warfare between scattered bands of resistance fighters and of English and Balliol Scots: the latter in castles which they fortified or rebuilt whenever possible, the former generally in

1   *Chron. Wyntoun* (Laing), ii, pp. 415-16; *Chron. Pluscarden*, i, p. 275. All the Scottish sources are very confused over the dating of events at this point, placing many of them in 1335. Nicholson, *Edward III*, pp. 163-6, places the risings in 1334, which makes good sense, despite the chroniclers!
2   *Chron. Lanercost*, p. 278; *Chron. Wyntoun* (Laing), ii, pp. 417-18; *Chron. Bower* (Goodall), ii, p. 317.
3   *Chron. Fordun*, i, pp. 360-2; *Chron. Bower* (Goodall), ii, p. 324.
4   *Chron. Wyntoun* (Laing), ii, pp. 460-2; *Chron. Bower* (Goodall), ii, pp. 333-4.
5   *Chron. Fordun*, i, pp. 361-2.
6   Ibid., p. 363.

hiding in woods and caves, apart from their two strong lowland fortresses of Dumbarton and Dunbar. Neither the chroniclers nor what we can piece together of events suggests any national movement. The evidence, as we would expect, indicates that most Scots were attempting to survive. The inhabitants of Lothian accepted the presence of English garrisons; when they could not avoid it they paid their dues to English sheriffs and when necessary paid protection money to the garrison of Dunbar and other guerillas as well.[1] The majority kept their heads down and became involved as little as they could, though they no doubt provided support unobtrusively when possible.

But if the Scots faced difficulties, Balliol with his English-supported government faced at least as many. At first, the situation had not seemed hopeless. The Bruce party appeared overthrown at Dupplin; Perth was entered and occupied; Balliol had support from his traditional family lands in Galloway; and Eustace Maxwell in particular was to be invaluable.[2] At least one Perthshire laird, Andrew Murray of Tullibardine, had helped Balliol by indicating the presence of a ford in the Earn, which enabled his army to cross and so did more than probably anyone else to bring about Balliol's victory at Dupplin[3] — a contribution for which Murray later paid with his life. Immediately after the battle, Duncan earl of Fife came over, bringing with him not only the possibility of enthronement, traditionally the prerogative of the earls of Fife, but the support of at least thirteen lairds from Fife and Perthshire: John de Bonneville, William of Cambo, John of Dunmore, William of Farsley, David Graham, John of Inchmartin, Alexander Lamberton, John de Laundells, Walter Lundie, Roger Morton, Michael Scot, and David and Michael Wemyss. In addition Balliol received backing from the bishop of Dunkeld and the abbots of Scone, Coupar Angus, Dunfermline and Inchaffray.[4] A document in the St Andrews Priory cartulary describes the consequences thus: 'there followed the lordship of Edward Balliol and Henry Beaumont and the other English', as a result of which the priory had not been able to levy the profits due to it from the church of Fordoun 'for fear of the said Lord Henry'.[5]

To an observer in Fife it must have seemed that way; but the support for Balliol was clearly limited to Galloway, Perthshire and Fife. Dupplin was won on 12 August; Balliol was crowned at Scone on 24 September. It is difficult to sort out what happened next. According to later Scottish chronicles, a Scottish force under James and Simon Fraser and Robert Keith recaptured Perth on 7 October.[6] Lanercost gives a more complex story,

1   *Chron. Lanercost*, p. 278.
2   Ibid., p. 269.
3   *Chron. Bower* (Goodall), ii, p. 307.
4   *Chron. Lanercost*, p. 269.
5   *St Andrews Lib.*, pp. 399-400. The whole document provides an interesting glimpse of the consequences of Dupplin and Halidon.
6   *Chron. Fordun*, i, p. 355; *Chron. Wyntoun* (Laing), ii, p. 394; *Chron. Bower* (Goodall), ii, p. 307.

according to which a Scottish force besieged Balliol in Perth; he was rescued by a force of Galwegians under Eustace Maxwell, in revenge for which Patrick earl of March, Thomas the new earl of Moray, Andrew Murray and Archibald Douglas ravaged Galloway, while Balliol fortified Perth and, leaving the earl of Fife in charge there, set off on a perambulation north of the Forth. In his absence, the Scots regained Perth, captured the earl by a stratagem, and burned the town. Balliol, on returning south, withdrew to Roxburgh on 14 October and took up quarters in Kelso.[1] There was probably just time for all this to have happened, if the sequence of sieges goes back to soon after Dupplin itself; but, in any event, the perilously narrow support for Balliol is clear, and the sequel is well known. On 17 December, he was overtaken by a strong force at Annan and sent in humiliating flight to England.[2] His regime, if it was to exist, had to be a Quisling one.

Edward III's overwhelming victory at Halidon made such a regime inevitable. It was the first of Edward's battle triumphs, and was greeted by the poet Laurence Minot as a fit revenge for Bannockburn.[3] For a time it seemed as if that victory had indeed destroyed the independence thought to have been won in 1314. Balliol's position was now clearly that of a subject king. Not only had he to pay homage, which he duly did in June 1334, having gone through the ritual of presenting excuses (no doubt genuine enough) when he was summoned to perform the ceremony at the York parliament in February; but he also had to cede to Edward a large part of southern Scotland into direct English rule, a rule of which the fact that Berwick is now in England is a reminder. Balliol was to rule the remainder of Scotland as Edward's vassal.[4]

It is not easy to be certain how effectively either part was governed. Naturally, there are English records of the 'English' sheriffdoms, which show administration operating within them.[5] Only one, Dumfries, had a native Scot as sheriff, the strong Balliol supporter Eustace Maxwell, who accounted from 15 October 1335. The revenues of Dumfries, Roxburgh and Berwick seem small, with more excuses than renders; but those of Edinburgh with the constabularies of Haddington and Linlithgow, were more ample, with receipts running at over £300 *per annum* for 1335-6 and 1336-7. Here English authority seems to have possessed some meaning. For Balliol's own government, however, we have only fragmentary evidence. At the end of 1333 or 1334, he appears to have appointed William Bullock as chamberlain, the key financial and thus administrative official in Scottish government. Bullock remained on Balliol's side, presumably active in

1   *Chron. Lanercost*, pp. 269-70.
2   Nicholson, *Edward III*, pp. 103-4.
3   *The Poems of Laurence Minot*, ed. J. Hall (Oxford, 1887), pp. 4-6.
4   Nicholson, *Edward III*, pp. 155-62.
5   *CDS*, iii, pp. 317-93.

administration and certainly holding Cupar castle for Balliol, until he was persuaded to rally to David II's side in 1339.[1] If he contrived to hold any Exchequer audits we have no evidence. We know only, as we have seen, that Balliol's government managed to prevent the holding of normal audits by their opponents. We also know, fortuitously, that Alan de Lisle was made sheriff of Bute and Cowal around the same time that Bullock became chamberlain[2] and held the office until killed in a local rising the following summer.[3] On this slender evidence, it looks as though Balliol was trying to set up the elements of an administration, probably on a much wider scale than we know of, for English records do not inform us of Balliol's activities, and later Scottish accounts had little interest in doing so. With hindsight, it is easy to dismiss Balliol's activities as contemptible; yet in 1333-4 we have evidence that 'Scottish' administration came to a standstill, and in 1335 there is much to suggest that the Scottish resistance was on the brink of collapse, with much negotiation going on with Edward III and very few left as continuing resisters. Only with Culblean and the emergence of Murray as an effective leader did the tide begin to turn.

Both Scottish and even sympathetic English accounts such as Lanercost refer to Balliol's supporters as 'English'. In this they were wrong. Even the 'English' garrisons at Edinburgh and Stirling included unquestionable Scots: William Ramsay, Alexander Craigie and three other knights in the Edinburgh garrison of 1335-6, and twenty-four of the men-at-arms in the same account. Many of the same men were still serving at Edinburgh in 1339-40, while the account for Stirling for the latter year includes nine men-at-arms (out of fifty-four) whose names suggest they were Scots.[4] William Ramsay, Alexander Craigie and some others received grants from Edward III to provide for their maintenance,[5] having presumably lost all by their consistent fidelity to Edward's cause. Among Balliol's supporters in Scotland it is possible even from the chronicles to list at least forty-six laymen who backed him for various periods, some indeed short, others much longer.[6] Most of these men, though not all, are of little note; that we can list them at all suggests that there were probably many more. Indeed, more could certainly be added from record sources. Yet the persistent references to them as English or at best 'Scoti Anglicati' reflect the fatal weakness: that this was not only a regime established with Edward III's support, but blatantly a Quisling one, and perceived as such *avant le mot*. Like more recent regimes of that type, however, it was tolerated by the mass

---

1   *Chron. Fordun*, i, p. 364; *Chron. Wyntoun* (Laing), ii, pp. 408, 451-2; *Chron. Bower* (Goodall), ii, pp. 313, 324, 330-1. These all date Bullock's appointment at the end of 1334; but, as we have seen (above, p. 228, n. 1), there is much confusion over chronology at this point, and one would expect Balliol to have appointed a chamberlain soon after he had been re-established on the throne.
2   *Chron. Bower* (Goodall), ii, p. 313.
3   *Chron. Wyntoun* (Laing), ii, p. 415; *Chron. Pluscarden*, i, p. 275.
4   *CDS*, iii, p. 360, and no. 1323.
5   *CDS*, iii, nos. 1351, 1367.
6   For details see the Appendix.

of the population. It did not simply collapse; it had to be pushed.

How then did the 'Scots' triumph? In the Second World War most occupied countries had to be liberated by external force. The only aid of that kind available to the supporters of David II was from France, but that hardly went beyond diplomacy. Philip VI's ambassadors always tried to keep in touch with the Scots in their efforts to resolve the differences with Edward III; but these efforts, and those of the pope, were in the end a total failure.[1] When the negotiations collapsed, there is no sign of active French intervention in Scotland. Philip had provided at Château Gaillard a retreat for the young David safe from English threats; he had always made clear that he regarded David as the rightful king. That was all he did but, as events proved, it was enough.

The moments of gravest danger were immediately after Halidon and perhaps again in the summer of 1335, when it seemed the Scottish resistance was divided and faltering. But 1336 saw Edward III's last personal effort in Scotland for some time, and culminating as it did in the death of his brother John of Eltham in Perth,[2] it may have left him discouraged and anxious to turn to other matters. Admittedly, he did not give up Scotland. Indeed, having established garrisons at Perth, Cupar, Stirling, Bothwell, Lochmaben, Edinburgh, Berwick and Roxburgh, and having demonstrated his authority in the north by sacking Aberdeen, Edward may well have thought it safe to leave Scotland to others — the earl of Warwick in 1337 and the earls of Arundel and Salisbury in 1338. The latter were specifically charged to reduce the castle of Dunbar, which was providing a base for resisters who could exact protection money from, or else ravage, the inhabitants of Lothian despite the garrison at Edinburgh. But when negotiations with the French finally collapsed in 1337, Edward transferred his personal concerns to the Continent, and in July 1338 he himself landed at Antwerp. Though he was to see Scotland again in the early 1340s and in 1356, it was now France that was inevitably his main concern. And without him, the English effort in Scotland lost much of its force.

Yet that effort still remained serious, and the Scottish resistance still had much to contend with, not least the readiness of the Scots themselves to acquiesce in English dominance when they had to. Lanercost's account of Edward III's own descent on Bothwell in 1336 is revealing: in October of that year, Edward, 'thinking, as all experts agreed, that the land of Scotland could be conquered only in winter, took himself and his army to the castle of Bothwell and the western parts. The men of that region, learning of his sudden and unexpected arrival, and being unable to resist him, gave

---

1   See E. Déprez, *Les Préliminaires de la Guerre de Cent Ans* (Paris, 1902), esp. pp. 109-19, 127-8, 132-3.
2   Fordun attributed John of Eltham's death to a violent assault by Edward III himself, provoked by John's having visited with fire and sword areas which Edward had taken into his peace: *Chron. Fordun*, i, p. 361.

themselves into his peace, more from fear than love; he accepted their allegiance, repaired the castle, formerly destroyed by the Scots, and left a garrison.'[1] The most that William Douglas of Lothian could do was lurk in hiding, and pick off a few of Edward's men when he had the opportunity. Without Edward's own presence, however, the situation improved for the Scots in 1337: Andrew Murray was able to recover and slight Bothwell, as well as seize St Andrews and Leuchars. But the essential problem of guerilla warfare against a well-based enemy remained, especially since that enemy still enjoyed at least acquiescence. Cupar in Fife remained in Balliol hands, held for him by William Bullock until Bullock came over in 1339; and attempts to capture Stirling and Edinburgh failed — in the case of Edinburgh 'because of English power backed by the deceit and guile of certain Scots'.[2]

Guerilla warfare does not usually eject an occupying force; its capacity lies in its ability to tie down and harass, in the hope that eventually the energy of the occupiers will be exhausted. For that it does not need the active and committed support of the population. It needs simply the failure of the population energetically to support the occupiers, and passive support of this nature clearly existed in Scotland in the late 1330s.

The turning-point was the failure of the siege of Dunbar in the spring of 1338. Alone, Murray's limited successes of 1337 might have been reversed, especially when his death early in 1338 deprived the resistance of the only energetic overall leadership it had had since 1332. The Stewart, as heir to the throne, was the inevitable successor as Guardian, but his record was not inspiring and it is indeed evident that Murray's death cost the Scots the initiative at a point when it could have been invaluable. Bower's comment on the Stewart is curious: 'though young in years, he showed himself an old man in his deeds, especially against the English'.[3] That can hardly be a compliment, but it is perhaps borne out by the fact that his first recorded action as Guardian was not until the siege of Perth in the summer of 1339. In the circumstances of 1338, the capture of Dunbar by the English might have changed everything: not only would it have removed a vital support for guerilla activities in Lothian, but it could have sapped the morale of the Scots and boosted that of the English at a critical moment. As it was, Black Agnes's successful defence seems to have broken the English effort. The occupation was by no means over: Perth, Stirling and Edinburgh were still in English hands in 1338, quite apart from Berwick, Roxburgh and Lochmaben. But most of Balliol's most important supporters had by now deserted him. There had been a steady trickle from 1335. The accounts of the sheriff of Edinburgh and the Scotch Rolls contain numerous references to forfeitures for rebellion and to losses by those who supported the English.

---

1   *Chron. Lanercost*, pp. 287-8.
2   *Chron. Fordun*, i, p. 362; *Chron. Bower* (Goodall), ii, p. 326.
3   Ibid., p. 328: 'qui, quamvis juvenis erat annis, senilem se gestis, potissime erga Anglos, exhibebat.'

For example, in November 1339 John Stirling, the constable of Edinburgh, who had received a grant of the baronies of Bathgate and Ratho and other lands as a personal reward in 1336, was given 200 marks *per annum* instead, as his lands were 'now occupied by the Scots through the war'.[1] Others, too, are described as having lost their properties through loyalty to Edward III.[2] The evidence is more and more of negotiations, particularly for the release by exchange and ransom of John Randolph, earl of Moray, who had been an English captive since 1335 (although it does not seem that these negotiations were completed until 1341).[3] From 1339, however, the key events were the reduction of the remaining English garrisons: Cupar was recovered on the defection of William Bullock at some point in the spring or summer of 1339 (at least according to Scottish accounts, though he was still apparently receiving wages as custodian from Edward in December of that year);[4] Perth surrendered to the Stewart in 1339,[5] and Edinburgh was taken by William Douglas and others in April 1341.[6]

By that time, too, the supporters of David were moving beyond the stage of mere resistance. At the end of 1337, as we have seen, it had been possible to hold an Exchequer audit at Aberdeen, at which those present included the holder of the regality of Moray on behalf of its captive earl, the sheriffs of Forres and Elgin, and Reginald More, chamberlain of David II.[7] In April 1340 there was a further session at Aberdeen, attended by the provosts of Aberdeen, Banff, Cullen, Kintore and Inverness, and by the custumars of Aberdeen, as well as by the chamberlain.[8] Clearly by this time there was a revival of Scottish administration in the name of David II, a revival — though still confined to the north-east — which made some effort to catch up at least by recording non-payments and requiring explanations. A much wider response is evident in the next series of accounts, rendered in May 1341, which included Dundee, Linlithgow and Haddington.[9] By that time, however, David II himself could return, to Inverbervie in the same north-eastern parts on 2 June 1341. He seems to have held court in the Mearns, issuing charters at Arbroath on 17 June and at Dundee on the following day, before going on to Kildrummy on 20 June.[10] Those in attendance, if we may believe the witness-lists, were the bishops of Aberdeen, Brechin and Ross, the abbots of Dunfermline, Coupar Angus and Lindores, and a handful of secular lords, Duncan earl of Fife, Robert Keith

1   *CDS*, iii, nos. 1209, 1319.
2   *CDS*, iii, nos. 1340 (Henry Ramsay, October 1340), 1370 (David Marshal, September 1341).
3   *CDS*, iii, nos. 1337, 1350, 1359; *Rot. Scot.*, i, pp. 599a, 609b.
4   *Chron. Fordun*, i, p. 364; *Chron. Wyntoun* (Laing), ii, pp. 451-2; *Chron. Pluscarden*, i, p. 288; *CDS*, iii, no. 1321.
5   *Chron. Fordun*, i, pp. 363-4.
6   Ibid., p. 365.
7   *ER*, i, pp. 440-54.
8   *ER*, i, pp. 455-68.
9   *ER*, i, pp. 469ff.
10   *RRS*, vi, nos. 25-30.

the marischal, David Hay the constable, Malcolm Fleming and Philip Meldrum, many of them from the north-east. In July David was at Stirling, though an English garrison still occupied the castle, under siege since 1339 and not to be surrendered until April 1342;[1] in August he was at Dumbarton, before moving on to hold a parliament at Scone in September, where naturally there was a wider attendance, including the Stewart, the earl of March, the earl of Ross, Maurice Murray and William Douglas. David was taking up effective government, and the supporters of Balliol were having to pull out. This had all come about since the failure of the siege of Dunbar in 1338, and was due in some measure to the continuing preoccupations of Edward III in France. In these three years, any basis for an occupation or a Quisling regime had gone. English control was limited briefly to Stirling, and more lengthily to Berwick; to a few outposts in the south-west (including Lochmaben, which was still held by the earl of Hereford who claimed Annandale, and whose successor was able in 1366 to divide its profits with the Scottish crown);[2] and to one or two bases, such as Hestan Island, still held by Balliol or his few remaining supporters. But David's government was secure and scarcely to be questioned, even during the difficulties of his captivity from 1346 to 1357.

The years of David's minority, however, presented a problem for the later writers on whom we have mainly to depend for a Scottish account of the period. On the one hand, the traumas, the devastation, the deceits and confusions of those years were obvious to them. Much of our evidence is in the accounts of Fordun, Wyntoun and especially Bower. Yet they knew that the Scottish kingdom and the Scottish identity did survive. Their accounts lead to a happy ending; and for all the emphasis on the disunity, the manifold treacheries and double-dealings of those years, they never seem to recognise that any other ending was possible. Nevertheless, on their own evidence it clearly was. For later Scottish writers, the Wars of Independence were won by Bruce and Randolph and Douglas, and ended with the treaty of Edinburgh/Northampton in 1328. This is the mythology of Barbour, and it has remained the Scottish national mythology ever since. How then to deal with the deviations of the nobles, many of them loyal supporters of Bruce? How, indeed, to deal with the deviations even of the future Robert II, whose weaknesses in this period are perfectly clear in the medieval accounts, but have somehow slipped out of many later descriptions, leaving a problem in explaining his inadequacies as king? From this problem comes another myth, implied in the medieval sources but never made explicit. The failings of the nobles were made to appear trivial; they were recognised but never given great weight, because after the successes of Bruce and his fellows their work could not be undone, however lacking in faith and honour were the puny nobles who survived. Wyntoun has a comment on the dark days of

1   *RRS*, vi, no. 31; *CDS*, iii, no. 1383.
2   *RRS*, vi, no. 363.

1335 which is apposite:

> Thus wes the kynryk off Scotland
> Sa hale in Inglis mennys hand,
> That nane durst thaim than wythsay ...
> Bot chyldyr, that na kyndly skyll
> Had to deme betwyx gud and iẅyll ...
> Qwhen men askyt qwhays men thai were,
> Thai rycht apertly wald awnsuere
> That thai war men to Kyng Dawy:
> Thus sayd thai all generaly
> That wes prenosticatyown
> That he suld joys efftyr the Crown.[1]

This is obvious exaggeration; but Wyntoun's point was that even if only the children remained loyal, Scotland herself was secure.

The truth is more mundane. Victory had not been finally secured by 1328; it was won in the 1330s by a few determined guerilla fighters; and they won not because of a national uprising but because most people remained, willingly or unwillingly, on the fence. So long as he had English support, Balliol could survive, and most Scots would let him, paying protection money to the guerillas if necessary. But they would do little to back him; and when English enthusiasm was diverted elsewhere, his fate was sealed. Edward III proved to be a good, perhaps even a great, general; but neither in Scotland nor in France did he find the secret of conquest. Yet his attempt in the 1330s was no mere aftermath to the triumph of Robert Bruce; it was an integral part of the Wars of Independence. Indeed, Edward may have come as close to success as did his grandfather. Scotland owes her survival to the guerillas of the 1330s, as well as to the far more often celebrated heroes of the wars of 1296-1328. It is time we paid them their due.[2]

1   *Chron. Wyntoun* (Laing), ii, p. 413.
2   I am grateful to Professor A.A.M. Duncan, who read this essay in typescript and made many very helpful suggestions; also to the editors for their patient and careful work to the great improvement of the final text. For the errors and misconceptions that remain, I am alone responsible.

Appendix

Balliol Supporters noted in the Chronicles
(omitting the chief 'Disinherited')

The list is confined to 'genuine' Scots, though due to lack of evidence some uncertainties remain. It is based on *Chron. Bower* (Goodall), ii, pp. 300-1, 306-8, 313, 315-17, 319, 321, 327-8, 330, 370; *Chron. Fordun*, i, pp. 355-6, 364; *Chron. Lanercost*, pp. 269, 271, 274-5, 277-8, 285, 290-1; *Chron. Pluscarden*, i, pp. 263, 266-7, 270, 272, 275-6, 278, 280, 282, 286, 288; *Chron. Wyntoun* (Laing), ii, pp. 381-2, 391-5, 403, 407-9, 415, 426-7, 432, 451-3; *Scalacronica*, p. 165; also *Rot. Scot.*, i, pp. 380-1, 463-4. With the few exceptions noted, the list is drawn mainly from Scottish chronicles; no attempt has been made to cover published or unpublished English records or English chronicles, from which it could certainly be considerably extended.

The list is divided into three sections, according to the dates at which the men concerned are mentioned as supporting Balliol or the English occupation. It is of course possible that some of them continued to give support after the dates recorded; for others the last mention records their return to David II's allegiance. It should not be assumed that those listed supported Balliol consistently between the dates shown. Many changed sides several times! Names are left in the form given in the source, except where there is an obvious modern spelling.

A. Those named as supporting Balliol only between the battles of Dupplin (12 August 1332) and Halidon (19 July 1333)

	First/Only Mention	*Last Mention*
Ralph de Baroun:	1332	—
John de Bonneville:	Aug./Sept. 1332	—
Alexander Bruce earl of Carrick:	Dec. 1332	17 Dec. 1332
W[illiam] of Cambo:	Aug./Sept. 1332	—
John of Dunmore:	Aug./Sept. 1332	—
William of Farsley:	Aug./Sept. 1332	—
David Graham:	Aug./Sept. 1332	—
Abbot of Inchaffray:	Aug./Sept. 1332	—
John of Inchmartin:	Aug./Sept. 1332	—
Richard Kirby:	17 Dec. 1332	—
Alexander Lamberton:	Aug./Sept. 1332	—
John de Laundells:	Aug./Sept. 1332	—
Twynie Laurison:	1332	—
Walter Lundie:	Aug./Sept. 1332	—
Roger Morton:	Aug./Sept. 1332	—
John Moubray:	1332	17 Dec. 1332
Andrew Murray of Tullibardine:	Aug. 1332	—
Abbot of Scone:	Aug./Sept. 1332	—
Michael Scot:	Aug./Sept. 1332	—
William Sinclair bishop of Dunkeld:	Aug./Sept. 1332	—
William Stodfort:	1332	—

B. Those named as supporting Balliol or the English occupation after the battle of Halidon (19 July 1333) and up to or including the battle of Culblean (30 November 1335), but not mentioned thereafter

	First/Only Mention	*Last Mention*
Michael Arnot:	1335	—
Robert Brade:	Nov. 1335	—
Thomas Brown:	Nov. 1335	—
Thomas Comyn:	Nov. 1335	—
Duncan earl of Fife:	Aug. 1332	Oct. 1335
John Gibson:	1335	—
Alan de Lisle:	Christmas 1333 or 1334	1334 or 1335
Thomas earl of Mar:	June 1334	—
Patrick earl of March:	July 1333	Michaelmas 1334
Richard Melville:	1335	—
Robert Menzies:	Nov. 1335	—
William Moubray:	1332	Nov. 1335
David Wemyss:	Aug./Sept. 1332	Lent 1335
Michael Wemyss:	Aug./Sept. 1332	Lent 1335
Thomas de Wollor:	1334	—

C. Those named as supporting Balliol or the English occupation after the battle of Culblean (30 November 1335)

	First/Only Mention	*Last Mention*
Laurence Abernethy:	no precise date	—
Abbot of Arbroath:	Aug./Sept. 1332	1336
Abbot of Balmerino:	1336	—
David Barclay:	1339	—
William Bullock:	Christmas 1333 or 1334	1339
Abbot of Coupar Angus:	Aug./Sept. 1332	1336
Abbot of Dunfermline:	Aug./Sept. 1332	1336
John of the Isles:	1335	—
Abbot of Lindores:	1336	—
Dungal Macdowall:	no precise date	—
Eustace Maxwell:	Oct. 1332	May/June 1337
Alexander Moubray:	1332	1337
Geoffrey Moubray:	1332	June 1337
Roger Moubray:	Oct. 1335	June 1337
Robert Prendergast:	1338 ?	—
Godfrey de Ros:	1335	Feb. 1336
Prior of St Andrews:	1336	—
William Spens:	1338	—

12

The 'Laws of Malcolm MacKenneth'

ARCHIBALD A.M. DUNCAN

Edward III's challenge to the Scottish kingdom in 1333 ended in failure, but seriously disrupted the processes of central government, not only in the 1330s but also after the capture of David II in 1346. On David's return in 1357 he faced a major task, the reassertion of royal authority in collecting revenues and enforcing the law in courts both royal and private. Yet Scotland still relied heavily upon oral record for the substantive law and its procedures, both of which had lost the coordinating and unifying effect of central judicatures: the king's council and his itinerant justiciars. When these were revived after 1357, there was little for them to look to as authoritative Scots law, little to inform a 'Scottish legal system' with practice and precedent, with old law and new reforms.

In the first volume of *The Acts of the Parliaments of Scotland* [hereafter *APS*] the minutes of parliaments and councils for 1357-99 contain decisions, but only the parliaments of 1318 and 1401 published sequences of statutes.[1] These alone were known to practising lawyers on the eve of James I's return in 1424, and survive as copied into lawyers' handbooks.[2] Of other printed laws and treatises, only a very few can be treated confidently as law-making by kings (William the Lion and Alexander II).

Legal practitioners in medieval Scotland had to communicate their mysteries to their apprentices by describing the law for this or that case in a brief paragraph; when in time the apprentice stood in court in his master's place, it was convenient to cite the law as he had received it, perhaps to add a few sentences relevant to the current case and to give it authority — the spurious authority of a king. How early such laws were ascribed to the good King David I is not known — certainly by the later thirteenth century,[3] but in the case of burgh law much earlier. That the need for a more systematic treatment did not stimulate the pen of an aspiring professor of private law

---

1 *APS*, i, pp. 466-74, 575-6. Individual statutes were published in other years, however, such as the 'statute of Stirling' of 1397 (so called when it was amended in 1399): *APS*, i, pp. 570, 572-3.
2 *APS*, i, p. 302, claims that the 1401 statutes are printed from the Haddington MS (an official register). They are not found there and must have been printed from one of the several law collections in which they occur.
3 The 1293 list of records handed back to King John included a 'roll of statutes of King Malcolm' (*APS*, i, pp. 116-17). The following printed words, *et regis Dauid*, are a guess by the editor. There is nothing to show which Malcolm was intended.

until the early fourteenth century must be attributed to lack of example and lack of materials. Of the great models which he might have known, the *Corpus Iuris Canonici* was built around papal letters, Glanville around royal writs, Bracton on cases from the plea rolls. Scotland, however, offered him no such materials, and could only borrow and adapt from England, as Robert I's clerks were manifestly doing in both form and content in the statutes of 1318.

The extent of this weakness is revealed in the attempt, made soon after 1318, to adapt Glanville's treatise by removing its structural girders, the writs, and tinkering with its walls; not surprisingly the task was abandoned when quarter-completed. But the adaptation was ascribed to the command of 'King David' [David I], for to be authoritative it had to be good old law, and David was already well established as the source of that; from the opening words of its prologue, it is known as *Regiam Majestatem* [hereafter *RM*].

A systematic account of the courts and the law which they administered was not written until the seventeenth century; the best the fourteenth century could manage was the treatise *Quoniam Attachiamenta* [hereafter *QA*] on procedure in baron courts which cites a clause (possibly two) from the 1318 legislation, but makes no claim to antiquity or royal authority.[1] Otherwise those practising in the courts had to make what they could from the guidelines and opinions cumulating in lawyers' handbooks, usually described as laws of Kings David, William and Alexander.

In his preface to *APS*, i, Cosmo Innes correctly ascribed *RM* and *QA* to the fourteenth century and concluded that the same period of fabrication, following upon the loss of genuine national records, produced the *Leges Malcolmi Makkenneth*, which Innes called 'unauthorised customs … absurdly distinguished by the name of Malcolm MacKenneth', and titled uniquely in black letter.[2] Here was a treatise (I shall call it *LMM*) which, like *RM*, was copied into almost all the handbooks, though it had little or no practical application, and which, like *RM*, because of its (bogus) antiquity and royal authorship, could claim to be what was so conspicuously lacking — an authority. It is the purpose of this essay to explore the origins of *LMM*.[3]

The first questions must be: whence came Cosmo Innes's text and how reliable is it? Some twelve pre-Skene manuscripts were used[4] to produce an eclectic version which follows the order of all but one of these, and with which subsequently discovered manuscripts are in agreement. In the

---

1   This treatise is fully discussed in T.D. Fergus, '*Quoniam Attachiamenta*, an edition' (Glasgow University Ph.D. thesis, 1988). He shows (i, pp. 182-3) that 1318 c. 19 is taken account of in *QA*, c. 40; there are no unambiguous references to *RM* in *QA*. At i, p. 196, he concludes that *QA* was probably compiled in the first half of the 14th century 'and that, within this period, an earlier rather than a later date is to be preferred'.
2   *APS*, i, pp. 38, 51.
3   *APS*, i, pp. 709-12.
4   *APS*, i, pp. 252-3, shows the MSS used. The only important MS to come to light subsequently is the Arbuthnott MS: NLS, MS 16497 (formerly MS Acc. 2006).

majority of these there are no accretions to *LMM*, unlike most of the other treatises, and a new edition would differ little from Innes's text.

Thus all manuscripts are agreed on two passages which contain what must be mistakes. In c. 11, a list of serious offenders is interrupted by words, here italicised: 'raptores de roboria et deforciatores mulierum et murdratores hominum *siue juste siue injuste non faciat sectam aperte* et combussores domorum ... comparebunt'. Only without these words (which I have not been able to trace to any other tract) does the text makes sense, and they must have crept in at a very early stage of the transmission.

Similarly in the middle of the last clause (c. 14), all manuscripts read: 'hoc agere consueuerunt pueros ac parentes dicti interfecti interficere ordinauerunt et tales malefactores fuerunt recepti ... de ... malefactoribus rex habebit sequelam'. In many MSS this nonsense was made more obscure by capitalising 'Et', where a new sentence cannot possibly begin. No more helpful was omission from the printed text of *ordinauerunt*, which as elsewhere introduces the 'legislation' which is the point of the clause; it is required here and should therefore be retained. A better resolution is the emendation of *et* to *si*: 'they ordained [that] if such malefactors were reset....'[1]

One Latin text, the Cromertie MS, differs significantly from the others both in the order of its clauses and in some words.[2] This manuscript is of the late fifteenth century, considerably later than, for example, the Bute MS which also has *LMM*, and it offers uniquely different and reordered texts of a number of other treatises — *RM*, *QA*, the Forest Laws — which, upon investigation have been shown to be the work of a gifted but idiosyncratic reviser.[3] The Cromertie text of *LMM* is of the same character, and for this discussion of the original treatise we may ignore its witness.

The manuscripts present us with a pretty well agreed text which for present purposes is adequate, and which shows that the exemplar lying behind all existing versions already contained misreadings. In c. 5 there is an obvious omission from the clerkships whose fees are described, for the text runs from the clerk of liverance to the clerk of the kitchen, omitting the clerks of the proof and wardrobe. I can see no reason why this should have been done deliberately, and suggest that at an early stage, possibly even in the compilation of *LMM*, the scribe's eye, seeking the second *clerici* in his exemplar, jumped by error to the fourth.

Cosmo Innes took the heading and c. 1 from the Cromertie MS: 'Secuntur

---

1   The printed vernacular translation from the Malcolm MS certainly knew this wording (they have 'ordanit gif sik misdoaris ...'), but I could find no Latin MS with 'ordinauerunt si'.

2   *APS*, i, pp. 183-5 (where too early a date of *c.*1470 is suggested), 252-3.

3   The Cromertie MS is NLS, MS Advocates 25.5.10. It is discussed in J. Buchanan, 'The Manuscripts of *Regiam Majestatem*', *Juridical Review*, xlix (1937), pp. 217-31; J.M. Gilbert, *Hunting and Hunting Reserves in Medieval Scotland* (Edinburgh, 1979), pp. 277-81; Fergus, *'Quoniam Attachiamenta'*, i, pp. 41-2, 146-58.

leges Malcolmi Makkeneth qui fuit rex victoriosissimus super omnibus nacionibus Anglie Wallie Ybernie et Norwagie'; fortunately, against the witness of Bute[1] and other manuscripts, which have 'virtuosissimus', the Arbuthnott MS (unknown to Innes) confirms 'victoriosissimus'. The word came from John of Fordun's account of this king in his *Chronica Gentis Scotorum*, the source too for c. 1 of *LMM* which narrates that Malcolm gave 'the whole land of Scotland to his men and retained nothing to himself in property except the royal dignity and the hill of pleading in the toun of Scone. And there all the barons granted to him the wardship and relief of the heir of any deceased baron, for the sustenance of the lord king'.

In his book IV, Fordun told a much more elaborate story of Malcolm's triumph over his foes.[2] Under King Grim, Malcolm won the nobility and people over to him; after he had slain Grim, Malcolm agreed to become king only if the laws allowed it. He confirmed the law of succession made in the time of Kenneth, his father, whereby each king should be succeeded by 'son or daughter, grandson or granddaughter, or brother or sister of collateral line' or another of close blood relationship.[3] This statement would have a contemporary meaning between 1346 and 1373 when the succession laid down under Robert I in 1318 and 1326 and following essentially this rule was both questioned and reaffirmed. Malcolm was victor over every neighbouring people ('gens') which challenged him, specifically Danes, an English earl and Norwegians, though not Irish and Welsh (as in *LMM*). Hence, says Fordun, he is always called 'the most victorious king' ('semper ... rex victoriosissimus appellatus'). This title shows beyond doubt that *LMM* was indebted to Fordun. He wrote of 'Malcolmus filius Kenethi'; *LMM*'s *mac* form[4] came from someone with a smattering of Gaelic.

Fordun described the many different ways land was granted, always for rents or feu-duty, then somewhat confusedly insisted that Malcolm improvidently gave away all the lands of his kingdom, keeping nothing from lands or rents for himself but the moot-hill at Scone. He therefore asked a general court for an allowance in lands, rents or yearly subsidy, provided that the poor were not burdened by a yearly contribution. This was granted. That again had contemporary relevance, for it reflected the taxation levied after 1357 to meet David II's ransom, which each estate had promised to pay.[5] Fordun's reticence about the income granted to Malcolm provided antiquity for the 'lands, rents and possessions [unspecified] which *ab antiquo* belonged to the royal crown and lordship', which in 1357 were to be resumed and held perpetually by the king without alienation;[6] Fordun

1   Now NLS, MS 21246.
2   *Chron. Fordun*, i, pp. 180-7.
3   Ibid., p. 172, where 'nepos aut neptis' must be translated 'grandson or granddaughter'; cf. ibid., ii, p. 165.
4   The spelling 'macKennet' is found in the rubric to *LMM* in the Bute MS. 'Mak', the almost invariable form, is typical of Scottophone scribes.
5   *CDS*, iii, nos. 1642-8, 1650-4, 1660-2.
6   *APS*, i, p. 492a.

provided a precedent for an improvident king, for an act of resumption and for the rejection of an annual contribution. The revocation of 1357 was being judged in Fordun's verdict that Malcolm was foolish to give away that which he had to ask to be returned.

According to Fordun, followed by *LMM*, Malcolm's nobles agreed that wardships, reliefs and marriages should be the king's. But in 1357 it was briefly provided that 'revoked lands and wardships shall not be alienated without mature counsel'. A deal must have been done subsequently over casualties, because while they were accounted for by the chamberlain in 1357-60, in 1361 he claimed that he had received nothing from them, 'because [he added in 1362] he had not intromitted with them'.[1] In 1363 he accounted for one relief sold, but in December 1364 made no mention of any casualty. In January 1366 and again in 1367 his failure to intromit was attributed to the king who 'disposed of them at the pleasure of his will'.[2] In a sense this was what Fordun and *LMM* prescribed, though it probably contravened the 1357 ordinance restricting disposal. A second, more drastic, revocation in September 1367 provided that all casualties and issues of the king's courts should remain for the sustenance of his household in the chamberlain's hands to be disposed of for the king's uses.[3] For the next four years the accounts are again silent about casualties and about why they are silent,[4] but in 1372 and thereafter sales of them returned to the chamberlain's account,[5] making it clear that David II had continued to enjoy free disposal of them rather than to make the restricted and responsible use intended in 1357 and 1367; in this matter Fordun and *LMM* put Malcolm II on David II's side.

The debt owed by *LMM* to Fordun is obvious, even if the words used are different (for example *nacio, defuncti* in *LMM, gens, decessum* in Fordun). *LMM* makes something of a nonsense of Fordun by omitting the landed revenues granted to Malcolm II and mentioning only the casualties, but this suggests a date in the 1360s for *LMM*, when David II was known to be disposing of wardships and reliefs personally. The reservation by Fordun's King Malcolm of 'the little hill (*monticulum*) of the royal seat of Scone where the kings ... are accustomed to give out judgements, laws and statutes to their subjects' explained the role of fourteenth-century Scone as the usual place for the holding of parliaments and framing of legislation.[6] But, in law-giving, *LMM*'s Malcolm was not Fordun's, for the latter made Kenneth (Malcolm's father) the legislator. Fordun's Malcolm did, however, found a new bishop's see at Mortlach, near the site of his supposed victory

---

1   *ER*, ii, pp. 73, 111.
2   *ER*, ii, pp. 163, 172, 220, 259.
3   *APS*, i, p. 502.
4   The relevant parts of the chamberlain's accounts are *ER*, ii, pp. 288, 304-5, 343, 355.
5   *ER*, ii, pp. 363, 430, 457, 498.
6   *Chron. Fordun*, i, p. 186; it is not clear why 'soleant' is in the subjunctive, but the present tense is undoubted — Fordun is referring to kings of his own time.

over the Norse, later transferred to Aberdeen. The author of *LMM* chose Malcolm to be his lawgiver, I suggest, as the pious tribute of an Aberdonian clerk.

This debt places the composition of *LMM* after Fordun finished the main work on his chronicle in 1363. That chronicle was not well known — manuscripts of it are few and late — and Fordun enjoyed no royal patronage such as was given to Barbour. Perhaps his work languished at Aberdeen. 1363 is a *terminus post quem* for the composition of *LMM*; a *terminus ante quem* is less obvious.

After the introduction, *LMM* deals with two separate matters: cc. 2-8 with the fees of a succession of royal officers (78 lines), the remainder, cc. 9-14, with a programme for the execution of the criminal law (62 lines). There is clumsy overlap in cc. 3 and 8 on the fines levied in courts. The two parts have a stylistic unity because the fiction of legislation was kept up throughout by the use of *ordinauerunt* (cc. 2, 3, 4, 5, 9, 10, 14). The king is almost always *dominus rex*, and the concern throughout with amercements and escheats gives a further unity, as well as indicating the concerns of the compiler.

But the treatment of justice, which becomes progressively didactic, is distinguishable from the material on fees, which is so factual that it has occasionally been treated as a reliable source on Scottish administration. That response to its simple clarity is, I suggest, a correct one; my view is that the fee-scale in *LMM* was based upon a genuine scale in a document used by the compiler of *LMM* and therefore earlier than *LMM* itself. Since c. 1 must have been written after 1363, the fees paid in the later fourteenth century are particularly relevant. *LMM* lists fees 'for the sustenance' of the following officials:

c. 2   Chancellor and the writing clerk
c. 3   Justiciar, crownar and their clerks
c. 4   Chamberlain
c. 5   Steward of the household, 13 named household servants and
         other 'lesser officials'
c. 6   Constable and marischal
c. 7   Sheriff, his clerk and serjeant
c. 8   Amercements to justiciar, crownar, clerk; sheriff, clerk,
         serjeant; burgh provost, bailies; chamberlain, clerk, serjeants.

LMM's information repays comparison with that recorded in the royal accounts, where fees should appear as allowances in the chamberlain's account and as receipts in the account of the official, if it survives. I shall deal with the offices in a rather different order to that of *LMM*.

The Chamberlain

The chamberlain's fee had been 250mk under Robert I and was first raised to 300mk in 1330; fees fell into disorder after 1332, and in 1342 a special fee of 150mk was allowed to this official.[1] The chamberlain from 1356 to March 1358 was the earl of Angus.[2] He was succeeded briefly by the earl of Mar (to November 1358), whose account does not survive and was probably never audited.[3] The history of the chamberlain's fee thereafter is instructive; in the following list all figures are converted to merks:

Walter Biggar

In Apr. 1359 (½ year as lieutenant of Mar):	100mk
In May 1360 (year):	150mk because he was unwilling to take more
In Jun. 1361 (3 terms):	300mk
In Aug. 1362 (year):	300mk
In Dec. 1364 (½ year to March 1363):	150mk plus 350mk arrears by inspection of accounts at king's command[4]

Robert Erskine

In Dec. 1364 (2 years):	600mk[5]

Walter Biggar

In Jan. 1366 (year):	300mk
In Jan. 1367 (year):	200mk 'this time'
In Jan. 1368 (year):	king assigned the issues of his ayre to his fee
In Jan. 1369 (year):	300mk[6]

Thereafter the fee remained at £200, or 300mk, *per annum* until 1424.[7]

It appears, therefore, that there was a purpose in 1358 to establish the chamberlain's fee at 300mk, the figure given in *LMM*, but it was held over, presumably as an economy, and was not paid till 1361 when ransom payments had been suspended; at some date in 1363, perhaps when Sir Robert Erskine was brought back to office for the negotiation of new terms with Edward III, the king honoured the fee of 300mk promised in 1358, by paying arrears, of 350mk, though calculation suggests that only 300mk was due.

LMM stated that the fee should be paid from the escheats of burghs, with other burgh revenues (including customs) going to the king. The issues of the chamberlain ayres, largely made up of escheats, were clearly rounded figures and hardly ever reached 300mk,[8] which may explain the reduced fee

1   *ER*, i, pp. 114, 285, 339, 511.
2   *ER*, i, pp. 595, 622, 624; ii, pp. 1-4.
3   *ER*, ii, p. cxxv and refs.
4   *ER*, ii, pp. 6, 52, 82, 118, 169.
5   *ER*, ii, p. 178.
6   *ER*, ii, pp. 223, 263, 288, 309.
7   *ER*, ii, pp. 349, 360, 366, etc.
8   *ER*, ii, pp. 48, 73, 111 (all 100mk), 220 (160mk), 259 (£140), 305 (£151), 343 (£180), 355 (£200), 363 (£190), 432 (£200) in 1373-4; thereafter the issues declined.

paid until 1361. When in 1367 the king assigned the issues of the ayre to the chamberlain's fee, he was carrying out part of the programme recorded in *LMM*.

Of the two other fees paid in 1358, that for the durward of the chapel was paid 'of grace' (in this and the next year), that is, his right to it was in doubt.[1] The ancient fee (1328-32) of the clerk of the rolls, £20, seems to have been halved from 1358 to 1363, when it was restored for a new appointee.[2] The fee of the clerk of audit (which had been £40) first appears in driblets in 1360-2 and was paid at £40 in 1363-4, when it was a charge on the customs of Montrose; it usually remained at that figure.[3] The fee of the clerk of the wardrobe, £10, first appears in January 1367, when arrears of £40 for four years were paid to the clerk's executors;[4] he had not seen a penny of it since his appointment. Thus these two offices were fee'd before 1362 uncertainly if at all, though the clerk of rolls was paid from 1358; in all three cases the fee ultimately restored was that of *c*.1330.

The Royal Household

The fees allocated to staff of the royal household in c. 5 are as follows:

Steward of household:	£40
Clerk of liverance:	£20
Clerk of kitchen:	£10
Panetar and butler:	£10 each
Baker, brewster, master cook, lardiner, durward of kitchen, janitor, durward of hall, durward of king's chamber:	100s. each
Maker of the fire in the hall:	40s.
Lesser servants:	40s., or as fixed by council

The Steward of the Household

Although Robert I and the young David II had stewards of their households, there is no trace of a fee to one in the accounts for 1327-33;[5] in the few accounts for the time David was in France and after his return (1334-46) the office is not mentioned and pretty certainly lapsed — household clerkships are mentioned. It was revived in 1357, with no mention of fee, for Sir William Vaus the first incumbent.[6] Sir Walter Moigne held office in 1359-60, without mention of fee, but £20 was allowed for part of his fee in May 1360-June 1361 and a further £20 in the year to August 1362, but expressly for the year 1361.[7] It seems that £20 was his

---

1   *ER*, ii, pp. 6, 52.
2   *ER*, ii, pp. 51, 114, 208, 285, 337.
3   *ER*, ii, pp. 82, 88, 142, 223, 261, 290, 359; cf. pp. 309, 347-8.
4   *ER*, ii, p. 263.
5   Sir Alexander Seton and Sir Adam More under Robert I (*RRS*, v, p. 143 and no. 333); Sir Malcolm Fleming under David II, see *ER*, i, index under his name and under 'Steward'.
6   *ER*, i, pp. 576, 578, 587, 606, 612, 616, 618, 625; ii, pp. 3, 76, 175 (for his death by 1365).
7   *ER*, ii, pp. 45, 49, 78, 82, 113.

fee for a term, equivalent to the £40 *per annum* given in *LMM*. This is the only record of a fee paid for holding the office of steward under David II. No fee is recorded for his successor, Sir William Dishington, nor for Master Gilbert Armstrong, steward in May 1363, who accompanied the king to England in October-December 1363.[1] However John Lisle, steward during the March 1364 parliament received £20 by the king's letters before December 1364, perhaps his fee as steward for a year.[2] Thereafter no steward and no fee is recorded for 1365-8, when ransom payments had been resumed. When the office was revived for Sir William Ramsay of Dalhousie, he received £20 'by the king's command' in 1369, and was paid £40 similarly in the year 1370; this was not said to be his fee and Ramsay in 1364 (when he was not steward), had been paid £80, 'receiving £40 annually by the king's letters'.[3] The payments in 1369 and 1370 may have been a revival of this personal fee — perhaps revived because he was steward. It is also possible, though not perhaps likely, that the £40 of 1370 silently included £20 of arrears from 1368.

When, under Robert II, the office passed to the king's half-brother, a fee (explicitly so) of £20 was paid; in 1373-4 he ceased to be called steward, but the payment continued for a few years.[4] No steward is named thereafter, though Sir Andrew Mercer 'stood in the office of steward of the household' in 1387-8, and is once called steward; from 1383 he enjoyed a heritable fee of 40mk from the revenues of Perth and this probably covered his services which were modest but real. He died about 1389.[5] The office is not mentioned under Robert III until June 1403, when it had evidently been revived because authority had passed to the duke of Albany. The incumbent appears in 1404-6 as clerk of liverance, factotum in a much depleted royal household.[6] The stewardship enjoyed a revival under James I, probably in conscious imitation of David II's time, but this is too late to have influenced the composition of *LMM*.

The only time when LMM's fee, £40, was paid to a steward was in 1360-1 and perhaps in 1370. Later the stewardship rated only £20 *per annum*, and was equal to the clerkship of liverance, not its superior as implied in *LMM*.

The Clerk of Liverance

He is assigned a fee of £20 by *LMM*. His annual fee in 1327-8 was 20mk, in 1328-9 £20, but reduced again to 20mk.[7] After 1357 the position is complex. John Leys had been appointed by November 1357, but was succeeded in 1358 by a Patrick, who lasted until some time in the account year June 1361-August 1362 when the clerk of audit acted 'in the name of

1   *ER*, ii, pp. 107, 112-13, 153, 156, 160, 164; pp. 130, 135.
2   *ER*, ii, pp. 135, 141, 173.
3   *ER*, ii, pp. 120, 348, 358.
4   *ER*, ii, pp. 365, 395, 436, 461, 501, 551.
5   *ER*, iii, pp. 170, 175-6, 178, 183, 218.
6   *ER*, iii, pp. 588, 647; cf. p. 536.
7   *ER*, i, pp. 114, 208, 286, 338; p. 510 for a part fee in 1341-2.

quondam Patrick clerk of liverance'. A period of uncertainty followed 'when there was not a clerk of liverance appointed *in certo*', and during which the steward fulfilled his functions. Robert Smailholm acted briefly, and then John Ross certainly from 23 October 1362 to the king's departure in October 1363.[1] Up to this time, no fee for the clerk is mentioned. From 11 March 1364 the clerkship was revived for John McKelly; his account shows a fee of £40 for two years paid in 1365.[2] The fee was held at the £20 level until 1399 when the king's authority was transferred to a Lieutenant, and then turns up as £20 for two years in 1405.[3] The clerk of liverance was certainly paid the fee given in *LMM* from 1364.

The Clerk of the Kitchen

He is known from the earliest surviving royal account, of 1326, and drew a fee of £10 as early as 1328-9; but by 1330-1 this had been reduced to 10mk, and was restored to £10 only on the king's return in 1341.[4] This clerkship would lapse in 1346, and there is no trace of its restoration after 1357; but its traditional fee, £10, was that given in *LMM*.

Other Household Staff

Those listed in *LMM* are very scarce in the accounts. In 1329 five men, including a durward and a durward of the kitchen, were paid £1 each *per annum*; the former might have been *LMM*'s durward of the king's chamber. In 1365 victuals were handed over to a granetar of Leith, baker, steward of wine and brewer, and in 1366 payment of some £61 in fees was made to servants 'as much in the pantry butlery and kitchen as in other households of officials [offices of the household?]'.[5] In 1368 the chamberlain paid 20mk at the king's command to 'officials of the king's household' as part of their fees. These may be the 'simple servants in the king's household' to whom the chamberlain paid a fee of £16 16s.4d. in one term of 1377-8.[6] The wine steward of 1366, Adam Page, received a fee of £16 jointly with his brother for 1361-4, which continued as 40s. for part of his (sole) fee in 1365 and 1366; in 1370 he was paid 5mk for his service and from 1371 to 1378 £5 — the only example I have found of this fee of *LMM* in the household.[7] This contrasts with an official not listed by *LMM*, the heritable durward of chancery and exchequer, who received 40s. annually in the 1360s in lieu of his robe.[8]

Various minor officials are listed in the earlier treatise on the King's Household [hereafter *KH*], datable 1290 × 1327. Much of *KH*'s account of the constable seems fanciful, but it claims that 'there shall be with [the

1   *ER*, ii, index under 'Liverance' (esp. pp. 104, 112, 183).
2   *ER*, ii, pp. 184, 228; cf. p. 224.
3   *ER*, ii, pp. 262, 310, 349, 366, etc.; iii, pp. 484, 647.
4   *ER*, i, pp. 55, 89, 202; pp. 195, 376; p. 286; p. 534.
5   *ER*, i, p. 185; ii, pp. 228-9, 253. The reading in this last is quite clear, but scarcely makes sense, and I suggest an error of transcription from the draft account.
6   *ER*, ii, pp. 307, 586.
7   *ER*, ii, index under 'Page, Adam'.
8   *ER*, ii, pp. 52, 118, 290, 309, 348, etc., to p. 622.

constable] in afforcement [of his court] the steward, marischal, panetar, butler, porter and other officers, if he can conveniently have them, with the other free men who shall be found in the court'. This catalogue may be compared with a later passage in *KH* on durwards 'of all houses of office of the king's court' (a phrase echoed in the account for 1366 cited above), who are to come to the petty audit with the officers, 'although the great and [*word missing*?] officers are heretofore of fee, like panetar, butler, lardiner, baker, naperer, chandler, waterer, and others such'.[1] There had been an honorific and heritable panetar and butler, but the accounts show no trace of active household deputes in these or the other listed functions, except a chandler who in 1373-4 drew a fee of 40s.,[2] that laid down in *LMM* for those of lowly status.

Nonetheless the functions of baker or naperer were as essential as those of wine steward, and it is inherently likely that such men did function within or on the edges of the royal household.[3] The accounts suggest that they were not heritable officers or even very regular in employment; that there was a shifting population of functionaries, perhaps taken on locally as the household moved about the country, each active in his own trade, receiving occasional handouts from the chamberlain or permanent clerks, and doubtless profiting from purchasing deals and from the sale of leftovers. The treatises represent attempts to make these posts permanent, to inflate their number and hence the importance of the offices in general. Anyone who has seen a government department prepare its estimates will be familiar with this form of functional self-importance, whereby the taxpayer is the helpless victim of bureaucratic expansion. Today 'new needs' justify more functionaries; in the fourteenth century, old precedents did so.

The Constable and the Marischal

There is no other evidence that the fees of £10 each prescribed by *LMM* were paid to these officers, nor is payment likely. This seems to be another example of bidding-up functional importance, perhaps seeking to bring these heritable officials within the household. But the verge of twelve leagues within which they are said to have jurisdiction was that laid down in *KH* for the constable's authority, and was probably real.[4]

The Chancellor

Clause 2 of *LMM*, dealing with chancery fees, correctly identifies 'the seal's fee' as the chancellor's remuneration; this is mentioned in *KH* and in the accounts for 1329, when, for some reason, the chancellor paid part of it

---

1   *SHS Misc.*, ii, pp. 33-4, 36; 'les grantz et plusours officiers' does not make sense. I suggest for 'plusours', 'plus' and a garbled word, meaning, say, 'important' or 'middling'.
2   *ER*, ii, p. 442.
3   See the macers, for example: *ER*, ii, pp. 124, 130, 135, 142, 150, 201, 361.
4   *SHS Misc.*, ii, pp. 34-5. *KH* describes a heritable constable and marischal, each of whom may have a knight as depute in the king's court; the constable's son was constable of the household under Robert I (*RRS*, v, nos. 293-4) and possibly early in David II's reign also (*ER*, iii, pp. 375, 402). The fees given in *LMM* may have been payable to these deputies.

to the chamberlain. It was probably increased later since the substantial annual fee paid by the chamberlain (£100 in 1290, 200mk in 1328-9, £100 in 1330-2) disappeared after 1332. An ancient fee (26mk and 10s.) from the mills of Perth was paid uncertainly until the 1360s[1] and is ignored by *LMM* c. 2, which offers a breakdown of the seal's fee, initially in diminishing order, and with a fee also for the writing clerk.

We can compare this scale with that of the 1363 fees for the durward of chancery and exchequer but in the order of *LMM*:

LMM	**Chancellor**	**Clerk**	**Durward equivalent[2]**	**Fee**
Charter of land				
worth £100+:	£10	2mk	Charter:	½mk
Letter of sasine:	1mk	2s.	—	
Letter of attorney			Letter patent	
or protection:	12d.	3d.	of course:	3d.
Brieve close:	6d.	4d.	Letter close:	1d.
Pardon:	£2	½mk	Remission, presentation,	
Presentation:	£2	½mk	warren or other	
			letter of grace:	2s.

The durward received 20s. as a fee for his exchequer work, the two clerks working there £4, raised to £5 in the 1360s.[3] If the durward's fees were roughly half those of a clerk, the two lists are not greatly out of line with each other. There are obvious correspondences, but also obvious omissions from *LMM*'s list, for example charters of land worth less than £100 annually and confirmations; if chancery knew the value of lands from retours, those stented at £100 or more must have been rare. This list seems best explained as a selection from a scale of fees in which items relevant to land (charter, sasine) were followed by those relating to pleading and offence (attorney, brieve, pardon); presentation interested every clerk!

I have reviewed elsewhere the evidence for the fees paid for royal charters and letters, but I missed there the fact that Durham Priory paid 15d. each for a letter of attorney and a *prohibicio* in 1329-30, corresponding exactly to what *LMM* laid down for such letters.[4] This is strong grounds for accepting the scale as a whole as accurate — and for suspecting that the durward was not always on duty!

The Sheriff

LMM c. 7 gives his fee as £10. The evidence in the accounts is scattered; I give references to the *Exchequer Rolls*:

---

1   *RRS*, v, pp. 200-1, 205; *SHS Misc.*, ii, pp. 31-2; *ER*, i, pp. 168, 264, 364, 485, 523-4. In 1358-9 the Perth fee was paid to the king; in 1360 the mills did not yield enough to allow payment; and a partial payment in 1366 seems to be the end of this fee: *ER*, i, p. 619; ii, pp. 26, 261.
2   *RRS*, vi, no. 306.
3   *ER*, ii, pp. 6, 52, 118, 178, 223, etc.
4   *RRS*, v, p. 213.

Aberdeen:	offered £85 [error for £95?] [for 19 terms]; refused, claiming this was only half of ancient fee of £20 *p.a.* (1359: i, p. 547,); £36 15s.4d. (= unspent balance in hand) for 3 years (1392: iii, p. 267)
Ayr:	£15 for 3 terms (1359: i, p. 558); £20 (1384: iii, p. 612)
Banff:	£5 (1359: i, p. 550)
Clackmannan:	£13 12s. [for 22 terms] (1359: i, p. 574); £5 (1374: ii, p. 421)
Edinburgh:	£20 (1372: ii, p. 364)
Fife:	£30 for 7 terms (1359: i, p. 563); £10 (1370: ii, p. 358; 1374: ii, p. 437; 1388: iii, p. 166)
Forfar:	£20 for 3 terms (1359: i, p. 593)
Kincardine:	£10 (1359: i, p. 587; 1391: iii, p. 265)
Kinross:	heritably 10mk (1359: i, p. 581); £5 (1374: ii, p. 437)
Lanark:	for life £20 (1375-81: ii, p. 463; iii, p. 82)
Perth:	£10 [for 6 terms] (1359: i, p. 559); anciently £10, doubled to £20 (1371-4: ii, p. 425); £20 (1381-3: iii, pp. 82, 667)
Peebles:	£5 (1359: i, p. 567)
Roxburgh:	total receipts, £5 18s. (1359: i, p. 568)
Stirling:	amounts below £10 'to complete the fee' (1372-4: ii, pp. 364, 394, 437); £10 (1391: iii, p. 271)

The fees were not haphazard and seem to have been based on £10 *per annum*, half for small or poor sheriffdoms, double for Edinburgh, probably because it included the constabularies of Haddington and Linlithgow; some allowance — very variable — was then made for the years before David II's return. Despite the lack of information for the 1360s it is striking that the heritable fee of the sheriff of Kinross was diminished after 1359, suggesting a strong impulse to rationalisation. Under Robert II there was some inflation of fees generally, as when that of the sheriff of Perth was doubled 'at present at the king's will'; the *LMM* fee of £10 was truest of the 1360s. But *LMM*'s claim that the fee was payable from escheats was also correct: in 1359 the sheriff of Stirling was allowed no fee 'because he presented nothing of the issues of his court'.[1]

The Justiciar

LMM c. 3 awards him no fee, but 100s. for each day of his ayre 'for sustentation'. The sheriffs' accounts show six 'ayres' (the word is used for a sitting at each place, not for the whole series) in 1358 (tabulated on the next page). This ayre had been carefully planned, with the justiciar sitting for a week in each major northern sheriffdom, and a rounded allowance for expenses was made to each sheriff. In December 1364 the chamberlain claimed to have paid to the justiciar north of Forth, as fee for the year,

---

1   *ER*, i, p. 577. For a discussion of the sheriff's fee see *Fife Court Bk.*, pp. l-liv.

Place	Date	Expenses	Reference
Inverness:	1 October	10mk	*ER*, i, p. 570
Aberdeen for sheriffdom of Banff:			
	8 October	16s.	*ER*, i, p. 550
[the sheriff of Aberdeen did not intromit			*ER*, i, p. 546]
Inverbervie for sheriffdom of Kincardine:			
	15 October	10mk	*ER*, i, p. 587
Forfar:	undated	17mk	*ER*, i, p. 591
	[prob. 22 October]		
Perth:	29 October	17mk	*ER*, i, p. 557
Cupar, Fife:	5 November	10mk	*ER*, i, p. 562
	for 'octave'		

£93 6s.8d., exactly 10mk short of £100, the 10mk shown as paid at Aberdeen from the issues of an ayre to a man who was presumably depute to Moigne, the sheriff. In 1374 the full fee of £100 was paid.[1] In 1372 the justiciar south of Forth was on a fee of £200; the chamberlain paid him £191 13s.4d. 'in complementum' — he had already received 12½mk.[2]

In later years a system can be traced whereby the justiciar uplifted some part of his fee from the sheriffs who accounted for this, and the chamberlain completed the fee during the following financial year;[3] that is probably the system represented by the expenses allowed to sheriffs in 1358, although the chamberlains' accounts do not reveal the *complementum*. In 1404 it was laid down that the fee was to be docked *proportionaliter* for failure to visit a sheriffdom; in 1406 Albany was paid £100 for five 'ayres' north of Forth, which, rather than £20 per sheriffdom, was, I suggest, the traditional fee with a suggestion that he had earned it.[4] Thus a standard fee of £100 for the northern justiciar probably persisted from 1358 and *LMM*'s extravagant provision of £5 per day for expenses (the 1358 ayre would have cost £180 with a six-day week) cannot be found in practice.

Fees from each case paid to the crownar and the justiciar's clerk do not figure in the royal accounts, but in two instances the clerk's fee is that found in the fifteenth-century *Ordo Justiciarie*,[5] and *LMM* is probably accurate in these matters.

The fact that fees and the administration of justice are expounded twice (cc. 3, 8) shows clumsy conflation in *LMM*. It is clear that the material on fees was based upon a real scale, not memorised (it is too detailed for that) but in written form. Although some of the fees could belong to *c.*1330, a sheriff's fee and a steward's fee seem to be unknown then, and the chancellor had a

---

1   *ER*, ii, pp. 176, 435.
2   *ER*, ii, p. 395.
3   *ER*, iii, pp. 30, 81, 316-17, 347, 376, 652.
4   A.A.M. Duncan, 'Councils-General, 1404-1423', *SHR*, xxxv (1956), pp. 135-8; *ER*, iii, p. 644.
5   *APS*, i, p. 707, c.11.

substantial annual fee unknown to *LMM*. For the date of this scale the chamberlain's fee suggests 1361 × ; the steward's fee is that of 1360-1 or just possibly of 1370; the clerk of liverance's fee suggests 1364 × , but the sheriff's fee points most clearly to 1357-71.

But there are discrepancies, especially the presence of the clerk of the kitchen, and the generous fee for the justiciar. It seems to me that these can be explained only as unfulfilled proposals, while the rest of the programme, a scheme for the household and other officials, was carried out hesitantly in the 1360s. When was such a scheme likely?

On 3 May 1371 Robert II (who had succeeded on 22 February) and his council drew up an ordinance 'on the estate or manner of living of [the king and queen] and on the ordering and governance of their households and on the keeping and maintenance of the castles', a document sadly lost with the register of temporary matters in which it was copied.[1] This ordinance could have contained the scale used by the author of *LMM*, but presumably it contained much more of which there is no trace in *LMM*. Moreover the restoration of the clerkship of the kitchen seems a very unlikely proposal for 1371, when the household was well accustomed to doing without one. Again, the steward of the household after 1371 seems to have been inactive on a fee which was less than that in *LMM*; the *LMM* fee for sheriffs was already usual in 1359, but was subject to modification in the years after 1371. For all these reasons, the scale is much more likely to belong to the last thirteen years of David II's reign, when there was a gradual achievement of many of the fees in *LMM*.

The provision (c. 4) that all the issues of the burghs and the customs were to go to the king after the chamberlain's fee had been paid anticipates a like determination on the return of James I in 1424[2] — and we know that James researched the records of David II for appropriate precedents, where, I suggest, he found this one. Then the silence of *LMM* about a queen's household may be significant, for Queen Joan left Scotland within months of David II's return, and he remarried in 1363 after her death. It is also noteworthy that c. 5 begins with the fee of the steward of the household; at the end 'the lord steward' and others of the king's council were to fix the fees of minor officials. 'Lord' is used nowhere else of an officer but only of the king. Examination shows the same feature in the accounts, except that the king's nephew and heir was 'Steward of Scotland', 'Steward', and twice 'lord Steward'.[3] *Dominus* therefore points firmly to a date after the return of David II, when Robert Stewart was not openly the object of the king's dislike, that is to the period 1357-*c*.1362. And the unfulfilled suggestion that there be a clerk of the kitchen fits 1357-8 much better than any other date.

To sum up, the scale of fees worked into *LMM* was, I suggest, drawn up

1   *APS*, i, p. 547a.
2   *APS*, ii, p. 4a, c. 8.
3   *ER*, ii, p. 111, and unpublished 17th-century transcripts from lost rolls of David II; SRO, Exchequer Records, E.38/1.

in 1357-8; it was based upon research, too, for the fees proposed often repeat those of 1329-32. And it was carried into effect gradually, or in some respects (clerk of the kitchen, justiciar's fee) not at all. Taken with the evidence of the accounts, it represents a remarkable recovery of government after the collapse permitted by Robert Stewart, and a determination that the king's officers were to be held to their functions by appropriate fees.

Unfortunately the matter in *LMM* cc. 8-14 is paralleled by few records which might show its accuracy. Clause 8 lays down fines before the justiciar, to which I shall return, and continues with those before the sheriff, burgh officers, chamberlain and their subordinates. How reliable is it ? The amercement before burgh officers, 8s., is stated in the Burgh Laws and confirmed in many burgh records.[1] There seem to be only two recorded amercements before the chamberlain, of 5mk in 1364 and 50s. in 1366-70;[2] the latter is *LMM*'s figure. I have not found any figure to compare with *LMM*'s amercement before the sheriff, which 'shall not exceed 16s.'; but the issues of the sheriff court were indeed modest. On the whole, these controls do suggest that *LMM* c. 8 reflects a real state of affairs in relation to these officers.

Whereas c. 3 deals with the justiciar *simpliciter*, c. 8 distinguishes amercements before the justiciar north, or 'on this side',[3] of the Scottish Sea, from those 'beyond the Scottish Sea in Lothian and in parts there from the Water of Tyne up to Forth'. Scocia, the thirteenth-century name for the northern justiciary, gave way to 'north of the Scottish Sea' or 'north of water of Forth' (indifferently) until the mid-fourteenth century, when the latter prevailed.[4] The southern justiciar, even more strikingly, was 'of Lothian' from *c*.1220, when the office first appeared, until 1360, and used a seal with that title in 1366;[5] in 1363, however, the office of clerk of justiciary 'south of the water of Forth' was conferred by the king, and from 1368 at least justiciars used that form invariably.[6] *LMM* represents a conflation of three usages: Scottish Sea, Lothian, and [water of] Forth. The first was used of the northern justiciar, and indicates the work of a northern clerk writing after *c*.1280, for he did not use Scocia; for the southern justiciary, he corrected himself to Lothian, and was perhaps therefore writing no later than the 1360s, but probably borrowing (as we shall see)

1   *APS*, i, p. 340, c.39.
2   *ER*, ii, pp. 150, 261.
3   Bute has 'citra mare Scocie', Arbuthnott 'ex parte boriali maris Scocie'. But all MSS describe the southern justiciary as 'vltra mare Scocie', so the northern origin of the treatise is not in doubt.
4   *RRS*, v, nos. 12, 140, 156, 202, 285, 311, 487; vi, nos. 3, 50, 70, 234, 462. I have not found any use of Scocia meaning 'north of Forth' after the accession of Robert I. I am much indebted to Dr Hector MacQueen for guidance on the question of the justiciars' titles.
5   *RRS*, vi, no. 237; Raine, *North Durham*, no. 326.
6   *RMS*, i, no. 100, a transcript which made Adam Forrester clerk of the rolls of custom — a misreading of 'iust'' as 'cust''; cf. *RMS*, i, app. II, no. 1461, where another mistake made him clerk north of Forth. He was clerk in the south: Raine, *North Durham*, nos. 147, 326. *RRS*, vi, no. 503; *ER*, ii, pp. 394, 462.

from an assize of 1245.

In the description of Lothian, the 'Tyne' can scarcely have been that of East Lothian, and Cosmo Innes may have been right to see this annexation of Northumberland as linking the treatise to the era of David I — not (as he suggested) by preserving an archaic fragment, but rather as deliberate and unauthentic archaising.[1] It seems more likely to me, however, that 'Tyne' in *LMM* had replaced a word which made sense, namely 'Tweed', perhaps because the Tweed valley had passed out of Scottish control after 1332. The need to define Lothian denotes a northern source; the use of 'Tweed [if I am right] to Forth' would suggest a source written between 1220 and 1332. If 'Tweed' did become 'Tyne', it probably happened in the early transmission of *LMM* itself.

In c. 8 the amercement of the justiciar north of Forth is given as eight cows, south of Forth as £10. Few as the sheriffs' accounts are, those for 1359 and 1392 are agreed that the standard amercement before the justiciar in northern sheriffdoms was 40s. for non-compearance of a suitor or an indicted person.[2] *QA*, a fourteenth-century treatise, explains that when a judgement of a barony court was challenged and falsed before the sheriff, a single amercement was due; but when that of the sheriff court was falsed before the justiciar 'each suitor [i.e. each member of the assize] before the sheriff should be amerced £10'. Elsewhere, in an obscure chapter, *QA* confirms that the £10 amercement for falsed judgement could be exacted only in the justiciar's court.[3]

The accounts rendered by sheriffs certainly show largish sums raised from justice ayres: in 1357-8, from Fife £88 13s.4d. and £45 16s.5d., from Forfar £187 6s.8d., from Inverness £186.[4] Only the last two could represent a falsed doom. All were probably sums raised in amercements of £10 from those convicted before the justiciar. The amount of this fine suggests that it had its origin in the ancient royal 'forfeiture' of £10, known since David I's reign, and which was still being levied in 1435 for breach of the king's protection.[5] Under Robert I and David II it was still the sanction for forest offences, but was also threatened in one of the earliest cases heard after the king's return in 1357. Arbroath Abbey complained successfully to the king in council of purpresture by an Aberdeen burgess on its lands; the sheriff was told that a recurrence would demand 'our full forfeiture of £10 for our needs'.[6]

But the twelfth-century royal forfeiture for non-payment of teinds in

1   *APS*, i, p. 51 (n. 5).
2   E.g., *ER*, i, pp. 559, 570, 587; iii, pp. 265-6, 268.
3   *APS*, i, pp. 649, c. 9; 651, c. 21. In 1385-7 the sheriff of Lanark was responsible for collecting £160 for 16 amercements 'in which Sir Robert Danielston was condemned by declaration of parliament for a falsed doom'; *ER*, iii, p. 164.
4   *ER*, i, pp. 561, 570, 590.
5   *ER*, iv, p. 670.
6   *RRS*, v; vi, indices under 'Forfeiture'; *RRS*, vi, no. 152.

Glasgow diocese was twelve cows, in Moray diocese eight cows;[1] an assize attributed to William I prescribed an eight-cow forfeiture for any *judex* of 'Scocia' (that is north of Forth) who left the king's court without leave, suggesting that it was an archaic penalty applied in traditional obligations to the king.[2] It may have survived in some thirteenth-century tract which became the source of *LMM*. Thus one of three clauses preserved now in the Bute MS among the 'statutes of King Alexander', on the penalties for failure in certain pleas, states that if the accused in a plea of the crown (which would be heard before the justiciar) falls in trial by battle, his sureties must answer to the king for 'ix cows and a colpindach'.[3] The rare word 'colpindach', a calf, links the clause to c. 8 in *LMM* where the crownar's amercement was a colpindach (north of Forth), a colpindach or 30d. (south of Forth), but the clerk, wherever he was, 2s. But there is also a link to a scale of forfeitures imposed in 1221 upon those who stayed away from the army: for a thane, nine cows and a calf ('iuvenca').[4] The otherwise bizarre calf in these clauses may be explained by *LMM* as the crownar's fee; it brings all three clauses into association and points to the reign of Alexander II as their period of origin — though this does not mean that *LMM* c. 8, or something like it, was necessarily written then.[5]

The colpindach is the only 'fact' to occur twice in *LMM*, for c. 3, which lists the sustentation of the justiciar and his clerk, is really about the crownar's rights to forfeitures and escheats:

> ... from each man amerced or sold [pardoned?] a colpindach or 30d. And for a man rightly cleared, nothing. And for a man condemned to death before the justiciar, the crownar shall have the unshod dantit [broken] horse, cattle up to 20, goats and sheep up to 10, the corn from broken stacks and ricks, and all utensils ben the house, that is under the crook hanging above the fire. Also, for a man not found at the time of attachment, the crownar shall stay at his house for a day and a night and he shall have sustenance for himself, two men of his own, two brought along as witnesses, and for his clerk; or else 2s. but no more. If he cannot find the indicted man, he should arrest all his movable and immovable goods under safe pledges to answer before the justiciar.

This detail suggests that the author of *LMM* was particularly concerned

---

1   *RRS*, ii, nos. 281, 507; *Glasgow Reg.*, i, no. 138. In the barony of Carnwath in the early 16th century a penalty of 'the best aucht' was laid down; the editor took this to mean 'the best eight oxen' (*Carnwath Court Bk.*, pp. cx, 100, 165), but the phrase means 'the most valuable possession'. For an escheat to the crown of 16 cows in 1434, *ER*, iv, p. 595.
2   *APS*, i, p. 379, c. 26.
3   *APS*, i, p. 402, c. 13.
4   *APS*, i, p. 398, c. 2.
5   The forfeiture of so many cows and a calf is also found in two clauses printed as cc. 14 and 15 of the Assizes of David I (*APS*, i, p. 320a). The vernacular translation uses 'colpindach' for 'calf'. I suspect that this vernacular word in *LMM* and Alexander II c.13 may be a consequence of early translation in a collection of laws made in the time of Alexander II. But the early laws will bear a great deal more textual study.

about the rights and duties of the crownar, the officer of the justiciar in each sheriffdom, charged with preparing for the ayre by securing attachments, pledges, and compearance, and with presenting the dittay (indictment) roll in court. By the 1360s the sheriff, not the crownar, was charged with collecting forfeitures and escheats (from which a profit might be made), and a crownar might be fee'd,[1] but legislation of 1487 and 1488 shows that the crownar was then taking a share of escheats, including dantit horse, very like those laid down in *LMM*.[2] It is unlikely that this right was established by *LMM*; rather *LMM*'s accuracy about dantit horse and other goods — but not the colpindach — is confirmed by it.

The crownar was one key to firm justice in the kingdom. In 1357 David II was adjured to appoint good and sufficient sheriffs and crownars and to hold the justice ayres in his own person, and the records suggest that he did indeed drive the ayres in person and by justiciars-depute with some energy.[3] But if the justice appeared and the defenders did not, little was achieved — and one ayre of David II was certainly abandoned from 'weakness'.[4] Possible obstacles to attachment, to the taking of pledges and to summons were many, for a defender could take refuge in a franchise, in a neighbouring sheriffdom, or in technical excuses (assonzies).

Clauses 3 and 8 are a transition to the third part of *LMM*, which continues the emphasis upon royal authority by attacking obstacles to the administration of justice before the justiciar, without mentioning that office. In a largely imaginary hierarchy of private courts (c. 9) — of baron, knight, sub-vassal, *armiger*, *subarmiger* — royal laws were to be administered by each 'taking' them to his court from the court above; each rank was to observe the law and enjoy half the amercement of the rank above, down to the *subarmiger* who 'shall observe common laws' and may take only a cow, a sheep or 3s.[5] This elaborates in a fanciful way the order of Robert I that the statutes of 1318 were to be proclaimed and observed in courts of the king and of 'prelates, earls, barons and others who have courts', to whom copies of the statutes were to be given.[6] The debt is confirmed by the use in *LMM* c. 9 of 'communes leges', for 'communis lex', not a phrase I have

---

1   In 1382 the crownar of Angus and Mearns had a fee of £4 annually (*RMS*, i, no. 735); but in 1359 the crownar had been exacting fees of the traditional kind from the regality of Arbroath, from which he was now excluded, and presumably from the royalty (*RRS*, vi, no. 223; *APS*, i, p. 525, no. 24); this regality included Tarves in Aberdeenshire, and exclusion would therefore affect the crownars of Angus and Aberdeen.
2   *APS*, ii, pp. 177, c. 7; 183, c. 18.
3   *APS*, i, p. 492a; *ER*, ii, p. 82, and index under 'Justice-ayre' and 'Justiciary'.
4   *ER*, i, p. 558.
5   Clause 9 gives to barons laws, amercements and fees as before the sheriff; knights have the same laws but not courts of life and limb, only of wrang and unlaw, 'and half amercement *sicut* in a baron's court'; sub-vassals have 'half amercement *sicut* in knights' courts'. This clause makes no sense if *sicut* be translated 'as', for that would make the baron's forfeiture the same as in the sheriff court, and also half of it. *Sicut* must be translated as 'of such as is': the knight has half the amercement as [the amercement is] in the baron's court.
6   *APS*, i, p. 466a; *RRS*, v, no.139 (p. 405).

found elsewhere, also occurred in the statutes of 1318.[1] But that is the sum of the debt, and other provisions of 1318 about the justiciar are missing from *LMM*, deliberately ignored by the author.

The last part of c. 9 provides that a thief taken in any of these jurisdictions was to be taken to and tried in the baron's court in the barony where he was taken, and the baron was to have his goods as escheat unless the lord of the toun where he was taken had that right by charter; the point seems to be to diminish the rights of knights' and other inferior courts. But according to *QA*, the thief's lord took his goods found in his lordship, provided he had a court in which escheats could be adjudged,[2] and *LMM* seems to be out of touch with reality in its picture of competing jurisdictions.

Other clauses are hostile to private justice: any baron or earl who protected a malefactor would lose his court (c. 10, a summary of a genuine assize of William I). There is a logical progression from c. 11 where the four pleas of the crown were to be tried before the justiciar — the king to have all escheats, the barons nothing — to c. 12, on traitors against the king, whose lands and goods the king was to have, 'unless special grace intervened', but c. 12 then bursts out against other criminals who hold of the king: 'the king shall have everything and the criminal shall be condemned to death without ransom', effectively a repetition of c. 11. Thus a traitor was offered the possibility of 'special grace', other criminals were hauled off to the gallows — yet surely there was no crime more serious than treason. The author of *LMM* was, in fact, more concerned about crime against the subject than about that against the king, as the tone of this outburst and of c. 14 shows. In c. 13 malefactors holding of barons are to be condemned in the same way as crown tenants, except for the four pleas of the crown, with which a baron had no power to intromit, the latter point already being dealt with in c. 11.

In all this material *LMM* shows no debt to *QA*, not even to *QA* cc. 14-15, which deal with homicide and treason (*prodicio*).[3] But there is certainly a debt to the assize of 1245 setting up a general ayre by the 'justiciar of Lothian' into '*murder, robbery* and similar felonies *pertaining to the lord king's crown*'; the chattels of these malefactors belong to the king. For *theft* and homicide, chattels belong to the lord where they have chattels; when convicted before the justiciar, they are to be handed over to the *baron* to do justice to them in their *baronies* 'without any *ransom* or other remedy, *unless* the *grace* of the lord king *intervenes*'.[4] The borrowings (italicised) of *LMM* are verbal, rather than of substance, but probably explain its reference to the justiciary of 'Lothian'.

1   *APS*, i, p. 467, c. 2; *RRS*, v, p. 406, c. 2: 'communis lex et communis iusticia fiant tam pauperibus quam diuitibus ...'.
2   *APS*, i, p. 649, c. 8.
3   *APS*, i, p. 650.
4   *APS*, i, p. 403, c. 14.

The denunciation in c. 14 of malefactors who slay a man, his children and friends because they covet his wife and possessions is highly rhetorical; it might relate to a specific occasion, but of several episodes from the period 1363-*c*.1400 where a woman was forced into marriage for her wealth, none was preceded by *LMM*'s slaying of the innocent. The clause deals with punishment of the slayers only indirectly:

> ... they have ordained that if such malefactors are reset and maintained by earls and barons of the realm unjustly against the law of God and the world (since, as it is said, *Who kills by the sword, by the sword shall he perish* [Rev. xiii, 10]), of all these malefactors the king shall have *sequela*, as much of the earls and barons as of those malefactors, because all the magnates who do this are false and perjured against the king and people of the realm.

The clause is a highly coloured version of the assize of William I (which also lies behind *LMM* c. 10), whereby bishops, earls, barons and thanes swore not to receive robbers, murderers, rapists or other malefactors, but to bring them to justice; anyone breaking this assize would lose his court for ever — that is, jurisdiction would pass to the king.[1] In *LMM* c. 14 the provision for loss of *sequela* defies translation, for the word meant astriction to something, as tenant to mill, or children to parent. *Sequela curie* was a fairly well-known expression for liability to suit of court,[2] but here the biblical quotation, implying that the slayer should be executed, justified the king having *sequela*, which must be his *secta*, duty to prosecute and, perhaps, jurisdiction.[3] As in other clauses of *LMM*, the royalist tone is accompanied by hostility not only to evil-doers but, even more strongly, to those magnates who reset them — and hence who defied the functioning of the crownar.

This survey of the contents of cc. 8-14 reveals their ignorance of *QA*, and some use of a few of the assizes and of the 1318 statutes, though in no case is anything quoted at length. The amercements of £10, eight cows and colpindach found in cc. 3,8 may well be penalties borrowed from assizes of William and Alexander II (such as the fine for a thane who has stayed away from the host) and woven into *LMM* to fit its supposed antiquity. Certainly the compiler of *LMM*, in using earlier material for phrases and even some substance, ignored much that was relevant to his concern with obstacles to justice. Thus in *LMM* a scale of royal fees was married to a much less systematic statement about forfeitures in the justiciary court (cc. 3, 8), and that to a tendentious account of other courts and penalties, which owed something to a collection of early assizes, but was otherwise an original composition. Why these disparate matters were brought together remains a

---

1   *APS*, i, p. 377, c. 20.
2   W.C. Dickinson, 'Sequels', *Juridical Review*, lii (1940), pp. 117-21.
3   'Secta' can mean suit of court or right to pursue action in court as in 1318, c. 20 (*APS*, i, p. 472a; *RRS*, v, p. 412); 'sequela' seems to be used here in the latter sense.

puzzle, though the justiciar seems to be the link between the first section, on fees, and the third, on penalties. Overall it seems to me that the first section was borrowed information designed to give verisimilitude to the hectoring of the third section, in which lay the purpose of the author. Although it makes no mention of the Highlands, *LMM* was clearly produced by a clerk familiar with the justice ayres north of Forth, a man angered by maintenance and its perversion of justice, anxious to bolster royal authority in the justiciary courts, and interested in the rights of the crownar.

There is, I believe, a link between *LMM* and a treatise on justiciary courts different from the later printed *Ordo*, which is preserved wholly in the Bute MS as *Ordo Justiciarie* in 12 chapters, and partly (the first five chapters) in the Arbuthnott MS as *Modus Tenendi Placita Justiciarie*; here I call it *OJ*. Cosmo Innes printed most of Bute's version, scattered through pages which he titled *Fragmenta Collecta* — an extraordinary description of dispersion! It is described below, using both MSS, in the Appendix.

Clauses 1-2B are a cursory treatment of indictment and judgement in the justiciary court, that is to say, the work of the crownar, the clerk and the justiciar; the rest of *OJ* is a melange whose common theme is that court. It reports cases heard there at Cupar and Perth; another heard at Forfar required the justiciar's intervention. Clauses 2B and 11 addressed a common problem but were not brought together, implying accretion of the later clauses. Clause 11 of *OJ* is likely to be earlier than legislation of 1384 dealing with the same matter.[1]

The text did not grow by haphazard accretion, for c. 1 refers to c. 3, and the apparent change in form between c. 4 and c. 5 is reversed when c. 6, a judgement in parliament, reverts to the form of c. 4. Was c. 3 real or apocryphal legislation? In the opening phrases, 'Ordinatum fuit coram rege in pleno parliamento apud Sconam per regem et communitatem', the words 'coram rege' cannot be paralleled. Sentences introduced by 'Nota quod' were usually glosses, but here each 'Nota' deals with a distinct matter: the clause is made up of notes of the gist of three separate provisions, which are likely to have been genuine legislation — c. 4 was a fourth such provision. The singular 'communitatem' points to parliaments held before 1357 perhaps in 1341-6, which left no surviving legislative record.

In c. 5 the abbot of 'L' [Lindores?] sought judgement against an earl of 'B' ['Buchan'], who alleged he need not answer a royal brieve of poinding for a debt, because he had from the king, on whose service he was absent, a letter of respite until his return. The abbot argued successfully that the debt had been confessed in court and could not be superseded as the case had been decided. The forum was not specified, but was probably the sheriff court, of which this may have been a record drawn up for an appeal to the justiciar or the king. Of the two possible dates, 1340 and 1360, the issuing

1   *APS*, i, p. 550b.

of royal brieves is improbable at the former, and the latter fits known circumstances. There was no earl of Buchan in 1340 or 1360; an initial 'M' was no doubt misread as 'B', for the earl must have been the impecunious Thomas earl of Mar who, at Kildrummy on 19 August 1359, promised to cease harassing the abbey of Lindores for homage, fealty and suit of court from its lands in Garioch; but he also promised them 8mk yearly in satisfaction for the harassment.[1] The action reported in *OJ* c. 5 must have arisen from the earl's reluctance to pay this 8mk; and his absence from Scotland reported in c. 5 is well attested. Mar is not found as witness to royal charters from 16 November 1358 until 26 October 1360, by which date he must have returned to Scotland.[2] The action raised in 1360 after his return is a *terminus ante quem* either for all 12 clauses of *OJ*, or for the first five only — but cc. 6-12 are unlikely to be much later for all 12 were surely the work of the one collector.

The materials in *OJ* deal with certain civil matters, and are much concerned with attachment also dealt with in legislation of 1357-89 and in *LMM* c. 3. But *OJ* also covers restrictions on baronial courts and the offence of reset, which were central interests of *LMM*, and this suggests that the compiler of *LMM* was also the collector of *OJ* and perhaps the author of its notes and summaries, which were made in or after — and not long after — 1360. *LMM* came from a pro-royalist environment familiar with northern justice ayres; *OJ* knew something of statute law and collected cases from northern sheriff and justiciary courts.

In whose circle might a clerk have composed these tracts? A connection with the northern justiciary seems clear, and at a date after the dimission of William earl of Ross, who held that office in 1339 and in parliament repelled a claim to it by Sir John Randolph in 1344. Ross was the king's cousin, but in 1346, at the muster for invasion of England, he murdered Ranald MacRuari and withdrew to the hills, thus escaping the debacle of Neville's Cross. He acted as justiciar during the king's captivity. In 1358 William Meldrum, who had been sheriff of Aberdeen, was depute justiciar, possibly for Ross, but in 1359 and 1367-8 Sir Robert Erskine was northern justiciar, and may have held office throughout the 1360s;[3] he was a frequent witness of royal charters, and chamberlain and justiciar of Lothian for a time. He married a daughter of the marischal, and acquired lands in the Garioch from the earl of Mar, but he was not active in northern society, and should be seen as close to the court; I do not see him as the inspirer of the two tracts.

The northern justiciar in 1370 was Sir William Dishington, lord of

---

1   *Lindores Cart.*, pp. 199, 205.
2   *RRS*, vi, nos. 199, 244.
3   *RRS*, vi, nos. 228, 230; *RMS*, ii, no. 3717; *APS*, i, p. 504, an appeal from Erskine as justiciar to the parliament of June 1368. Sir Hugh Eglinton and Erskine were 'justiciars' and 'justiciars of Scocia' in 1360 (*RRS*, vi, no. 228; Fraser, *Menteith*, ii, no. 29). 'Scocia' here meant, I believe, 'Scotland', and Eglinton was justiciar south of Forth.

Ardross in Fife, who from August 1364[1] until the end of David II's reign witnessed 40 out of 94 royal charters, almost always appearing last among the knights. Dishington had briefly been steward of the household in 1361-2, and succeeded Sir David Wemyss as sheriff of Fife by 1362, holding office until David II's death.[2] From 1362 until 1370 he was also responsible for building work at the new kirk of St Monans, endowed by the king after escaping from drowning in the Forth.[3] He served in parliament and on its committees in the late 1360s,[4] put his seal to the tailzies of the throne in 1371 and 1373, but was conspicuously absent from the court of Robert II, who replaced him as sheriff of Fife and justiciar;[5] he was David II's man. His duties as steward and justiciar, and the three reports in *OJ* from cases heard at Cupar, point to authorship in his circle. But he had no connection with Aberdeen and was remote from the problem of reset of criminals which afflicted the Grampian region.

The strongest case seems to me to be that for a man who was both sheriff and crownar, and could well have acted as depute justiciar to Erskine: Sir Walter Moigne, a frequent witness of charters of Thomas earl of Mar.

Thomas succeeded to his earldom after 1347, in the aftermath of Neville's Cross at which two men of Mar, Sir Laurence Gillibrand and Sir Walter Moigne, were taken prisoner. In 1350 an embassy of Mar's men, including Gillibrand and Master John Cromdale, went to England to negotiate David II's release.[6] Earl Thomas was clearly hard up — in August 1357 at Bruges he sold the barony of Foveran for 1,000 écus,[7] and for the next six years he disposed of assets at a striking rate. But he was in Flanders at the same time as his clerk John Cromdale, who was there on David II's business, raising money for the king's ransom.[8] Perhaps to help Mar's finances, perhaps as a reward for labours in securing the king's release, Thomas received the lordship of Garioch from David II in January 1358. About this time he became the king's chamberlain,[9] acting through a lieutenant from November 1358 to April 1359 when he was furth of Scotland.[10] He issued charters at Kildrummy on 9 July 1358, and on 19 August and 2-15 September 1359,[11] but between these dates he was at Westminster on 24 February 1359, when he made an indenture of service

1   *RRS*, vi, nos. 327, 462.
2   *ER*, ii, p. 358.
3   *ER*, ii, index under 'Dishington'.
4   *APS*, i, pp. 506b, 507a, 534a.
5   *APS*, i, pp. 546, 549b; *ER*, ii, p. 366. Alexander Lindsay was justiciar north of Forth before 1373; *ER*, ii, p. 435.
6   *Rot. Scot.*, i, pp. 678b, 736a.
7   SRO, Mar and Kellie Muniments, GD.124/1/107, badly calendared in HMC, *Mar and Kellie Report*, p. 6.
8   *Rot. Scot.*, i, p. 807b.
9   *RRS*, vi, no. 176.
10   *ER*, ii, p. cxxv and refs. For Walter Biggar as acting chamberlain, *ER*, ii, pp. 4-6.
11   HMC, *Mar and Kellie*, pp. 3, 6; *Aberdeen-Banff Ills.*, iv, pp. 716-19.

with Edward III,[1] whom he had gone to visit on behalf of David II.

Mar undertook to serve Edward III in his wars against all men except David Bruce, in return for £400 annually or a wife of equal value. While it is true that the indenture provided for compensation should Mar lose his Scottish lands, this could arise only with a breach of the Anglo-Scottish truce, and the undertaking was not really hostile to David II. It showed that Edward III expected war with France to break out again, with a consequent need for captains and colonels, and that Mar, like David II, needed money. He reported home in the summer of 1359, then returned to England to earn his wages, with a safe-conduct to go to France.[2] But peace with France broke out, neither captain nor colonel was needed, and by 28 May 1361 the earl was back at Kildrummy.[3] He had done nothing to earn King David's wrath, and the appointment of Gillibrand as steward of Queen Joan's household in 1357 and of Moigne as steward of King David's household late in 1359 suggest that both before and after his chamberlainship Thomas had considerable influence with the king, in whose release he had taken a marked interest.

The claim that he fell from favour because of his agreement with Edward III is not found in any early comment. According to Bower the king went north in 1361 to escape a recurrence of the plague, and, because of an unspecified disagreement with Earl Thomas, besieged Kildrummy, which was surrendered. He handed it over to the keeping of Sir Walter Moigne and Ingram Winton. Earl Thomas left Scotland but in due course returned, was restored to favour and regained his possessions; the king spent Christmas in Kinloss.[4] Andrew Wyntoun claims Mar was out of Scotland when the siege took place and unambiguously placed these events in the year following (a word Bower failed to include) 1361, that is between 25 March 1362 and 24 March 1363. Sir Thomas Gray's *Scalacronica*, written very shortly after the event and an independent source, also places the siege after Christmas 1362.[5] David issued acts at Kildrummy and Aberdeen in September-November 1362, when Aberdeen Cathedral business was done; Mar was definitely present on 1 November 1362. King David was at Kinloss on 24 December; on his way south he made grants dated at Aberdeen on 15-20 January 1363.[6] This is the month, four years after Mar's Westminster indenture, when the king must have besieged Kildrummy and secured its surrender. Why did he do so?

Gray first reports a cause: extortions by Mar and his men on neighbouring people 'as was alleged against him by the king'. Then he

1   *Rot. Scot.*, i, p. 836; *Aberdeen-Banff Ills.*, iv, pp. 156-7.
2   *Rot. Scot.*, i, p. 842. *Aberdeen-Banff Ills.*, iv, p. 715, lists all safeconducts for Mar from *Rot. Scot.*
3   *Aberdeen Reg.*, i, p. 89.
4   *Chron. Bower* (Goodall), ii, pp. 364-5.
5   *Chron. Wyntoun*, vi, pp. 249-51 (Wyntoun has the bishop of St Andrews pass Christmas at Kinloss, the king at Elgin); *Scalacronica*, p. 202. *Aberdeen-Banff Ills.*, iv, pp. 719-20, quotes Bower, Wyntoun and Gray on this episode.
6   *RRS*, vi, nos. 275-86; *Aberdeen Reg.*, i, pp. 96-102.

claims that the castle was surrendered and restored to the earl for £1,000 to be paid to the king at the end of five years on pain of losing castle and earldom; this *'mouvement'* arose 'chiefly' from an appeal to battle by Sir William Keith (the marischal) against Mar in the king's court. They appeared in the lists at Edinburgh, and the case was settled by the king, who showed greater favour to Keith.

We can only conjecture what was afoot between the king and Mar in 1362-3. The chamberlain's account for August 1362-March 1363 shows that the earl received 1,000mk, the king 500mk. I suggest that this was a loan by the crown of 1,000mk to Mar, repayable after five years (Gray's period) as £1,000; 500mk of this was interest which the king drew immediately from the chamberlain for his own uses. The loan was secured by 'an agreement about the lands of the earldom of Mar',[1] but was to enable the earl to go into an exile to which (I suggest) he had been sentenced by the king in the lists after combat with Keith. All these events would have occurred in the autumn of 1362, before the taking of Kildrummy in January 1363, itself attributable to Mar's evasion of the king's will by remaining in Scotland to extort further funds from the countryside.[2]

Perhaps the king seized Kildrummy as creditor's surety. Certainly Bower and Wyntoun are correct in saying that he entrusted it to Moigne who was sheriff of Aberdeen and is found provisioning it in 1363-4;[3] Gray may also be right in saying that it was restored to Mar — but after a year or so. Mar became a pensioner of Edward III,[4] but he arranged in December 1364 for his dependants to bring cattle from Mar to be sold to Newcastle merchants,[5] and so he still held his lands and was still hard up.

Gray appreciated the effects of this punishment: 'soon after that, there arose disagreement between King David and William earl of Douglas, *who had the sister of the earl of Mar as wife* [my italics], because of divers matters wherein it appeared to the said earl that the king had not shown him such fair lordship as he would have liked. So [Douglas] made a conspiracy ...'[6] — the conspiracy in which he was joined by Robert the Steward and the earl of March, and which David so successfully overcame. Mar and David II were barren and each had an heir presumptive, Mar's being Douglas, the king's being the Steward. If Mar were to fail in his repayment under the agreement of 1362-3, the king could foreclose on the earldom of Mar, and Douglas would lose his wife's inheritance.

But there was more to it than this, for the treaty of Berwick in 1357 made explicit provision for a default by the Scots in ransom payments, such

1   *ER*, ii, p. 164.
2   The king seems to have gone to Kildrummy about 13 October 1362: *RMS*, i, no. 108.
3   *ER*, ii, p. 166; Ingram Winton was presumably appointed constable in immediate charge.
4   He was paid 100mk in November 1363, and February 1364, 125mk in June 1365 and (probably) 25mk a little later; PRO, E.403/417, m. 20, 32; E.403/422 mm. 20, 29.
5   *Rot. Scot.*, i, p. 896a.
6   *Scalacronica*, pp. 202-3.

as had now occurred for three years: King David was to return to captivity or was to send two earls or two of the Steward, Douglas (not yet earls) and Sir Thomas Murray of Bothwell in his place.[1] In 1363 he effectively sent Mar, and the threat to the others was plain. The rebels' rhetoric was the king's mishandling of the ransom funds; their real agenda was a defence of lineage and heritage against the dangers of exile or even forfeiture. But Mar played no part in this rebellion and thereby may have won favour with the king. Thus Master Gilbert Armstrong, who had been a canon of Aberdeen since 1343, was a frequent witness of Earl Thomas's charters and was of his council, had, by May 1363, been appointed steward of the king's household and, at the king's nomination, provost of St Mary-on-the-Rock.[2] Evidently he had joined the king's household and affinity by the time the Steward's rebellion collapsed.

In June 1363 Sir William More of Abercorn granted his lands of Craigforth in Stirlingshire for life to Sir Robert Erskine (chamberlain when the Steward was regent and again in 1363-4), so that Erskine might mediate between More and the Steward, Mar, Keith, Sir Archibald Douglas or the men or relations (within the third or fourth degree) of Erskine.[3] This grouping of Mar with the Steward and with supporters of the king warns us against a simplistic 'two-party' view of politics in 1363; it suggests that Mar was still in Scotland in June when David II was embarking upon a high enterprise: a new treaty with England to cancel the ransom and make a Plantagenet heir presumptive to the Scottish throne. This proposal had first surfaced when the embassy of Mar's men went to London in 1350; now Mar, recipient of a loan from the king, went to London a few months before the king himself, and stayed there while terms were debated in 1364 and 1365, returning in 1365-6 after a new ransom treaty had been sealed. Mar had promised to serve Edward III in 1359 and was clearly on good terms with him; but David II may well have known of this and approved, for he was not averse to promising military help himself in return for cancellation of the ransom. Mar may have been his friend at the English court in 1363-5, and the taking of Kildrummy a brief fall from favour, after which David compromised by entrusting the castle to Mar's ally, the king's sheriff, Sir Walter Moigne.

The only family of his name recorded in Scotland in the early fourteenth century was associated with Berwick. But David earl of Huntingdon and lord of Garioch (d. 1219) had a household knight, Sir Bartholomew le Moyne, in whom the ancestry of Sir Walter is most likely to be found;[4] presumably the family was endowed in the Garioch, although I have found

1   *RRS*, vi, p. 187.
2   D.E.R. Watt, *A Biographical Dictionary of Scottish Graduates to A.D. 1410* (Oxford, 1977), p. 17.
3   SRO, GD.124/1/156, calendared after a fashion in HMC, *Mar and Kellie Supplementary Report*, p. 8; More's mediation was envisaged in friction between Erskine and the Steward or More's men and relations only.
4   K.J. Stringer, *Earl David of Huntingdon, 1152-1219: A Study in Anglo-Scottish History* (Edinburgh, 1985), p. 166.

no trace of their heritage. David II granted the Park of Drum in the Mearns
to Walter Moigne apparently between 1341 and 1346, in which year both
were taken prisoner at Neville's Cross.[1] In 1356 Moigne, now released,
proposed to go on crusade to Prussia with Sir Thomas Bisset and the
brothers Norman and Walter Leslie, an intention abandoned when the return
of David II became a probability.[2] Moigne witnessed most charters of the
earl of Mar in 1356 and 1358-62[3] and was of his council along with Sir
William Liddell (sheriff of Aberdeen in 1358), Gillibrand, Master Gilbert
Armstrong and others in 1358 × 1362.[4] He is found witnessing charters of
the marischal in 1351 and 1378,[5] but not in between; they were frequent co-
witnesses of Mar's charters and farmed Aboyne jointly from the crown,[6]
evidence of at least one common interest.

Moigne witnessed royal charters on 2 July 1359 and nine times between
December 1361 and March 1363, five of these charters being dated at
Aberdeen; four were witnessed also by Keith, but none by Mar, who was
abroad for part of the time.[7] In 1360 the king exempted his grant of Drum
from the 1357 revocation of earlier gifts, describing Moigne as his 'dear
bachelor', a status above that of a simple knight.[8] Like others who held the
office, Moigne did not witness in 1360-1 when steward of the king's
household, a sign of the modest importance of an office which he is most
likely to have obtained through Mar's influence. He replaced William
Liddell as sheriff of Aberdeen between 1359 and 1361 when the king gave
him that office and the crownarship for life;[9] unfortunately I can find no
evidence of the name of the sheriff at the time of *The Abbot of Lindores v.
The Earl of Mar*. Moigne accounted for the sheriffdom for the period
between March 1363 and December 1364[10] and held it certainly until 1366,
perhaps until 1368 or 1369, when he had been replaced by Sir Alexander
Fraser.

Moigne witnessed one further royal charter, at Dumbarton in February
1367, was exporting wool through Aberdeen in 1368, had a royal charter in
1370 but after the accession of Robert II appears only as witness to an
Aberdeenshire charter in 1378.[11] In the June 1368 parliament he was
appointed to a committee of royal servants enjoined to visit four key castles
and prescribe what was needed to make them defensible, to be carried out

1   *RMS*, i, app. II, no. 874; *Rot. Scot.*, i, p. 678.
2   *Rot. Scot.*, i, p. 797.
3   *RRS*, vi, nos. 294, 323; *RMS*, i, app. I, no. 128; *Aberdeen-Banff Ills.*, iv, p. 159;
Aberdeen-Banff Colls., pp. 537-9.
4   *Melrose Lib.*, ii, no. 528; *ER*, i, p. 551.
5   *RRS*, vi, no. 135; *Aberdeen Reg.*, i, pp. 121-4.
6   *RMS*, i, app. II, no. 1362; *Aberdeen-Banff Ills.*, iii, p. 293.
7   *RRS*, vi, nos. 220, 266, 289.
8   *RRS*, vi, no. 236.
9   *ER*, i, 586; *RMS*, i, app. II, no. 1387; for the sheriffs of Aberdeen, see *Records of the
Sheriff Court of Aberdeenshire*, i, ed. D. Littlejohn (New Spalding Club, 1904), pp. 407-10.
10   *Aberdeen Reg.*, i, p. 160.
11   *RRS*, vi, no. 500; *ER*, ii, p. 299; *RMS*, i, no. 368; *Aberdeen Reg.*, i, p. 124.

by the king.[1] The accounts do not suggest that any visitation took place, but the plan nonetheless attests to Moigne's presence in parliament and his acceptability to the estates and the king. His sudden loss of life-office and virtual disappearance from 1369 are probably to be explained, therefore, by personal circumstances, such as a stroke or other disabling illness.

Mar had come back from England by May 1366.[2] He had used attorneys in his litigation in 1360; probably he would now have a pardon, would require sasine of his earldom, surely worth more than £100, and might seek a charter confirming it — just those royal letters listed in *LMM*. He witnessed royal charters on 26 July 1366 and 22 June 1368 and was appointed to parliament's committee in September 1367 when he lost the lordship of Garioch under the act of revocation, the king holding it through 1368.[3] He had thought to visit Compostella in mid-1367 (and a year later Amiens)[4] probably to intercede for a child by his second wife, and was one of the many nobles who acted as commissioners for David II in renewing the truce in London in June 1369; he was back at Kildrummy on 20 August 1369,[5] though whether he paid off his debt of £1,000 or was excused is unknown.

During these years there was a litany of complaint about the responsibility of a small group of men for serious disorder. A provision of 1366 that rebels in the provinces of Atholl, Argyll, Badenoch, Lochaber and Ross were to be arrested and to pay specially to taxation as punishment[6] suggests that many 'rebels' had simply refused payment of the fairly heavy taxation of these years and sought reset in franchises, thus incurring the condemnation of parliament. In the June 1368 parliament the king publicly commanded the Steward (whose lordship of Badenoch, but not, it seems, earldom of Strathearn, had been under consideration for resumption in January), his two oldest sons and the earl of Mar to swear that the public would not be hurt by the inhabitants of their lordships, and that malefactors would not be allowed to cross them, nor be reset within them.[7]

This demand was in vain. The Steward was detained in Loch Leven castle late in 1368, and one of his his younger sons, Alexander, remained there for some three months until the March 1369 parliament, probably as surety for compliance with the statute.[8] In that parliament the Steward and his sons were again singled out, being required to do justice to malefactors

1   *APS*, i, p. 504a.
2   HMC, *Mar and Kellie*, i, p. 6.
3   *RRS*, vi, nos. 353, 400, 404; *APS*, i, p. 501b. The revenues from Garioch were £27 5s.7d. plus a belated £32 for Martinmas 1367; £56 plus a belated £16 for Martinmas and Whitsun 1368 (*ER*, ii, pp. 288, 302, 341). The first figure may be actual rents, but the others, all multiples of £8, suggest that the chamberlain had farmed the revenues out. Garioch must have yielded Mar between £60 and £100 annually.
4   *Rot. Scot.*, i, pp. 915b, 924b.
5   *RRS*, vi, p. 470; *Rot. Scot.*, i, p. 938; *Aberdeen-Banff Ills.*, iv, p. 722.
6   *APS*, i, pp. 498-9; 'ad contribucionem specialiter exsoluendum'.
7   *APS*, i, pp. 503a, 528a, 529a.
8   *ER*, ii, pp. 309, 347.

in their earldoms and other Highland possessions, not to reset them there, to ensure that taxes were paid, and to swear to fulfil this undertaking under penalties. This undertaking was to be extended to the earls of Mar and Ross, John of Lorn, Archibald Campbell and others,[1] evidently two groups from east and west — Ross, also lord of Skye and of lands in Buchan, could fall into either group.

The Steward must have failed again, for he lost his style as earl of Strathearn between July 1369 and April 1370.[2] But Mar was also in trouble once more, for he was imprisoned on the Bass Rock in 1370, or perhaps even 1369.[3] Little is heard of Ross after 1357; he became a creditor of David II, who granted him the casualties of the sheriffdom of Aberdeen 'up to a certain sum' and resumed them for a reckoning in 1368.[4] These revenues, which *LMM* reserved to the king, would otherwise have been collected by the sheriff, Moigne.

Earl William of Ross later described graphically how David II gave to Sir Walter Leslie that part of the earldom of Buchan which Ross should have inherited; when Ross sought from the chancellor letters making Robert Stewart, Mar, Keith and William Meldrum (a former sheriff of Aberdeen) his attorneys in an action for recovery, his clerk was arrested and held to ransom. Ross had to surrender lands at Forfar, and the king bamboozled him with questions based on Roman law, forced him to confirm the Buchan lands to Leslie at Inverness, and insisted he entail the earldom on his daughter and heiress and her husband, Leslie.[5] Earl William, the killer of Ranald MacRuari in 1346, comes closest to that lord of *LMM* c. 14 who slays a man and his relatives in pursuit of wife and heritage, and against whom the author feels such hostility as an enemy of king and people. Ross did not marry an heiress to MacRuari,[6] but the demand of *LMM* c. 14 that the king shall have the *sequela* of all slayers and resetters thereof, bears a remarkable resemblance to the treatment of Ross when his daughter was married to Leslie, for although *sequela* means more here, in common use the main 'sequel' of a man was his daughter, on whose marriage the lord had a right to merchet.

The writer who took so much interest in justiciar and crownar in *LMM*

1   *APS*, i, pp. 506b-7a.
2   *RRS*, vi, nos. 444, 447, 464, 466.
3   *ER*, ii, p. 357.
4   *APS*, i, p. 529b.
5   SRO, J. and F. Anderson Muniments, GD.297/193; accurately translated in *Ane Account of the family of Innes*, ed. C. Innes (Spalding Club, 1864), pp. 70-2; printed in abbreviated and badly garbled form in *Aberdeen-Banff Ills.*, ii, pp. 387-9. The marriage of Ross's daughter to Leslie took place in 1366 (*Scots Peerage*, vii, p. 240); the king was at Inverness late in 1369; *RMS*, i, no. 296. Ross's uncle had married one of the Comyn co-heiresses to the earldom of Buchan and secured an entail of all his lands to his older brother, Earl William's father (*RMS*, i, app. I, no. 11; app. II, no. 49; see also Alan Young's essay, 'The Earls and Earldom of Buchan in the Thirteenth Century', above, pp. 182-3).
6   But it is worth adding that MacRuari's sister and heiress, Amy, was married to John MacDonald Lord of the Isles — whose sister was the wife of William earl of Ross; thanks to the killing, the MacRuari inheritance actually went to Ross's brother-in-law (*Scots Peerage*, v, pp. 37-9; vii, p. 237).

could have been the clerk of justiciary, to whom it allows 4d. for each extract from the rolls, while *OJ* includes examples of such extracts. In October 1369, at Aberdeen, David II gave the office of clerk of the rolls of justiciary north of the Forth to a William de Camera, to be held at the king's pleasure;[1] William pretty certainly belonged to the de Camera family of Aberdeen, and was perhaps an illegitimate son of his namesake who had a charter of East Ruthven in Cromar from Thomas earl of Mar in 1356 witnessed by Keith and Moigne.[2] He remained in office until the 1390s, but in 1380 was also secretary to Alexander Stewart lord of Badenoch and northern justiciar,[3] a leading disturber of northern peace and leader of caterans, whose reported misdeeds in the 1390s occurred too late to have inspired the revulsion of the author of *LMM*. In fact, trouble had begun in Badenoch even during the king's captivity, for in September 1365, when at Kildrummy, perhaps for the restoration of Earl Thomas, David II had acknowledged that 'at the times of our absence out of the kingdom, both through the expulsion of certain tyrants who were then reigning and by defect of justice, the exercise of which lapsed at that time', tenants of the bishop of Moray in Strathspey and Badenoch had committed many crimes, punishment for which before the justiciars would cause devastation to ecclesiastical lands; the bishop was to appoint a bailie to try them and ensure redress to those who had been wronged and justiciars and crownars were forbidden to interfere.[4] This may be criticism of Ross as justiciar in the 1350s, but if so it is muted, blaming the times rather than the man.[5] When the king renewed the grant in 1367, his letters were directed impartially to justiciars of the crown and of the regality of Moray — so that the king was reinforcing the kind of franchise criticised in *LMM*.[6] By 1370 Alexander Stewart had apparently become lord of Badenoch, for he then promised to be protector of the men and lands of the bishop of Moray there.[7] Under Robert II he also became justiciar and king's lieutenant in the north, but his dealings with the bishop, who sought the exemption of his lands in Badenoch from Alexander's regality, were still conducted within the framework of the law in the 1380s. If anyone in this situation inspired the author of *LMM*, it was surely the bishop and his unruly tenants; but the bishop fits no other part of the bill and it is unlikely that *LMM* would have been composed by Alexander Stewart's secretary. However, since the clerkship of justiciary fell vacant in 1368-9, it is possible that de Camera's predecessor in office had also served Moigne or knew him well and was the author of *LMM* and collector of *OJ*.

---

1   *RMS*, i, no. 295.
2   *RRS*, vi, no. 294.
3   Watt, *Scottish Graduates*, p. 77, under '(a) William de Camera'.
4   *RRS*, vi, no. 348.
5   The use of the plural 'tirannorum' shows that something more than the 'reign' of Edward Balliol was meant.
6   *RRS*, vi, no. 374; see also no. 456.
7   *Moray Reg.*, no. 154.

In short, the disorders of Moray, well-documented in the cathedral's register, are less likely to be the background to *LMM* and *OJ* than the more obscure problems of the sheriffdoms of Aberdeen and Banff such as the feud between two lairds for whom Mar and the marischal went surety in June 1368.[1] The tone and burden of *LMM* and *OJ* belong to the environment of the late 1360s, or possibly 1370. *LMM* denounced magnates who reset, those answerable for the four pleas of the crown, and false traitors against the king, a phrase which particularly recalls the exiling of Mar and the risings of Douglas, the Steward and March in 1363, but which also had a particular relevance to the position of William earl of Ross and lord of half the earldom of Buchan. The treatise was written, I suggest, by a clerk connected with Sir Walter Moigne, formerly steward of the king's household, sheriff and crownar of Aberdeen, and ally and counsellor of the earl of Mar. The clerk had some familiarity with Fordun's chronicle and a devotion to the see of Aberdeen and its supposed founder, Malcolm II.

More important than the individual who wrote or inspired these treatises is the legal and political climate which they represent, a climate best summarised as 'reconstruction' but driven by an ambitious king with a programme for firm administration of law and order and accountable administrators. They show the franchise courts in flourishing condition (as the surviving regality and barony court records of the sixteenth century do not), but under heavy criticism as refuges for offenders and needing to observe the 'king's laws' — an early demand for that uniformity which James I's parliament was to lay down. As in many other respects, the brief thirteen years of David II's personal rule, although racked by political controversy, showed a will to rebuild royal authority without regard to the inherited privileges of an uncertain aristocracy.

1   *APS*, i, p. 506a.

Appendix

The Ordo Justiciarie

This Appendix provides texts, based upon the Bute and Arbuthnott MSS, of part of *OJ* and a guide to the rest as it is printed in *APS*, i. The Bute MS (NLS, MS 21246) = *B*; the Arbuthnott MS (NLS, MS 16497) = *Arb*.

c. 1 (based on *B* and *Arb*).

In primis vocandus est rotulus de plegiagiis attachiatorum[a] vt intrent attachiatos de quorum pena si non veniant[b] inferius[c] loco suo dicetur. Postea vero[d] vocandi sunt quatuor vel quinque[e] vel sex ponendi in assisam, et tunc querendum est ab eorum plegiis si velint[f] esse plegii quousque fiat judicium. Et si sic, ponendi sunt[g] attachiati[h] in curia, sin autem[i] custodiri debent firmiter[j] per coronatorem et ministros curie. Deinde eligenda est assisa et iurare debent magno sacramento. Deinde culpandi sunt per rotulum in presentia[k] assise. Et si ponant se in assisam tunc querendum est ab eis vtrum[l] consentiunt in personas assise an non, et remoueantur illi quos racio[m] exigit amoueri. Et nota quod causa sanguinis remoueantur ab assisa filii amulorum amicarum et infra.[n] Et cum venerit[o] assisa, recitanda[p] est inculpacio eorum, ad quemlibet inculpacionem assisa dicat veredictum suum.

Item clerici indictati et attachiati ac in curia presentati inculpandi sunt; si qui sint[q] tales quod priuilegium clericale gaudere debent et illud allegauerint,[r] liberentur diocesano[s] vel aliis ordinariis ad hoc potestatem habentibus sub saluis et securis plegiis, et inculpacio nominari debet et mitti ad diocesanum[t] vel alium potestatem habentem sub sigillo justiciarii.[u]

c. 2 (= *APS*, i, p. 727, c. 36).

Indicted burgesses and men of a regality may be repledged to the burgh or regality court. [*Arb* varies only in using singular (burgensis, homo, etc.).]

c. 2B (from *Arb*).

Vbi attachiatus debet pati judicium.

Item notandum est quod vbi aliquis attachiatus sit ad sectam partis vbi delictum siue contractus[v] commissum fuerit ibi debet pati iudicium licet alibi inueniat plegios sicut in diuersis commitatibus accidit. Sed tamen per statutum regis Roberti primi indictatus potest essoniari per seruicium regis et per infirmitatem lecti.[w] Item per statutum regis Roberti tercii secunda dies erit terminus peremptorius et 2a et etc.[x]

c. 3 (= *APS*, i, p. 737, c. 4).

In parliament at Scone it was ordained that assonzies for an indicted attached offender were not to be heard, but all done on the first day which used to be done on the fourth. Note that the penalty (referred to in c. 1 as 'inferius loco suo') for not entering an attached person who had been repledged is 12mk for avoidance of the statute. Note also that an action for wrong should not stop the king's indictment, even though the defender in the matter of wrong goes quit by exception or by assize, but if [unless ?] the action touches life and limb otherwise he shall not suffer judgement in the king's court or elsewhere. *Arb* reads near the end: non pacietur iudicium.

c. 4 (= *APS*, i, p. 737, c. 5).
In the same place it was adjudged that the malefactor must be tried before the person accused of reset of him is tried.

c. 5 (= *APS*, i, p. 735, c. 8; the following text is based on *B* and *Arb*).
De curia domini regis tenta apud Abirden.[y]

Memorandum quod in curia domini regis tenta apud Abirdene anno etc M CCC sexagesimo[z] comparuit abbas de L et peciit[aa] iudicium super quodam plegio inuento ex parte comitis de Buchan quod non teneretur respondere ad breue domini regis de districto impetratum[bb] per dictum abbatem super eodem comite[cc] de quodam debito. Comparuerunt P de L et T de A attornati dicti comitis recognoscentes dictum debitum et allegantes dominum suum non debere ad hoc compelli ibidem,[dd] pro eo quod rex concessit eidem comiti, qui in seruicio suo alibi occupatus fuit,[ee] litteram de omnibus querelis motis seu mouendis ponendis in respectu vsque ad redditum suum, que quidem littera ibidem[ff] per eosdem attornatos fuit ostensa; replicante parte dicti abbatis per dominum R de K senescallum suum[gg] quod postquam debitum suum fuit confessatum in curia supercessio non debet fieri in hoc casu, cum non fuit querela placitanda sed determinata per confessionem.[hh] Et iudicatum fuit pro parte replicante quod, non obstante littera illa, pro debito debet distringi.

B remains the only authority for cc. 6-12. Its text is not always intelligible.

c. 6 (= *APS*, i, p. 752, c. 25): Scone parliament.
Anyone alleging a privilege should not be compelled to pay or be judged, but if he finds a pledge should have a day to show [his privilege].

c. 7 (= *APS*, i, p. 731, c. 10): Cupar, justiciary court.
As between half-sisters, who should succeed a deceased brother?

c. 8 (= *APS*, i, p. 737, c. 6): [*ibidem*].
An accused of rape before the justiciar alleged that in a baron's court he had proved that he was under age, although he was more than fifty. Responded that this corrupt law of old men alleging boyhood had been quashed by the king in parliament at Scone, but the accused claimed that this could not be made retrospective to publication.

c. 9 (= *APS*, i, p. 732, c. 11): [*ibidem*].
Should the widow of a son who predeceased his father have a terce when the grandson succeeds?

c. 10 (= *APS*, i, p. 732, c. 11): [*ibidem*].
Can a husband infeft his wife or children after marriage.

c. 11 (= *APS*, i, p. 738, c. 7): Forfar.
The extradition of an attached person from one sheriffdom to another by letter of the justiciar.

c. 12 (= *APS*, i, p. 738, c. 8): Perth, justiciary court.
One accused of rape alleged he had been cleared in the court of the bishop of St Andrews. The king's doomsman or bailie should be present in the baron's court for a good judgement.

NOTES. [a] attachiamentis, *B*. [b] venerunt, *Arb*. [c] *B* omits inferius. [d] Postea si, *B*. [e] *Arb* omits vel quinque, perhaps correctly. [f] volunt, *B*. [g] sint, *B*. [h] attachiamenta, *Arb*. [i] sue assise for sin autem, *B*. [j] *B* omits firmiter. [k] principio, *B*. [l] *Arb* reads Deinde eligenda est assisa et si ponant se in assisa tunc inculpandi sunt per rotulum in presentia assise et querendum vtrum. [m] illi qui rotulo, *B*. [n] This sentence is found only in *B* and is clearly a gloss which has crept into the text. It is nonsense and I am unable to suggest an emendation. Sanguinis fits with MS fil' (filii), but et infra suggests (say) viginti annorum et infra. I suspect that amulorum amicarum represents two attempts to read the same word. [o] venit, *B*. [p] Both MSS have reintrand' but recitanda seems a justifiable emendation. [q] qui si sint, *Arb*. [r] *B* reads Item clerici indictati et attachiati et in curia presentati inculpandi sunt qui allegauerunt. [s] decano, *Arb*, *B*. [t] decanum, *B*. [u] A version of this second paragraph is printed in *APS* i, p. 738, c. 13, from the Cromertie MS version of *QA*. [v] contractus must be a misreading, but no emendation occurs to me. [w] *APS*, i, p. 468, c. 6 (1318). [x] *SHR*, xxxv (1956), p. 135, c. 4 (1404), and ibid. pp. 137-8. This sentence is an accretion, bearing on the next clause not this one; the 1404 statutes are the text before the treatise *OJ*. [y] De debito distringendo, *B*. [z] M CCC xl (sic), *Arb*. [aa] petit, *Arb*. [bb] tenetur respondere breui domini regis impetrato, *B*. [cc] *B* has communitas (in appropriate case) for comes throughout. [dd] *B* omits ibidem. [ee] fuerat, *Arb*. [ff] *Arb* omits ibidem. [gg] domini abbatis R de L et ... suum, *B*. [hh] cum fuit querela determinata per confessionem, *B*; placita, *Arb*, emended here to placitanda.

13

The Kin of Kennedy, 'Kenkynnol' and the Common Law

HECTOR L. MACQUEEN

The administration of the Scottish common law was primarily dependent upon local courts linked to central government by itinerant justiciars, all acting in the king's name and by royal authority; as such, it neatly illustrates that interplay between the crown and the localities which was the principal characteristic of the later medieval Scottish polity. The law itself was the product of diverse influences, reflecting something of the development of the kingdom. Most conspicuous was the Anglo-French or feudal element, but Celtic laws and customs still found their place. The contrast, even opposition, sometimes said to exist between them parallels the division usually held to have grown up in the later Middle Ages between the Scots and the Gaelic-speaking peoples of Scotland. Yet the law, both in content and action, in fact illustrates that such views are at best simplistic.[1]

This essay seeks to demonstrate the point with an examination of two disputes involving Gilbert, Lord Kennedy of Dunure in Gaelic-speaking Carrick. He straddled the Scots and Gaelic worlds of fifteenth-century Scotland. As justiciar, lord of parliament and guardian of the young James III, he was an important figure in royal government and the administration of the common law.[2] But the earlier of the two disputes concerned the office of 'kenkynnol', a word derived from the Gaelic *cenn ceneóil*, later *ceann cineil*, meaning 'head of the kindred' and denoting a form of lordship stemming from the Celtic legal tradition.[3] The kindred in question was the

---

1   See W.D.H. Sellar, 'The common law of Scotland and the common law of England', in *The British Isles 1100-1500: Comparisons, Contrasts and Connections*, ed. R.R. Davies (Edinburgh, 1988); H.L. MacQueen 'Scots law under Alexander III', in *Scotland in the Reign of Alexander III, 1249-1286*, ed. N.H. Reid (Edinburgh, 1990); and J. Bannerman, 'The Scots Language and the Kin-based Society', in *Gaelic and Scots in Harmony: Proceedings of the Second International Conference on the Languages of Scotland*, ed. D.S. Thomson (Glasgow, 1990).
2   Details of Gilbert's career emerge from the works on his brother, Bishop James: see A.I. Dunlop, *The Life and Times of James Kennedy Bishop of St Andrews* (Edinburgh, 1950); N.A.T. Macdougall, 'Bishop James Kennedy of St Andrews: a reassessment of his political career', in *Church, Politics and Society: Scotland 1408-1929*, ed. N.A.T. Macdougall (Edinburgh, 1983). See also *Scots Peerage*, ii, pp. 452-4. For Gilbert as justiciar, see HMC, *4th report*, p. 507, and *RMS*, ii, no. 812; as justiciar of Galloway, *ER*, vi, p. 574.
3   See W.D.H. Sellar, 'Celtic law: survival and integration', *Scottish Studies*, xxix (1989), p. 8; also H.L. MacQueen, 'The laws of Galloway: a preliminary survey', in *Galloway: Land and Lordship*, ed. R.D. Oram and G.P. Stell (Edinburgh, 1991), p. 131, and Bannerman, 'Scots Language and Kin-based Society', pp. 5-6.

prolific one of Kennedy, the numerous branches of which dominated Carrick society. The Gaelic character of this dispute contrasts with the other, later dispute, a legal action brought by Gilbert against Robert, Lord Fleming by brieve of mortancestor in a justiciary court held in the tolbooth at Dumbarton on 15 April 1466.[1]

This later case will be considered initially. It seems much more intimately connected with the Scottish side of Gilbert's life, as well as illustrating both the combination of centre and locality in legal administration, and a form of action derived from the Anglo-French legal tradition. The litigants were two of the leading figures of the day.[2] There were two justiciars specially constituted for the action, and they presided over a gathering of the suitors of the sheriffdom of Dumbarton. Also present were many royal councillors, including the bishop of Glasgow, the chancellor, Andrew, Lord Avandale, and a number of lords of parliament. After a legal debate determined by the justiciars with the advice of the barons and freeholders of the court, an assize of twenty-five good and worthy men was appointed to decide the case. Gilbert was in court, but was represented by his prelocutor, Master David Guthrie, who was armed with his *Regiam Majestatem* and *Quoniam Attachiamenta* and was himself a rising star in royal government. Fleming, who was absent, had two procurators, his son Malcolm and David Reid.[3] Events were formally recorded by Alexander Foulis, one of the foremost notaries of the time; he acted for Kennedy, while another notary, Andrew Hynde, took instruments on behalf of Fleming.[4] Almost certainly there would in addition have been present a crowd of curious locals, drawn to the court by such a gathering of notables, and there may have been yet others whose motive in attendance was less innocent. Through his procurators, Fleming explained his absence by saying that he had come to the court with his ordinary attenders, but had then heard that Gilbert, Lord Kennedy had come to Dumbarton with an army; hence he did not compear personally.[5] The atmosphere may therefore have been a charged one, of the kind sought to be avoided by contemporary legislation against men coming to court 'with multitude of folkis [and] armys'.[6]

The brieve of mortancestor was an action for the recovery of lands on the grounds that an ancestor of the pursuer had died vest and saised of them as of fee, that the pursuer was his ancestor's nearest heir, and that the defender was unjustly withholding them from him.[7] The lands put in issue

1   For the description which follows see SRO, Ailsa Muniments, GD.25/1/102.
2   For Fleming see *Scots Peerage*, viii, pp. 532-3.
3   On Guthrie see A. Borthwick and H.L. MacQueen, 'Three Fifteenth-Century Cases', *Juridical Review*, xxxi (1986), p. 149; on Reid see H.L. MacQueen, *Common Law and Feudal Society in Medieval Scotland* (Edinburgh, 1993, forthcoming).
4   See the instruments in the Fleming of Wigtown Papers: NLS, Charters 15566-71 (= *Wigtown Charter Chest*, nos. 38-41). On Foulis see Borthwick and MacQueen, '15th-cent. Cases', pp. 150-1.
5   NLS, Ch. 15566; noted *Wigtown Charter Chest*, no. 38.
6   *APS*, ii, pp. 16, c. 10; 51, c. 29; 176, c. 3.
7   See MacQueen, *Common Law and Feudal Society*, ch. 6.

by Gilbert's brieve, Easter and Wester Mains, Shirva, Bar, Bedcow and Wester Gartshore, were collectively known as the forty-merk lands of Kirkintilloch, and were part of Lord Fleming's barony of Lenzie.[1] Indeed, at an earlier stage of the proceedings, there had been a possibility that the assize might be held in Fleming's baron court.[2] Gilbert's claim was that these lands had been held of Fleming's predecessor by his grandfather, also Gilbert. The litigation began with legal debate on dilatory pleas proponed by Fleming's procurators against the court,[3] and an argument that the action should not proceed in the absence of the defender. This was met by Guthrie with reference to an indenture between the parties agreeing to submit themselves to an assize at the justice ayre of Dumbarton; then he cited *Regiam Majestatem* and *Quoniam Attachiamenta* to the effect that no essonzies or absence could prevent an assize going ahead once the parties had submitted to it.[4] The court decreed that the brieve should proceed at once to the assize. The assize was then 'fully advised by both sides', probably meaning that evidence was led, before finally holding that Gilbert should be infeft in the disputed lands. Two days later, he was given sasine at the lands by the sheriff of Dumbarton[5] and, a little after, the Flemings carried out a formal ceremony of protest, symbolically breaking the sasine as proceeding upon an illegal sentence.[6] At the end of the month, the saga was completed when the king granted a confirmation of what appears to have been the key document in the Kennedy case, a charter issued by Malcolm Fleming of Biggar in 1384, confirming a tailzie of the forty-merk lands of Kirkintilloch by John Kennedy of Dunure in favour of his son, Sir Gilbert Kennedy and his heirs male.[7] It was this Gilbert who was the grandfather of the 1466 claimant.

There we lose sight of the matter. That is unfortunate, as it appears that

1   Most of these lands are readily identifiable today as farms or settlements within a mile or two of the town of Kirkintilloch (see Ordnance Survey map grid reference NS, between eastings 65 and 72 and northings 72 and 76). Easter and Wester Mains, which cannot now be identified, were probably contiguous to the castle or 'peel' of Kirkintilloch, the caput of the barony of Lenzie (see NLS, Ch. 15556); memories survive today in the street names of Eastermains, in a council estate on the eastern edge of Kirkintilloch, and Westermains Avenue, a row of 1960s bungalows on the western edge of the town. The earliest Kennedy document relating to them refers to the lands as the 'villa' of Kirkintilloch (SRO, GD.25/1/9, pre-1384). There are several candidates in the hilly country around the town for the 'Bar' of the documents. For a convenient map of the barony, see T. Watson, *Kirkintilloch: Town and Parish* (Glasgow, 1894), facing p. 46. See also *The Court Book of the Burgh of Kirkintilloch, 1658-1694*, ed. G.S. Pryde (SHS, 1963), pp. 41 (n.), 52 (n.), 139 (n.). The barony was a detached part of the sheriffdom of Dumbarton because the Flemings had held the sheriffship in the 14th century.
2   SRO, GD.25/1/97; NLS, Ch. 15561.
3   NLS, Chs. 15567-70 (= *Wigtown Charter Chest*, nos. 39, 40); SRO, GD.25/1/102. For an account of Scottish medieval pleading see H.L. MacQueen, 'Pleadable brieves, pleading and the development of Scots law', *Law and History Review*, iv (1986).
4   For the indenture see SRO, GD.25/1/97, and NLS, Ch. 15561 (= *Wigtown Charter Chest* no. 34). The *Regiam* references are to IV, 49 and 51 (*APS*, i, pp. 639-40); the *Quoniam* one to ch. 35 (*APS*, i, p. 654).
5   SRO, GD.25/1/103.
6   NLS, Chs. 15572-5 (= *Wigtown Charter Chest*, no. 42).
7   *RMS*, ii, no. 874; SRO, GD.25/1/104.

eventually the Flemings regained the lands, and it would be interesting to know why, when and how.[1] There is no evidence of further legal procedure in the case, and it is tempting to see a part for politics rather than law in the reversal of the verdict of the assize. Within a few months, Gilbert Lord Kennedy had dropped out of the political limelight as a result of the Boyd coup in July 1466. While Fleming generally does not seem to have been a beneficiary of this political upheaval,[2] it may be that Kennedy became more concerned with consolidating his position in the ancestral territories of Carrick than with lands of relatively small value and far away. In a much weaker bargaining position than hitherto, he may have sought a settlement of the dispute and made concessions to the Flemings.

The main interest of the events at Dumbarton lies with what had gone before, however, rather than with what happened later. Clearly there was a problem about the Kennedy inheritance which had prevented Gilbert from taking up title to the forty-merk lands of Kirkintilloch until relatively late in life. That problem arose after the death of his grandfather and paralleled the difficulty which also occurred in respect of the 'kenkynnol'. Indeed, the two are tied together by a crucial document of 1454 by which Gilbert received a quitclaim from one Gibboun Kennedy in respect of the Kirkintilloch lands and the 'kenkynnol'.[3] These inheritance difficulties have already been much referred to in Kennedy family histories.[4] The purpose of the present analysis is distinct in that the genealogy is set against the background of forms of social organisation in south-western Scotland, methods of dispute settlement, and the processes of medieval Scots law — all matters on which the writing of Geoffrey Barrow has given us new and firmer insights.[5] Gilbert's

1   See NLS, Chs. 15583 (instrument of sasine in favour of David Fleming, Robert, Lord Fleming's grandson and apparent heir, dated 2 December 1480, also recorded in SRO, Darow Protocol Book, B.66/1/1/1, pp. 272-3) and 15586 (service of John Fleming as heir to David Fleming, 18 April 1497) (= *Wigtown Charter Chest*, nos. 50, 53). In both documents the 40-merk lands of Kirkintilloch are amongst the properties transferred. See also Watson, *Kirkintilloch*, pp. 151-5.
2   See N.A.T. Macdougall, *James III: A Political Study* (Edinburgh, 1982), pp. 70-4. Note that the well-known bond of manrent between Sir Alexander Boyd of Drumcoll, Gilbert, Lord Kennedy and Robert, Lord Fleming (NLS, Ch. 15560) discussed by Macdougall (p. 71) was entered on the same day, 10 February 1466, that Kennedy and Fleming agreed to submit the Kirkintilloch dispute to an assize. While therefore it certainly had a political significance, it also represents Kennedy's concession in return for Fleming's agreeing to undergo the assize. For further comment on the bond see J. Wormald, *Lords and Men in Scotland: Bonds of Manrent 1442-1603* (Edinburgh, 1985), pp. 145-6.
3   SRO, GD.25/1/59.
4   See in particular R. Pitcairn, *Historical and Genealogical Account of the principal families of the name of Kennedy* (Edinburgh, 1830); D. Cowan, *Historical Account of the Noble Family of Kennedy* (Edinburgh, 1849); J. Paterson, *History of the County of Ayr* (Ayr and Paisley, 1847-52), ii, pp. 272-88; A. Carrick, *Some Account of the Ancient Earldom of Carric* (Edinburgh, 1857); *Scots Peerage*, ii, pp. 444-54; J. Fergusson, *The Kennedys* (Johnston's Clan Histories, 1958), pp. 5-8.
5.   See particularly the first three chapters of *The Kingdom of the Scots* (London, 1973); *The Anglo-Norman Era in Scottish History* (Oxford, 1980), pp. 145-68; *Kingship and Unity: Scotland 1000-1306* (London, 1981), pp. 11-13; 'Northern English society in the twelfth and thirteenth centuries', *Northern History*, iv (1969), pp. 22-3; 'The pattern of lordship and feudal settlement in Cumbria', *Journal of Medieval History*, i (1975), p. 117; 'Popular courts

problems also show the intertwining of Scots and Gaelic elements in Scottish society. We will also see how litigation, the importance of which has sometimes been under-rated by comparison with less formal methods of dispute settlement, could play a significant role in individuals' resolutions of their problems.

In 1372 Robert II issued three confirmations, all apparently in favour of John Kennedy of Dunure.[1] By the first of these he confirmed a confirmation by Alexander III, dated January 1276, of the gift by Neil earl of Carrick to Roland (a Latinised form of Lachlann) of Carrick that he, and his heirs after him, should be head of all his kindred both in *calumpnie* and in the other matters pertaining to 'kenkynnol', together with the office of bailie of Carrick and the leading of the men of that district, all under whomsoever should be the earl of Carrick. The second confirmation was of Earl Neil's original charter to Roland of Carrick, while the third confirmed a remission of Robert I forgiving Gilbert of Carrick for letting 'our castle of Loch Doon' pass into the hands of the English. It is clear that the purpose of these transactions was to establish Kennedy's titles to the offices of 'kenkynnol', bailie of Carrick, and keeper of the castle of Loch Doon, a major fortification in south-eastern Carrick.

The political and social history of south-west Scotland and especially of the province of Carrick assists understanding of these confirmations.[2] At some point late in the reign of William I (1165-1214), Duncan son of Gilbert son of Fergus, a descendant of the kings of Galloway, became lord of Carrick, in consequence, it is thought, of a royal grant compensating him for giving up his claim to Galloway. Duncan appears as earl of Carrick, although his designations vary throughout his long career, and he survived until 1250. He had several sons: Neil, Colin, John, Alexander and two Alans.[3] His successor as earl was Neil, who died in 1256. The earldom then passed to Neil's daughter Marjorie, who married Robert Bruce the

---

in early medieval Scotland: some suggested place-name evidence', *Scottish Studies*, xxv (1981); 'Popular courts in early medieval Scotland: some suggested place-name evidence — additional note', ibid., xxvii (1983); *Robert Bruce and the Scottish Identity* (Saltire Soc., 1986), pp. 15-18; and 'The lost Gàidhealtachd of medieval Scotland', in *Gaelic and Scotland*, ed. W. Gillies (Edinburgh, 1989).

1   See *RMS*, i, nos. 508-10.

2   For what follows, see A.A.M. Duncan, *Scotland: The Making of the Kingdom* (Edinburgh, 1975), pp. 184, 585; G.W.S. Barrow, *Robert Bruce and the Community of the Realm of Scotland* (3rd edn, Edinburgh, 1988), pp. 276-7; *Scots Peerage*, ii, pp. 421-39.

3   For Colin see W.D.H. Sellar, 'The earliest Campbells: Norman, Briton or Gael?', *Scottish Studies*, xvii (1973), p. 116. Dr Kenneth Nicholls has drawn my attention to an entry in the plea rolls for Trinity term 32 Edward I in the Public Record Office, Dublin (RC.7/10, p. 241), which refers to Duncan son of Nicholas son of Alexander of Carrick suing Ralph Pendelu for a messuage and two carucates in Drumaliss, Co. Antrim, of which his ancestor Duncan (i.e. the earl) was seized in the time of King John, from whom it descended to Alexander as son and heir, from him to Nicholas as son and heir, and from him to Duncan the claimant as son and heir. For the Carrick lands in Ireland, see R. Greeves, 'The Galloway lands in Ulster', *TDGNHAS*, 3rd ser., xxxvi (1957-8).

younger of Annandale in *c*.1272. Their eldest son, also Robert, became earl of Carrick in 1292 and, in due course, king of Scots. Thus the earldom became linked with the kingship. It was held by Robert I's brother Edward (d. 1318), by the future David II in 1328-9, and by Edward's illegitimate son Alexander (d. 1333); by the end of the fourteenth century it had become customary for the king's eldest son to hold the title.

The history of the earldom of Carrick provides a classic example of the transformation of a territorial dignity into an honorific title; but while the institution of the earldom changed, Carrick society did not change with it. In the earlier medieval period it had formed part of Scottish Cumbria, that south-western region of Scotland which has been described as 'a true melting pot of races and language'.[1] Place-name evidence suggests that the earliest known settlers spoke Cumbric, closely related to Welsh, and that they were gradually absorbed by Gaelic-speakers. Cumbric elements still survived in the twelfth century, but Gaelic was to be the dominant language until the seventeenth century.[2] The witness-lists of Carrick charters are full of Gaelic names[3] while it seems highly likely that the fifteenth-century Kennedies still spoke Gaelic, if William Dunbar's famous description of Gilbert's son Walter in his poem 'The Flyting of Dunbar and Kennedie' can be taken as evidence on the point. Dunbar calls Kennedy an 'Iersche brybour baird' ('a Gaelic-speaking wandering bard') and an 'Ersche katherene' ('a Gaelic-speaking reiver').[4] He also questions Kennedy's sense of 'Inglis' speech:

> Sic eloquence as thay in Erschry use,
> In sic is sett thy thraward appetyte.
> Thow hes full littill feill of fair indyte:
> I tak on me ane pair of Lowthiane hippis
> Sall fairer Inglis mak, and mair parfyte,
> Than thow can blabbar with thy Carrik lippis.[5]

Later in the poem Kennedy defends the use of 'Ersche' or Gaelic as a patriotic motif:

> Thou lufis nane Irisch, elf, I understand,
> Bot it suld be all trew Scottis mennis lede;
> It was the gud langage of this land,
> And Scota it causit to multiply and sprede ...[6]

In many respects offices and customs found in Carrick until a very late

---

1   Barrow, *Kingship and Unity*, p. 12.
2   See generally W.L. Lorimer, 'The persistence of Gaelic in Galloway and Carrick', *Scottish Gaelic Studies*, vi (1949), and J. MacQueen, 'The Gaelic speakers of Galloway and Carrick', *Scottish Studies*, xvii (1973).
3   For Gaelic witnesses to Earl Duncan's charters see mainly *RMS*, ii, no. 142, *Melrose Lib.*, i, nos. 30, 32, and *North Berwick Chrs.*, nos. 1, 13, 14; also Duncan, *Making of the Kingdom*, p. 450; for 14th-century Carrick charters, see *Laing Chrs.*, nos. 40, 64, 69.
4   *The Poems of William Dunbar*, ed. J. Kinsley (Oxford, 1979), no. 23, ll. 49, 145.
5   Ibid., ll. 107-12.
6   Ibid., ll. 345-8.

period reflect the Gaelic nature of its society. For example, Carrick had its *judices* and toiseachdeors, and the men of the province were under obligation to render common army service to the king under the earl.[1] The custom of fosterage, disapproved of in the canon law but widespread in the Celtic areas of Britain for many centuries, was used in Carrick, most famously in the case of Robert Bruce, later king of Scots; it is also mentioned in the sources of Kennedy family history.[2] There is evidence for kin-groups and their heads or captains in Carrick in addition to the 'kenkynnol' of the kindred of Kennedy: for example, the undertaking made to the king some time before 1282 by the captains and freeholders of Carrick;[3] references to kin-groups who brought themselves under the captaincy of the Kennedies at different times in the fourteenth and fifteenth centuries;[4] and parliament's abolition in 1490 of the custom of taking the tribute of calp 'usit be heddis of kin' in Carrick.[5]

As already noted, 'kenkynnol' is a Gaelic form meaning 'head of the kindred'. The form of words has apparently no exact parallel in Irish Gaelic, although the idea of the head of the kindred (*cenn fine, agae fine, conn fine*) was of course familiar.[6] Geoffrey Barrow has pointed out, however, that 'kenkynnol' is cognate with the Welsh *pencenedl*, which has precisely the same form and meaning.[7] It would be dangerous to make too much of this point with reference to the survival in Carrick of Cumbric custom, but it does suggest that Welsh as well as Irish parallels may be relevant to an understanding of the institution. The early Irish and Welsh texts themselves seem to be dealing with broadly similar ideas.[8] Thus the *cenn fine* 'is chosen — presumably by election among the kin-members — on the basis of his superior wealth, rank and good sense', while the office of *pencenedl* was not hereditary and should not be held by lesser men. The Welsh texts generalise that 'it is right for [the *pencenedl*] to intervene with his kinsman in every need which comes upon him', while the Irish laws list various obligations which the *cenn fine* incurs on behalf of the kin. Both, for instance, have an important role to play in the marriages of kinswomen. It is important to remember, however, that these texts are early, and that there may have been development in the nature of the offices. Indeed, with respect to the *pencenedl*, attention has been drawn to a fourteenth-century example of the office passing by virtue of inheritance rather than election.[9] In any event,

---

1   For references see MacQueen, 'Laws of Galloway', pp. 136-7.
2   Ibid., p. 138.
3   *APS*, i, p. 109.
4   *RMS*, i, app. II, no. 914; SRO, GD.25/1/63.
5   *APS*, ii, pp. 214, c. 5; 222, cc. 19, 20.
6   For these phrases see F. Kelly, *A Guide to Early Irish Law* (Dublin, 1988), pp. 13-14.
7   Barrow, *Kingship and Unity*, p. 12.
8   For the Irish texts see Kelly, *Early Irish Law*, p. 14; for the Welsh texts, D. Jenkins, *The Law of Hywel Dda* (Aberystwyth, 1986), p. 123. For discussion of the early medieval Welsh kin-group, with brief reference to the *pencenedl*, see W. Davies, *Wales in the Early Middle Ages* (Leicester, 1982), pp. 71-81.
9   Jenkins, *Law of Hywel Dda*, p. 371. See also for changing ways of choosing the king in Ireland, K. Simms, *From Kings to Warlords: The Changing Political Structure of Gaelic Ireland in the Later Middle Ages* (Woodbridge, 1987), pp. 41-9.

the position in Carrick, or, indeed, anywhere else in Scotland, may never have been that described, somewhat obliquely, in the Welsh and Irish sources.

Our earliest piece of evidence, Earl Neil's charter granting the 'kenkynnol' to Lachlann of Carrick, must have been made before 1256, the year of the earl's death. It is usually said to have been occasioned by the requirement that the head of a kin-group be male while Neil himself had only female issue — his daughter Marjorie, who succeeded him as countess of Carrick in accordance with the feudal rules of inheritance which had become the common law of Scotland by the mid-thirteenth century. Since, however, the Welsh and Irish texts hint at a process of election for the head of kindred, the possibility that Lachlann was simply designated 'kenkynnol' and that the matter had nothing to do with conscious avoidance of the feudal rules of inheritance should not be dismissed. Perhaps, as John Bannerman has suggested above, it should be seen as an instance of the accommodations between two systems of which there are other examples at this period:[1] that is, the king could take the profits of Marjorie's ward and marriage, while the kindred continued to deal with the internal order of its society.

The legislation of 1490 shows that the taking of tribute in the form of calp was another important aspect of 'kenkynnol', and that it was still a sufficiently significant part of Carrick life to cause resentment amongst the inhabitants. The *calumpnie* mentioned in Earl Neil's charter as pertaining to the 'kenkynnol' were the accusations of sergeants or peace officers acting under him in the protection of the kindred. There is evidence for such sergeants in thirteenth-century Carrick. In 1225 the earl of Carrick agreed that the clergy in his earldom should not be liable to give hospitality to his sergeants. In 1285 Robert Bruce, as earl of Carrick by virtue of his marriage with Countess Marjorie, exempted the Carrick tenants of Melrose Abbey from the *superdictus* or accusation 'of our sergeants', which suggests that Bruce was exercising the lordship of 'kenkynnol' at this time. Despite the hostility which their activities aroused, it is quite likely that these sergeants remained a feature well after 1300.[2] The protection of the kindred is also evident in Walter Kennedy's reference in the 'Flyting' to the king of the underworld as Dunbar's head of kin: 'Pluto thy hede of kyn and protectour/To hell to lede the on lycht day and leme'.[3]

All this shows that in Carrick, Gaelic speech, social organisation and customs remained dominant until late in the fifteenth century, and quite possibly at a later period as well. This is the milieu in which we must first place John Kennedy of Dunure, the beneficiary of the 1372 confirmations discussed above. It is apparent that by 1372 he was the leading member of the already numerous Kennedy kindred, and the greatest landowner in

---

1   See John Bannerman's essay, 'MacDuff of Fife', above, pp. 32-3.
2   See MacQueen, 'Laws of Galloway', pp. 133-5, for references and discussion.
3   *Poems of William Dunbar*, no. 23, ll. 535-6.

Carrick.[1] He was married to a lady named Mary, of whom next to nothing else is known,[2] and he appears as steward of the earl of Carrick in 1367[3] (although the earldom was in the king's hands at that time). This probably establishes his descent from the Kennedies of Carrick, who are found in the thirteenth century as part of the entourage of Earl Duncan. In two charters, one dated 1243, there is mention of a Gillescopewyn Mackenedi, steward of Carrick;[4] he is probably the same person as 'Gillascop steward of Carrick' who witnessed another charter of Duncan of Carrick.[5] Gilchrist son of Kennedy was one of a number of persons with Gaelic names who, some time before 1223, witnessed a charter of Thomas de Colville 'named the Scot' to Melrose Abbey, in which the leading witness was Duncan of Carrick.[6] Other members of the Kennedy family were associated with Duncan's sons. Murdoch MacKenede witnessed a charter of Colin son of Duncan of Carrick in favour of the church of St Cuthbert at Maybole.[7] Probably it was the same Murdoch who, along with Samuel and Henry mac Kenedy and several others, formed an inquest in 1260 to draw up an extent of the earldom of Carrick for the king.[8] Murdoch and Henry also appear in another Carrick inquest about the same time,[9] while one Murdoch the steward, who witnessed a charter of Earl Duncan, can probably be identified with Murdoch MacKennedy.[10] It is clear that the Kennedies were leading members of Duncan's menzie who had accompanied him into Carrick after the defeat of his lineage in Galloway in the 1180s. The first mention of a Kennedy is in fact in the twelfth century, fighting against Roland of Galloway,[11] Earl Duncan being the son of Gilbert whom Roland overthrew in Galloway in 1185; while some time probably shortly after 1197 Gilbert Mackenede appears in the witness-list of a charter alongside Duncan.[12]

A second point at which we can link John Kennedy of Dunure with thirteenth-century Kennedies is in his ownership of the forty-merk lands of

---

1   See *Laing Chrs.*, nos. 64, 68-74, for Kennedies of Dunure, Blairquhan, Knockdolian, Dalmorton and Gyltre between 1370 and 1400; and, for acquisitions of property, *RMS*, i, no. 135, and *RRS*, vi, no. 301 (Cassillis); *RMS*, i, nos. 381, 529, 660 (Dalrymple). For colourful accounts of Kennedy's acquisition of these properties, see Pitcairn, *Account of principal Kennedy families*, p. 3.
2   For Mary see *RMS*, i, nos. 378, 428. Note that in the 'Historie of the Kennedyis' (Pitcairn, *Account of principal Kennedy families*, p. 3) John is said to have obtained Cassillis by marriage to its heiress, who, doubtless worn out by the vigorous and contested courtship there described, died shortly after the wedding.
3   *ER*, ii, p. 293.
4   *Melrose Lib.*, i, nos. 190-1.
5   *North Berwick Chrs.*, no. 14.
6   *Melrose Lib.*, i, no. 192. For Thomas de Colville the Scot, see Barrow, *Anglo-Norman Era*, pp. 31-2.
7   *North Berwick Chrs.*, no. 15.
8   I.A. Milne, 'An extent of Carrick in 1260', *SHR*, xxxiv (1955), pp. 48-9. See also the additional notes by A. McKerral and J. Fergusson, ibid., pp. 189-90.
9   *CDS*, i, no. 2674.
10   *North Berwick Chrs.*, no. 1; Fergusson, *Kennedys*, p. 5.
11   *Chron. Bower* (Goodall), i, p. 490.
12   *Melrose Lib.*, i, no. 36. This can be dated no earlier than 1197, since it refers to the sheriff of the new castle of Ayr, founded in that year.

Kirkintilloch in the barony of Lenzie. At some date before 1384, John gave the forty-merk lands of Kirkintilloch to his heir Gilbert [hereafter Gilbert I]; we know this because a 1384 confirmation of the grant by Malcolm Fleming, the baron of Lenzie, survives.[1] Now in the 1260s (probably) an 'open court of Lenzie' held by its lord John Comyn of Badenoch was attended, amongst others, by 'Fergus Kennedy our steward and John his brother'.[2] Fergus the steward almost certainly received lands in Lenzie as the fee for his office; and by 1296 Hugh Kennedy certainly held the lands of Kirkintilloch.[3] Probably Fergus is to be identified with Fergus Makennedy who in 1266 rendered accounts at the king's exchequer as attorney for William Comyn of Kilbride, sheriff of Ayr.[4] The link with the Comyn family must have arisen as a result of its activity in the south-west from the mid-thirteenth century, perhaps especially after the king had farmed the wardship of the earldom of Carrick to Alexander Comyn earl of Buchan in c.1260. Moreover, John Comyn of Badenoch had been justiciar of Galloway since 1258.[5]

The Wars of Independence led to the fall of the Comyn family in Scotland and the redistribution of their estates amongst the supporters of Robert I. Sir Malcolm Fleming, head of the Lanarkshire family of Biggar, received the barony of Lenzie.[6] The position of the family of Kennedy during the Wars is unknown. The Carrick connection would suggest adherence to the Bruce side, but it is possible that the Comyn link led to some ambiguities and uncertainties, particularly after the death of Robert I in 1329. In 1358 David II granted to Malcolm Fleming of Biggar 'all those lands with the pertinents which were of the late John Kennedy within the barony of Lenzie, pertaining to us by reason of the forfeiture of the same late John'.[7] Presumably the late John Kennedy had been guilty of treason. So far as we can tell, the Kennedy holdings in Lenzie had been held as tenants of the baron (after 1314, of the Flemings) rather than of the king. But a traitor's lands were forfeit to the king whether or not they were held of him,[8] and it must have been this which entitled David II to make the grants he did to the Flemings. It would be interesting to know more about what John Kennedy had done, as chronicle sources tell of a John Kennedy fighting for David II in Carrick in 1346.[9] It is likely that the lands in Lenzie forfeited by John Kennedy were those held by his family from the mid-thirteenth century. It is significant that John Kennedy of Dunure is described as the son of John Kennedy;[10] he would seem therefore to have succeeded in

1   *RMS*, i, no. 466; SRO, GD.25/1/20.
2   *Cambuskenneth Reg.*, pp. xxxi-xxxii.
3   *Rot. Scot.*, i, p. 29b; *Ragman Rolls 1291-1296* (Bannatyne Club, 1834), p. 143.
4   *ER*, i, p. 34
5   For these Comyns see *ER*, i, p. 28, and G.W.S. Barrow, 'The justiciar', in his *Kingdom of the Scots*, p. 107.
6   *RMS*, i, no. 80.
7   *RRS*, vi, nos. 160, 270.
8   *Regiam Majestatem*, II, 49-50 (*APS*, i, pp. 618-19); IV, 1 (*APS*, i, p. 632).
9   *Chron. Bower* (Goodall), ii, p. 347; *Chron. Wyntoun* (Laing), ii, p. 477.
10   *RRS*, vi, no. 148, at p. 175; no. 150 at p 187; both dated 1357.

re-establishing his family's interests within the Fleming barony.

We can now return to the three confirmations of 1372, to consider the basis for the claim by John Kennedy of Dunure to the offices of 'kenkynnol', bailie of Carrick, and keeper of Loch Doon castle. The first point to emerge from the confirmations is that Neil earl of Carrick had held the office of 'kenkynnol' in the 1250s and that it related to his kindred — the kindred of the family of Carrick. Lachlann was the son of one of Neil's many brothers and the appointment must have been to the headship of the Carrick kindred. The royal confirmation of the earl's grant in 1276 is most

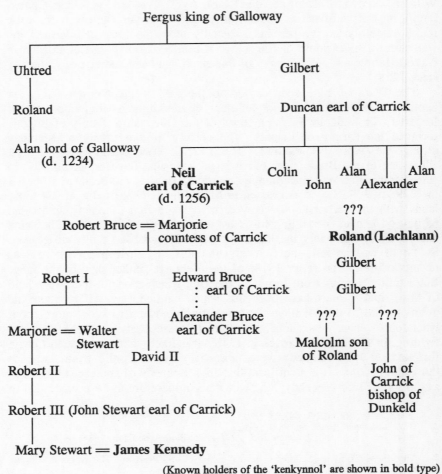

(Known holders of the 'kenkynnol' are shown in bold type)

The Carrick Kindred

probably to be explained by Lachlann's recent death, although it is tempting to speculate on the possibility that it was also a response to the threat posed by the arrival of the Bruces upon the Carrick scene.[1] Lachlann was succeeded by his son Gilbert, who as keeper of Loch Doon castle incurred the wrath of Robert I and was then reconciled, as recorded in the document confirmed in 1372. In turn Gilbert had a son, also called Gilbert, who received a grant of lands in the Lennox in 1315 × 1333, and was made crownar for life of the territory between the rivers Ayr and Doon by David II in c.1342.[2] The last known members of the family of Carrick are Malcolm son of Roland of Carrick, who made a grant of lands to John Kennedy of Dunure in c.1369, and (perhaps) John of Carrick, bishop of Dunkeld and chancellor of Scotland from 1370 to 1377.[3]

What was the connection between these successive members of the kin of Carrick and John Kennedy of Dunure? The answer usually given is to say that John was the heir of line of the Carrick family, or that the Carricks changed their name to Kennedy. From the material here surveyed, however, it is clear that this view cannot be accepted. While there may have been some fairly close link of cousinship between the Kennedies and the Carricks in the twelfth century, it had become distant at best in the late fourteenth century. The line of descent which leads to John Kennedy of Dunure may not be clear in all its details, but it is certainly distinct from that of the Carricks. On the other hand, the main line of Carricks comes to an end just at the time when John Kennedy became really prominent. The family may well have become extinct either because there was no issue, or because there was female issue only and by marriage the line became subsumed in that of another family. This latter, I suggest, is what happened: that Gilbert of Carrick left only a daughter who had married a Kennedy, and that it was as a result of this marriage that John Kennedy could make a claim to the Carrick inheritance. The evidence apart from the confirmations of 1372 is strong for this inference. First, the Kennedies of Dunure took over the armorial bearings of the Carricks;[4] second, the Carrick lands in Lennox were in 1393 held by John Kennedy's heir;[5] third, amongst John's descendants there appears the name Gilbert, which had powerful Carrick family associations going back to its twelfth-century progenitor in Galloway;[6] and fourth, the Kennedies came to be described as cousins of

---

1 Note in this connection the 1285 Bruce charter concerning 'surdit de sergaunt' (*Melrose Lib.*, i, no. 316, and above, p. 281.
2 *Lennox Cart.*, pp. 43-4; *RMS*, i, app. II, no. 844.
3 SRO, GD.25/1/8. Malcolm of Carrick was possibly a brother of the second Gilbert (*Scots Peerage*, ii, p. 425). For Bishop John of Carrick, see D.E.R. Watt, *A Biographical Dictionary of Scottish Graduates to A.D. 1410* (Oxford, 1977), pp. 89-90 (though no suggestions about his family are made there).
4 For discussion of this point see Cowan, *Family of Kennedy*, p. 10 (n.).
5 See *Lennox Cart.*, pp. 44-5.
6 J. Goody, *The Development of the Family and Marriage in Europe* (Cambridge, 1983), p. 201, makes the interesting point that often the first child of a marriage would be named after the maternal grandfather. It cannot yet be said whether such customs were general in medieval Scotland.

the royal house.[1] The Carricks were close relatives to the Bruces and the Stewarts, but there is no evidence to show that this was true of the Kennedies. Cousinship could have arisen only by virtue of marriage into the Carricks. Was John's wife Mary the last of the Carricks in the main line?

Finally, however, it is possible that inheritance has nothing to do with 'kenkynnol', and that John Kennedy was either elected to the position or received a submission from the Carricks. It may also be no coincidence that John received his confirmations the year after the house of Bruce had itself become extinct in the male line with the death of David II in 1371. As already noted, it is possible that as early as the 1270s there was some tension between the Carricks and the Bruces as earls of Carrick concerning the 'kenkynnol', and this may have been heightened by the *contretemps* over Loch Doon castle. In addition, there does not always seem to have been an easy relationship between David II and the Kennedies. David retained the earldom in his own hands after the death of Alexander Bruce in 1333, although Alexander's widow Eleanor survived into the 1370s and continued to be styled countess of Carrick. William Cunningham was granted the lordship of the earldom in 1362 but had lost it before 1368. During all this time, therefore, David may have laid claim to the 'kenkynnol' himself, as heir to the Carrick titles. Nor would he have deprived himself of this claim when in 1368 he granted the earldom to his great-nephew John Stewart, son of Robert the Steward.[2] When David died, Robert became king, but although he was a descendant of Neil earl of Carrick, the descent was through his mother and ultimately, of course, from Neil's daughter, which may have weakened his claim, at least in Carrick eyes. In any event, Robert II may not have been much interested in the 'kenkynnol' (especially given the complication that his son was earl of Carrick), and John Kennedy certainly appears to have been favoured by him. Just conceivably, therefore, the death of David II provided an opportune moment for John to push forward a claim, the details of which were hazy even to contemporaries but which no-one, given the demise of the Carricks, was in a position to resist.

We have seen that John Kennedy of Dunure recovered what were probably his ancestral lands of Kirkintilloch and granted them to his son and heir, Gilbert I. The charter to Gilbert I was one of tailzie, which specified as the first line of heirs the heirs male lawfully begotten of the bodies of Gilbert I and his wife, Agnes Maxwell. The second line commenced with another Gilbert (Gilbert II), the first-born son of Gilbert I, and his lawful heirs male, whom failing Gilbert II's brother german John and his lawful heirs male, whom failing Roland, the brother german of Gilbert II and John, and

1   Gilbert, son of John Kennedy of Dunure and Mary, is the first so described (*RMS*, ii, no. 379, confirmation of charter to him of 1406, copied into the Ailsa Muniments as SRO, GD.25/1/29).
2   For the earldom from 1306 to 1368 see A. Grant, 'Earls and earldoms in late medieval Scotland (*c*.1310-1460)', in *Essays Presented to Michael Roberts*, ed. J. Bossy and P. Jupp (Belfast, 1976), pp. 28, 30-1.

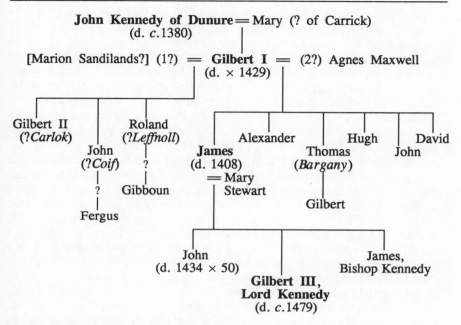

(Known holders of the 'kenkynnol' are shown in bold type)

The Kin of Kennedy

his lawful heirs male. If all these lines failed, then the heirs whomsoever of John Kennedy lord of Dunure would inherit.[1] It was this charter which appears to have been the key document in establishing the title to the Kirkintilloch lands of Gilbert, Lord Kennedy (Gilbert III) in the litigation of 1466, judging by its royal confirmation immediately after the case.

What then are the implications of this tailzie? First, it is apparent that Gilbert I had some fairly long-lasting liaison before his marriage with Agnes Maxwell, accounting for his 'first-born' son Gilbert II and his brothers german (that is, full brothers). Had these sons been born to Gilbert I and Agnes unlawfully (that is, prior to marriage), the doctrine of legitimation by subsequent marriage would have applied to enable them to be counted amongst its lawful issue.[2] Two interpretations of the evidence seem possible. Either Gilbert I had an earlier canonical marriage, in which case this tailzie was to a large extent an act of disinheritance; or we are dealing with an

1   SRO, GD.25/1/20.
2   A.E. Anton, 'Parent and child', in *Introduction to Scottish Legal History* (Stair Soc., 1958), pp. 117-18.

instance of 'Celtic secular marriage', a subject recently illuminated by the researches of W.D.H. Sellar.[1] With particular reference to the West Highlands and Islands, Sellar has argued for the late survival of customs allowing polygamy, concubinage and divorce, and recognising the legitimacy of the offspring of polygamous and concubinatory unions, despite the disapproval of canon and feudal law. Such customs may have been found in Galloway, and it is possible that the earlier liaison of Gilbert I would have been recognised as a marriage under Carrick custom but not under canon law. Kennedy tradition holds that Gilbert I had another wife besides Agnes Maxwell, called Marion Sandilands, but is inconclusive as to which of these possibilities should be preferred;[2] indeed no historical evidence of either Marion or her marriage to Gilbert I has been found.

In 1392 Gilbert I gave Kirkintilloch to James Kennedy, a son of his marriage with Agnes, although significantly the document recording the gift does not explain its terms and speaks of the possibility of the heirs of Gilbert I raising actions against James and revoking the gifts made to him.[3] One possible interpretation of this is that James was not the eldest son of his parents' marriage; but perhaps a better explanation is the claim of the sons of the earlier union of Gilbert I. By November 1408 James was dead, but his parents survived him. What now happened to the forty-merk lands of Kirkintilloch was the central disputed issue in the 1466 litigation at Dumbarton. James had had sons from his marriage of c.1404 with Princess Mary, daughter of Robert III: his heir John, then Gilbert (Gilbert III, the pursuer of 1466), and James, later to be the famous Bishop Kennedy.[4] In 1408 John cannot have been much more than a baby, but if his father had held the fee of Kirkintilloch then he was entitled to succeed. The pleading of Robert, Lord Fleming in 1466 was that John had indeed been vested in the lands, and had then forfeited them.[5] It is a matter of record that John was arrested and incarcerated in Stirling castle in 1431,[6] but the reasons for this have been mysterious. Fleming's pleadings state, however, that John had committed felony against the king and therefore, according to *Regiam Majestatem*, his lands were forfeit to the king for a year and a day, after which they returned to the overlord, that is, Fleming.[7]

---

1   W.D.H. Sellar, 'Marriage, divorce and concubinage in Gaelic Scotland', *Transactions of the Gaelic Society of Inverness*, li (1978-80). See also Barrow, 'The lost Gàidhealtachd', pp. 72-3, and MacQueen, 'Laws of Galloway', p. 137. The best modern treatment of the development of the canon law of marriage is J.A. Brundage, *Law, Sex and Christian Society in Medieval Europe* (Chicago, 1987).
2   See further below, pp. 290-1.
3   SRO, GD.25/1/24.
4   SRO, GD.25/1/31, records the death and offspring of James.
5   See NLS, Ch. 16632. Attention was first drawn to this document in A.R. Borthwick, 'The King, Council and Councillors in Scotland, c.1430-1460' (Edinburgh University Ph.D. thesis, 1989), pp. 22, 48.
6   For John's arrest see *Chron. Bower* (Watt), viii, pp. 264-5.
7   NLS, Ch. 16632. The *Regiam Majestatem* reference is II, 49 (*APS*, i, p. 618). Both Fleming and Kennedy had reason to be familiar with these rules: Fleming because his father's lands might have been forfeited under them in 1441, Kennedy because as justiciar he had presided over at least one treason trial in 1464 (*RMS*, ii, no. 812).

Gilbert III's case in 1466 was that his grandfather Gilbert I had died vested as of fee in the lands. It follows that, by his argument, John had never had sasine. Otherwise the action could have been based on John's title, either by a brieve of mortancestor referring to John's death vested as of fee, or, supposing John to have been served heir and then dispossessed without ever regaining sasine, by brieve of right.[1] It is also of interest that Gilbert III made no reference to the title of his father. If James had had a heritable title, why was it not the basis of the claim in 1466? Either Gilbert I had retained the fee in making his 1392 gift to James, or he had somehow recovered it following James's death in 1408. Such a recovery would have been a departure from the laws of succession and, if made, would have constituted a fatal weakness at the heart of Gilbert III's case. His argument must have been that neither James nor John had ever had heritable sasine.

It seems clear that Gilbert I remained infeft in the principal Kennedy estates in Carrick until his death. The date of this is unknown, but in 1429 John was designed 'of Cassillis' and 'of Dunure',[2] strong evidence that he had succeeded to the principal Kennedy estates and that Gilbert I was dead by this time. Gilbert III became entitled to the family lands because John died without issue at an unknown date. He was apparently still in custody in Stirling in 1434; but according to the *Liber Pluscardensis*, written around 1461, he subsequently 'escaped and exiled himself beyond return'.[3] It seems that he had never been put on trial; Fleming's pleadings in 1466 develop an elaborate argument, based on the 'imperial (that is, Roman) law', that the crime of lese majesty 'is sa odiouse that a man may be accusit of it efter his deide'.[4]

This analysis of the descent of the Kennedies to Gilbert III throws much light on a series of documents concerning the office of 'kenkynnol' from 1406 to 1455. The first document is a 1406 royal confirmation of a resignation of the office of 'kenkynnol' by Gilbert I in favour of his son James.[5] The confirmation used the formula first set out in Earl Neil's

---

1 On the brieve of right see MacQueen, *Common Law and Feudal Society*, ch. 7.
2 *RMS*, ii, nos. 128-9. Dunure appears as 'Dunovin' in the printed volume (no. 129) and a check with the MS Register (SRO, C.2/3, no. 40) suggests that this is a correct reading. The error may have been that of the clerk; compare the lettering for other early forms of Dunure, e.g. Donnouire, Dounouire. It is not difficult to misread '-ouire' as '-ovin'. The original of no. 128 survives in the Bargany Muniments, SRO, GD.109/267.
3 *ER*, iv, p. 591; *Chron. Pluscarden*, i, p. 377; ii, p. 284.
4 The 'imperial law' in question, a constitution of Marcus Aurelius to be found in Justinian's Code (C.9.8.6.2), is quoted as follows in the plea: 'Hoc iure uti, cepimus ut etiam post mortem nocentium tam hoc crimen lese maiestatis inchoari, possit [post mortem *interlined*] ut convicto motus eius memoria dampnitur et bone eius succession eius eripiantur'. This would probably have been an extension of the Scots law of treason, although trials of the recently dead were known, e.g., of Sir Roger Mowbray in 1320, and Archibald Douglas earl of Moray in 1455. Treason law was being expanded in the mid-15th century, however: see R. Nicholson, *Scotland: The Later Middle Ages* (Edinburgh, 1975), p. 373, and A. Grant, review of Macdougall, *James III*, in *SHR*, lxii (1983), p. 170. Compare English law: J.G. Bellamy, *The Law of Treason in England in the Later Middle Ages* (Cambridge, 1970), pp. 183-95.
5 SRO, GD.25/1/29.

charter of the mid-thirteenth century: James was to be head of his kindred both in *calumpnie* and in other matters pertaining to 'kenkynnol', as well as bailie of Carrick with responsibility for leading its men. In addition, the holding of weapon showings was mentioned as a further duty. It is not clear what happened when James predeceased Gilbert I, or how the office was held, until a whole series of documents dating from 1450 on becomes available. In that year Gilbert III had a notarial transumpt made of the 1406 confirmation and obtained a further royal confirmation.[1] In February 1451, and again in the following May, Gilbert resigned the office of 'kenkynnol' to the king and received it back again.[2] Then in October 1454 Gilbert resigned the 'kenkynnol' yet again to the king, this time, however, in favour of his son John.[3] The king granted the office to John in November, with reservation of a liferent to the father,[4] so that Gilbert would continue to exercise the office until his death, whereupon John would assume it. However, the king, Gilbert and John appear to have gone through all this again one year later, in October 1455, when yet another royal charter narrated Gilbert's resignation in favour of John, the grant to John, and the reservation of the liferent to Gilbert.[5] It is clear that all these are all distinct transactions because the location as well as the date of each differs.

Alongside this series we must place an unpublished document in the Ailsa Muniments dated October 1454 (that is, at the same period as Gilbert III was making the arrangements in favour of his son John). It narrates in Scots a quitclaim to Gilbert III by Gibboun Kennedy son and heir of Roland Kennedy of Leffnoll, promising on pain of £6,000 'to make na pursuit nor claim nor questioun, nor to pursue, to vex, disturb or inquiet in the law or by the law' in relation to a number of lands and offices. These form the main Kennedy properties mentioned in this essay, including the forty-merk lands of Kirkintilloch in the barony of Lenzie and, finally, 'the office of the hed of kynschip'.[6]

The explanation of this curious series of documents must lie in the question of the state of relations between the many sons of Gilbert I, including Gilbert II, John and Roland, the offspring of his liaison with, perhaps, Marion Sandilands prior to his marriage with Agnes Maxwell. The documents may be linked with a Kennedy tradition that Gilbert I disinherited his eldest sons. According to the sixteenth-century *Historie of the Kennedyis*:

> The nixt Laird of Donour [Gilbert I] had tua wyffis. On his first wyff he begatt ane sone. Scho was the Laird of Cadderis dochter and Sandylandis to name. And aftir hir decease he mareyitt the Laird of Caderwodis dochter, Maxwell, quha buir him ane sone also. It is

---

1   SRO, GD.25/1/40; *RMS*, ii, no. 379.
2   *RMS*, ii, no. 414; SRO, GD.25/1/52.
3   SRO, GD.25/1/58.
4   SRO, GD.25/1/60.
5   SRO, GD.25/1/66. The reason for this transaction was probably the completion of the king's *quadriennium utile*: see Borthwick, 'King, Council and Councillors', pp. 190-1.
6   SRO, GD.25/1/120, a notarial transumpt of 1475.

now affermitt be the neme of Kennedy that King James the First send
ane of his dochters to this Laird of Donour to foster, quha remaynit
with him quhill scho was ane woman. At the quhilk tyme, the ladyis
awin sone, haffing mair credeitt in his moderis house nor hir
stepsone, he being in luiff with this young lady, gettis hir with
bairne. The King hir fader, being far offenditt thairatt, could find na
better way nor to caus him mairie hir. And sa the Laird of Donour
disereist his eldest sone and maid his secund sone Laird. And his
eldest sone, he gaiff him the landis of Carloik.[1]

There are obvious flaws in the history here, but the theme of two wives and
disinheritance of the first wife's son in favour of the second wife's because
the latter's son had married a daughter of the king clearly ties up with some
of what we know about Gilbert I and his family. The documents hint that
James, although not the heir thereto, received the Kennedy patrimony, and
he was married to a daughter of Robert III. Some parallel is apparent with
the preference for the family born of the royal spouse shown earlier by John
first Lord of the Isles; after his death Donald, his eldest son by his second
marriage to Margaret Stewart, daughter of Robert II, succeeded to the
Lordship, rather than any of his sons by his first marriage to Amy
MacRuari.[2] Finally, there is some evidence later in the fifteenth century for
a family of Kennedies of Carlok,[3] which would tie in with the later Kennedy
tradition.

There is also a good deal of indirect evidence for dissension amongst the
various branches of the Kennedies in the first half of the fifteenth century.
Family tradition has it that James was slain by the disinherited Gilbert II,[4]
which left the aging Gilbert I, the three sons of his first liaison, the
surviving sons of his marriage with Agnes Maxwell (Alexander, Hugh,
John, Thomas and David), and James's three sons, none of whom can have
been more than a few years old and who were presumably with foster
parents. Scattered references over the next forty years show parts of the
Dunure patrimony in other hands. In 1434, for example, one Fergus
Kennedy appears as keeper of Loch Doon castle.[5] Fergus was certainly not
one of the sons of James Kennedy and it seems most plausible to suppose
that he was a relative of the John Kennedy of Coif who in 1450 resigned the
keepership of Loch Doon castle to Gilbert III.[6] In 1434 Thomas Kennedy
rendered his accounts at the exchequer as bailie of Carrick,[7] and he still held
the office in 1438.[8] He has been identified as one of the sons of Gilbert I

1   Pitcairn, *Account of principal Kennedy families*, p. 6.
2   See *Acts of the Lords of the Isles*, pp. 286-99.
3   *The Acts of the Lords Auditors of Causes and Complaints*, ed. T. Thomson (Edinburgh,
1839), p. 141 (1489/90).
4   *Scots Peerage*, ii, p. 449. I have not been able to trace a source for this statement.
5   *ER*, iv, p. 596.
6   *RMS*, ii, no. 412.
7   *ER*, iv, p. 594.
8   *ER*, v, p. 25.

and Agnes Maxwell. It would be interesting to know more about the relations between the other sons of Gilbert I and Agnes Maxwell in this period, and between them and their nephews. Family legend has it that one son, Alexander, laid claim to the leading position within the family and was eventually slain by his brothers.[1] Entries in the Register of the Great Seal show Hugh, Thomas and David entering into rearrangements of their landholdings, including property which had belonged to Alexander, in 1429.[2] The destinations are, however, in favour of their nephews, which does not suggest feud or division at this time. Nonetheless, for Thomas later to exercise the bailieship of Carrick was a diversion of the inheritance of Gilbert I. Since Earl Neil's pre-1256 charter it had always been linked with the 'kenkynnol'. We might infer from this the possibility that Thomas also exercised the position of 'kenkynnol'. There is a hint of feud between Thomas's descendants and Gilbert III, perhaps associated with these events. This emerges from much later evidence showing Gilbert III entering an indenture in September 1465 with Thomas's son, Gilbert Kennedy of Bargany, whereby there were to be various marriages between the two families; a classic technique for the settlement of feud.[3] Earlier indentures, bonds of manrent, and letters of retinue between Gilbert III and other Kennedies found in the Ailsa Muniments also suggest the settlement of various family feuds arising during Gilbert's career.[4]

If this interpretation of the evidence is correct, then the quitclaim by Gibboun Kennedy of Leffnoll in 1454 fits into a pattern of family dispute over its patrimony, including the 'kenkynnol'. There is no evidence as to the basis of his claim. The *Scots Peerage* speculates that he and the Kennedies of Coif were descendants of one of Gilbert I's first-born sons,[5] and, while the Kennedies of Carlok should not be forgotten, it is a plausible hypothesis in the context of the idea that these sons had been disinherited. A continuing claim from this source also fits in well with what we know of Carrick society. If the 'kenkynnol' retained real significance, then there were many problems following the death of James in c.1408, the death of Gilbert I at some uncertain point between then and 1429, and the incarceration and exile of John after 1431. These events must have opened up a sequence of vacuums at the head of the Kennedy family in an already tense situation, and provided many opportunities for the assertion of claims and counter-claims

---

1   Pitcairn, *Account of principal Kennedy families*, pp. 5-6
2   *RMS*, ii, nos. 128-9. The property in question was Ardstinchar, which Alexander owned in 1416 (SRO, GD.25/1/32).
3   See SRO, GD.25/1/91. For the use of intermarriage as a technique for the settlement of feud, see Wormald, *Lords and Men*, p. 79, where she suggests that 'the marriage contract was the weakest form of alliance', and that bonds would usually be required in addition 'to bolster up and strengthen the marriage contract ... for all it achieved in practice was the bringing into juxtaposition of two distinct kin groups. It did not automatically impose obligations of kinship'. But here the marriage was linking members of the same kin-group: see further ibid., p. 86.
4   SRO, GD.25/1/34, 35, 54.
5   *Scots Peerage*, ii, p. 452.

to the family offices and properties. It looks as though these opportunities were exploited.

Gilbert III emerges from obscurity in the mid-1440s.[1] Apart from destinations in tailzies, many of which have been mentioned already, his only appearance in record before this time is as holding some of the king's lands in Carrick in 1434.[2] By 1448, however, he was the bailie of Carrick, a position which he continued to hold until 1469.[3] There was a series of royal confirmations in 1450-1, which included not only the keepership of Loch Doon castle and headship of his kindred, but also other Kennedy lands.[4] It may have taken longer to confirm the position of 'kenkynnol' with the other members of his family, and this could explain the series of transactions about it up to 1455 and the feud with the Kennedies of Bargany, resolved only in 1465.[5] By the time he embarked upon this process of recovery Gilbert III was a mature figure in his forties, late in the day to be asserting his hereditary claims. Obviously he was favoured by his cousin, James II, who only assumed active authority at the head of government in 1449 and may have had in mind the need to counter-balance the Douglas dominance of the south-west; but another reason for Gilbert's late emergence may have been the eventual death of his elder brother John, since it was only on this occurrence that Gilbert would have had any rights by inheritance to be supported by the king.

This essay began by referring to the apparently dual aspects of Gilbert, Lord Kennedy: on the one hand, royal servant and follower of the forms of the common law, on the other, holder of an ancient form of Gaelic lordship. The duality may, however, may be more apparent to modern eyes, aware of the ultimate fall of Gaeldom before the modernising forces of central government and law, than it would have been in medieval times. Clearly, kings from Alexander III to James II saw no contradiction in dealing with the 'kenkynnol' and employing a characteristically feudal form of document, the confirmation, to do so. Indeed 'kenkynnol' fitted quite well into a society such as that of fifteenth-century Scotland, in which kinship played such an important stabilising role.[6] The continuing relevance of this form of

1   See SRO, GD.25/1/34 (1444), 35 (1447).
2   *ER*, iv, p. 596.
3   *ER*, v, pp. 328, 356, 416, 522, 678; vi, pp. 73, 236, 341, 633; vii, pp. 26, 260, 387, 450, 562.
4   See *RMS*, ii, nos. 381 (August 1450), 415, 416 (February 1451). Presumably the royal confirmations in August 1450 of the charters in favour of Gilbert I Kennedy of Dunure and of his son James Kennedy (*RMS*, ii, nos. 378-80) were obtained at the instance of Gilbert III as well. Notice a further royal confirmation of Cassillis in August 1453 (SRO, GD.25/1/56).
5   See also J.M. Brown (now Wormald), 'The exercise of power', in *Scottish Society in the Fifteenth Century*, ed. J.M. Brown (London, 1977), discussing the royal confirmations of 'kenkynnol' at p. 58; and her *Lords and Men*, p. 86. See also C. McGladdery, *James II* (Edinburgh, 1990), p. 57.
6   See in particular on this the works of Jenny Wormald already cited; and in addition her 'Bloodfeud, kindred and government in early modern Scotland', *Past and Present*, 87 (1980), and *Court, Kirk and Community* (London, 1981), pp. 28-32.

lordship well illustrates Sellar's recent argument about the persistence of Celtic law and custom into the medieval period in Scotland generally.[1]

At the same time, however, it is possible to see change taking place; the 'kenkynnol' exercised by Gilbert Kennedy is likely to have been quite distinct from that which pertained to Earl Neil in the thirteenth century. It seems unlikely that Neil required royal support to enable him to wield his lordship over his kindred and there is no hint that it was required when the time came for the transfer to Lachlann. The essence of 'kenkynnol' must have been the recognition of the headship by the kindred. But if the suggestion that the confirmation by Alexander III was linked to the arrival in Carrick of the Bruces is correct, it is also possible to see how change was forced by contact with the world outside the kindred. By the time the Kennedies assumed the office in the fourteenth century, it was said to be held of the king or his heir as earl of Carrick, and in the 1450s Gilbert Kennedy and the king employed the conveyancing process of resignation and regrant to ensure the former's title. The impact of feudal or common law form could hardly be clearer than in these transactions.

Another possible change closely linked to this is the heritability of the 'kenkynnol'; that is to say, the way in which its transfer between successive heads of kin came to parallel the way in which land was transferred under the common law. Under the Kennedies, the guiding principle was clearly that the current holder would be succeeded by his common law heir, usually his first-born legitimate son. But in Earl Neil's time and before, if Welsh and Irish comparisons are relevant, some form of election or choice by the kindred would have taken place. It is interesting that on at least two occasions, a Kennedy transferred the 'kenkynnol' *inter vivos* to his heir: Gilbert I to James, and Gilbert III to his son John. Perhaps this was to ensure the son's position with the kindred. On the other hand, grants of land to heirs in this fashion became common in the fourteenth century, and Gilbert III's grant of 'kenkynnol' to his son employs the characteristic technique of reserving a liferent to himself.[2] So the documentation, again, most probably reflects the effect of conveyancing methods which assumed heritability and primogeniture and were designed to minimise inheritance formalities.

Documentary and legal forms do not, of course, necessarily reflect social reality, and it need not follow that the impact of 'kenkynnol' upon those subject to it had changed much by Gilbert III's time: for example, calp was clearly still exigible. But, if our inference of strife within the kin of

---

1   Sellar, 'Celtic law: survival and integration', and his 'Common law of Scotland and of England'; and also Bannerman, 'Scots Language and Kin-based Society', pp. 11-14 (all as cited on p. 274 above).
2   See in particular W.C. Dickinson, 'Freehold in Scots law', *Juridical Review*, lvii (1946). Two instances which antedate Dickinson's earliest example of the technique of granting away the fee of lands while reserving a liferent (*liberum tenementum*) are *RRS*, vi, nos. 266, 269. See also A. Grant, 'The Higher Nobility in Scotland and their Estates, c.1371-1424' (Oxford University D.Phil. thesis, 1975), pp. 197-211.

Kennedy after the death of James in *c.*1408 is correct, then clearly the peace-keeping role of the 'kenkynnol' had broken down. Although Henry Kennedy of Bennane still held the office of sergeant of Carrick in 1450,[1] we hear nothing else besides about the peace sergeants; instead, the bond of manrent and similar devices had become the established techniques for the settlement of disputes and feud. While the head of the kindred probably retained a critical role in the achievement and maintenance of these settlements, the ways in which this was done were changing.

Such changes were, however, still at a comparatively early stage in the time of Gilbert III, and it may be suggested that part of the conflict amongst the Kennedies in the first half of the fifteenth century is to be explained by the collision of the old ideas about 'kenkynnol' with the new ones emerging from contact with a world dominated by principles derived from feudal and canon law. Again, however, it would be mistaken to draw a picture of, as it were, conflicting ideologies, with Gilbert III standing in one camp and his kin in another. Rather there was a power struggle and both sides drew their support from whatever source would help them. After all, the quitclaim of 1455 by Gibboun Kennedy of Leffnoll referred to his pursuit of the 'kenkynnol' 'in the law or by the law'; presumably he would not have hesitated to use legal process had he thought that would serve his purposes.[2]

The same point applies to Gilbert III's own case in 1466 against Robert, Lord Fleming, when he sought to recover the lands of Kirkintilloch. The litigation illustrates much about late medieval forms of process, representation of the parties, pleading and legal sources.[3] But what happened at Dumbarton was also the culmination of at least five years of direct dispute between the two parties, during which they had not only employed the courts but had also endeavoured to resolve matters through an arbitration at Edinburgh, as well as by making use of bonds of manrent and contracting for marriage amongst their offspring.[4] It is in other words a microcosm showing the many ways in which the nobility of the period settled their conflicts and preserved the stability of their society. Litigation was the last resort, but no less important or significant for that. The apparent collapse of

---

1   *RMS*, ii, no. 413 (another instance of the king confirming a grant of the fee which reserved a liferent to the grantor). John Kennedy of Bennane appears as bailie depute of Carrick for Gilbert III in a notarial instrument of 1454 (SRO, GD.109/3). The witnesses have Gaelic names and the notary is Eugene Makyllan.
2   SRO, GD.25/1/120.
3   In addition to the matters noted elsewhere in this essay, see the falsing of the doom, and the comparison of the brieves of mortancestor and novel dissasine probably drawn from *Regiam Majestatem*, IV, 40 (*APS*, i, p. 638) (mortancestor touches fee and freehold where novel dissasine touches only freehold), in NLS, Chs. 15563-5.
4   NLS, Chs. 15559, 15562 (the arbitration); 15560 (bond of manrent); 15559 (marriage contract). Notice that the day after the arbitration was due to take place Gilbert, Lord Kennedy had a notarial transumpt made of the 1384 Kirkintilloch charter (SRO, GD.25/1/88). For evidence showing the dispute going on in 1462 see *Wigtown Charter Chest*, no. 31; also Raehills (Annandale) MSS, Bundle 58 (two notarial instruments of 1461 and 1464 respectively, showing the Flemings taking sasine of the 40-merk land of Kirkintilloch). I owe this last reference to Dr Alan Borthwick.

Kennedy's success at Dumbarton, resulting from his political eclipse months later, is the final element in the complex picture. Ultimately, therefore, Gilbert III represents not two contrasting societies, the local Celtic world of Carrick and the feudal world of the Scottish kingdom, but rather a single, very complicated and changing world. Comparatively late in his career he gained the opportunity to achieve prominence on both local and national stages, and he used all the tools available to him to those ends. No doubt he in turn was used, by James II against the Douglases, and by Robert, Lord Boyd as he prepared his coup of July 1466 — but such was the nature of the world in which Gilbert played out his life.[1]

1   I am deeply indebted to my colleague David Sellar, with whom I have discussed the subject matter of this paper over many years, for reading and re-reading various drafts in a critical yet encouraging way. Dr Alan Borthwick guided me to some of the more obscure corners of Register House, and like Dr John Bannerman gave me generous access to his unpublished work. Others too numerous to mention have commented helpfully at various stages of the research and writing. My thanks to them all. Responsibility for what is now presented is of course mine alone.

A Bibliography of the published works
of G.W.S. Barrow, to the end of 1992

JULIA BARROW

1950 'A Scottish master at Paris: Master Adam de Gulyne', *Recherches de théologie, ancienne et mediévale*, xvii, pp. 126-7.

1951 'A twelfth-century Newbattle document', *SHR*, xxx, pp. 41-9.

1952 'The cathedral chapter at St Andrews and the Culdees in the twelfth and thirteenth centuries', *Journal of Ecclesiastical History*, iii, pp. 23-39.
'A Scottish collection at Canterbury', *SHR*, xxxi, pp. 16-28.

1953 'The earls of Fife in the twelfth century', *PSAS*, lxxxvii, pp. 51-62.
'Scottish rulers and the religious orders, 1070-1153', *Transactions of the Royal Historical Society*, 5th ser., iii, pp. 77-100 (Alexander Prize Essay).
'Early Stewarts at Canterbury', *The Stewarts*, ix, pt. 3, pp. 230-3.
'Dr Stuart and the "Gilleserfs" of Clackmannan', *Scottish Gaelic Studies*, vii, pp. 193-5.

1954 'A bogus tax-collector in Lothian', *The Stewarts*, ix, pt. 4, pp. 323-8.
'Caerlaverock Castle', *SHR*, xxxiii, pp. 175-6.

1955 'King David I and the honour of Lancaster', *Eng. Hist. Rev.*, lxix, pp. 85-9.
Review: R. Kirk, *St Andrews* (London, 1954), in *SHR*, xxxiv, pp. 177-8.

1956 *Feudal Britain* (London, Edward Arnold).
'The beginnings of feudalism in Scotland', *Bulletin of the Institute of Historical Research*, xxix, pp. 1-31.
'The earliest Stewart fief', *The Stewarts*, x, pt. 2, pp. 162-78.
'A letter of Pope Innocent III concerning Ecclefechan', *TDGNHAS*, 3rd ser., xxxiii, pp. 84-90.
'The Culdees of St Andrews', in *Common Errors in Scottish History*, ed. G. Donaldson (London, Historical Association Pamphlets, General Series, G32), pp. 7-8.

1957 'A writ of Henry II for Dunfermline abbey', *SHR*, xxxvi, pp.
138-43.
Reviews: *Felix's Life of St Guthlac*, ed. and trans. B. Colgrave
(Cambridge, 1956), in *History*, xlii, p. 213; *Regesta Regum Anglo-
Normannorum*, vol. ii (Henry I, 1100-35), ed. C. Johnson and
H.A. Cronne (Oxford, 1956), in *SHR*, xxxvi, pp. 59-62.

1958 (Ed.) *Regesta Regum Scottorum: Handlist of the acts of William the
Lion, 1165-1214*, compiled by G.W.S. Barrow and W.W. Scott
(reproduced from typescript, Edinburgh).
'The Gilbertine house at Dalmelling', *Ayrshire Archaeological and
Natural History Society Collections*, 2nd ser., iv, pp. 50-67.
'A note on the date of document no. 942 in *Registrum Antiquissimum
of the Cathedral Church of Lincoln*, ed. C.W. Foster, iii (1935)', in
Registrum Antiquissimum of the Cathedral Church of Lincoln, ed.
K. Major (Lincoln, Lincoln Record Society), viii, pp. xxi-xxiii.
Review: *Medieval Religious Houses, Scotland*, ed. D.E. Easson
(London, 1957), in *History*, xliii, p. 223.

1959 J. Harvey and G.W.S. Barrow, *The Normans* (London, BBC).
'Treverlen, Duddingston and Arthur's Seat', *Book of the Old
Edinburgh Club*, xxx, pp. 1-9.
Review: *Studies in the Early British Church*, ed. N.K. Chadwick
(Cambridge, 1958), in *History*, xliv, pp. 249-51.

1960 (Ed.) *Regesta Regum Scottorum*, vol. i: *The Acts of Malcolm IV, King
of Scots* (Edinburgh, Edinburgh University Press).
'From Queen Margaret to David I: Benedictines and Tironians',
Innes Review, xi, pp. 22-38.
'Early East Lothian charters', *Transactions of the East Lothian
Antiquarian and Field Naturalists' Society*, viii, pp. 1-7.
Reviews: P.F. Anson, *A Monastery in Moray: The Story of Pluscarden
Abbey, 1230-1948* (London, 1959), in *History*, xlv, p. 194; A.
Ashley, *The Church in the Isle of Man* (Borthwick Papers, London
and York, 1958), in *Journ. Eccles. Hist.*, xi, p. 66; C. D'Olivier
Farran, *The Principles of Scots and English Land Law: A Historical
Comparison* (Edinburgh, 1958), in *SHR*, xxxix, pp. 148-51.

1961 Review: G. Donaldson, *Scotland: Church and Nation through Sixteen
Centuries* (London, 1960), in *History*, xlvi, pp. 186-7.

1962 *The Border*, Inaugural Lecture, King's College, Newcastle upon
Tyne (Durham, Durham University).
'The Scottish clergy in the War of Independence', *SHR*, xli, pp.
1-22.
'Rural settlement in central and eastern Scotland: the medieval
evidence', *Scottish Studies*, vi, pt. 2, pp. 123-44.

'Medieval England', bibliographical section in *Handbook for History Teachers*, ed. W.H. Burston and C.W. Green (London, Methuen), pp. 443-51.

Reviews: *Adomnan's Life of Columba*, ed. and trans. A.O. Anderson and M.O. Anderson (Edinburgh, 1961), in *History*, xlvii, pp. 53-4; W.C. Dickinson, *Scotland from the Earliest Times to 1603* (Edinburgh, 1961), in *History*, xlvii, pp. 295-6.

1963 'The Highlands of Scotland in the lifetime of Robert the Bruce', *The Stewarts*, xii, pt. 1, pp. 26-46.

Review: *Celt and Saxon: Studies in the Early British Border*, ed. N.K. Chadwick (Cambridge, 1963), in *History*, xlviii, pp. 350-1.

1964 'James the Stewart of Scotland', *The Stewarts*, xii, pt. 2, pp. 77-91.

Reviews: P.H. Blair, *Roman Britain and Early England, 55 B.C.-A.D. 871* (Edinburgh, 1963), in *SHR*, xliii, pp. 153-4; J. Bulloch, *The Life of the Celtic Church* (Edinburgh, 1963), and N.K. Chadwick, *Celtic Britain* (London, 1963), in *History*, xlix, pp. 205-6; J.H. Burns, *Scottish Churchmen and the Council of Basle* (Glasgow, 1962), in *Journ. Eccles. Hist.*, xv, pp. 109-10; V.H. Galbraith, *An Introduction to the Study of History* (London, 1964), in *SHR*, xliii, p. 65; R.W. Southern, *St Anselm and his Biographer* (Cambridge, 1963), and *The Life of St Anselm, Archbishop of Canterbury, by Eadmer*, ed. and trans. R.W. Southern (Edinburgh, 1962), in *SHR*, xliii, pp. 155-6.

1965 *Robert Bruce and the Community of the Realm of Scotland* (London, Eyre and Spottiswoode; Berkeley and Los Angeles, University of California Press).

'Les familles normandes d'Ecosse', *Annales de Normandie*, xv, pp. 493-515.

'Simon de Montfort's Parliament', *The Times*, 19 January, p. 11.

'The enigma of Malcolm the Maiden', *The Glasgow Herald*, 18 December, p. 8.

Reviews: M. McLaren, *If Freedom Fail* (London, 1964), and *The Bruce: An epic poem written around the year A.D.1375 by John Barbour, archdeacon of Aberdeen*, ed. and trans. A.A.H. Douglas (Glasgow, 1964), in *SHR*, xliv, pp. 66-7; R. Nicholson, *Edward III and the Scots* (Oxford, 1965), in *History*, l, p. 352.

1966 'The Anglo-Scottish Border', *Northern History*, i, pp. 21-42.

'The Scottish judex in the twelfth and thirteenth centuries', *SHR*, xlv, pp. 16-26.

'1066: the end of the Age of the Vikings', *The Glasgow Herald*, 14 October, p. 8.

1967 'Looking ahead in Scottish History', in *Scottish History Society 1886-1966: A Commemorative Record* (Edinburgh, printed for the Society by T. and A. Constable Ltd.), pp. 29-39.

'The wood of Stronkalter', *SHR*, xlvi, pp. 77-9.

Reviews: H.R. Trevor-Roper, *George Buchanan and the Ancient Scottish Constitution* (London, 1966), in *Annali della Fondazione italiana per la storia amministrativa*, iv, pp. 653-5; R.H.C. Davis, *King Stephen* (London, 1967), in *History*, lii, pp. 309-10; *Anglo-Scottish Relations 1174-1328*, ed. E.L.G. Stones (London and Edinburgh, 1965), in *SHR*, xlvi, pp. 58-61.

1968   Reviews: *Regesta Regum Anglo-Normannorum*, vol. iii (Stephen and Matilda, 1135-54), ed. H.A. Cronne and R.H.C. Davis (Oxford, 1968), in *SHR*, xlvii, pp. 169-70; *Ancient Petitions Relating to Northumberland*, ed. C.M. Fraser (Surtees Society, 1966), in *Northern History*, iii, pp. 226-8; T.I. Rae, *The Administration of the Scottish Frontier, 1513-1603* (Edinburgh, 1966), in *SHR*, xlvii, pp. 171-3.

1969   'Northern English society in the twelfth and thirteenth centuries', *Northern History*, iv, pp. 1-28.

'The reign of William the Lion, King of Scotland', in *Historical Studies: Papers Read before the Irish Conference of Historians*, vol. vii, ed. J.C. Beckett (London, Routledge and Kegan Paul), pp. 21-44.

'The Anglo-Scottish union of 1707', *History Today*, xix, pp. 534-42.

'The bearded revolutionary', *History Today*, xix, pp. 679-87.

'The Park in the Middle Ages', in *Northumberland: National Park Guide*, no. 7, ed. J. Philipson (London, HMSO), pp. 57-61.

Review: W.P. Hedley, *Northumberland Families*, vol. i (Society of Antiquaries of Newcastle upon Tyne, 1968), in *Archaeologia Aeliana*, 4th ser., xlvii, pp. 189-91.

1970   Patricia M. Barnes and G.W.S. Barrow, 'The movements of Robert Bruce between September 1307 and May 1308', *SHR*, xlix, pp. 46-59.

'Angerius Brito, Cathensis episcopus', *Traditio*, xxvi, p. 351.

Reviews: A. Squire, *Aelred of Rievaulx* (London, 1969), in *New Blackfriars*, li, no. 607, pp. 577-9; A.J. Otway-Ruthven, *A History of Medieval Ireland* (London, 1968), in *Welsh History Review*, v (1970-1), pp. 84-6.

1971   (Ed., with the collaboration of W.W. Scott) *Regesta Regum Scottorum*, vol. ii: *The Acts of William I, King of Scots* (Edinburgh, Edinburgh University Press).

Feudal Britain: revised paperback edition (London, Edward Arnold).

'The Scottish justiciar in the twelfth and thirteenth centuries', *Juridical Review*, 1971, pp. 97-148.

'The early charters of the family of Kinninmonth of that Ilk', in *The Study of Medieval Records: Essays in Honour of Kathleen Major*, ed.

D.A. Bullough and R.L. Storey (Oxford, Oxford University Press), pp. 107-31.

'Anglo-French influences', in *Who are the Scots?*, ed. G. Menzies (London, BBC), pp. 114-26.

Reviews: *Regesta Regum Anglo-Normannorum*, vol. iv, ed. H.A. Cronne and R.H.C. Davis (Oxford, 1969), in *SHR*, l, pp. 164-5; J. Ferguson, *The Declaration of Arbroath* (Edinburgh, 1970), in *Scottish Studies*, xv, pp. 158-60; M. Gelling, W.F.H. Nicolaisen and M. Richards, *The Names of Towns and Cities in Britain* (London, 1970), and H. Marwick, *The Place Names of Birsay*, ed. W.F.H. Nicolaisen (Aberdeen, 1970), in *SHR*, l, pp. 79-80; *Fasti Ecclesiae Scoticanae Medii Aevi ad annum 1638*, 2nd draft, ed. D.E.R. Watt (Scottish Record Society, 1969), in *Journ. Eccles. Hist.*, xxii, pp. 145-6.

1972 'Wars of Independence', in *The Scottish Nation*, ed. G. Menzies (London, BBC), pp. 15-29.

Reviews: *Edinburgh Studies in English and Scots*, ed. A.J. Aitken, A. McIntosh and H. Pálsson (London, 1971), in *SHR*, li, pp. 200-2; *Scottish Historical Documents*, ed. G. Donaldson (Edinburgh, 1970), in *History*, lvii, pp. 101-2.

1973 *The Kingdom of the Scots: Government, Church and Society from the eleventh to the fourteenth century* (London, Edward Arnold).

Reviews: *The Agrarian History of England and Wales*, vol. i, pt. 2, *A.D. 43-1042*, ed. H.P.R. Finberg (Cambridge, 1972), in *SHR*, lii, pp. 195-6; K. Nicholls, *Gaelic and Gaelicized Ireland in the Middle Ages* (Dublin and London, 1972), in *History*, lviii, pp. 257-8; *Glamorgan County History*, vol. iii: *The Middle Ages*, ed. T.B. Pugh (Cardiff, 1971), in *Welsh History Review*, vi, pp. 475-8.

1974 (Ed.) *The Scottish Tradition: Essays in honour of Ronald Gordon Cant* (Edinburgh, Scottish Academic Press).

'Some East Fife documents of the twelfth and thirteenth centuries', in *The Scottish Tradition* (above), pp. 23-43.

'A note on Falstone', *Archaeologia Aeliana*, 5th ser., ii, pp. 149-52.

Reviews: M. Dolley, *Anglo-Norman Ireland* (Dublin, 1972), and J. Watt, *The Church in Medieval Ireland* (Dublin, 1972), in *History*, lix, pp. 451-3; G. Fellows Jensen, *Scandinavian Settlement Names in Yorkshire* (Copenhagen, 1972), in *Northern History*, ix, pp. 170-1; R.A. Griffiths, *The Principality of Wales in the Later Middle Ages*, vol. i: *South Wales, 1277-1536* (Cardiff, 1972), in *SHR*, liii, pp. 80-2.

1975 'The pattern of lordship and feudal settlement in Cumbria', *Journal of Medieval History*, i, pp. 117-38.

'Macbeth and other mormaers of Moray', in *The Hub of the Highlands: The Book of Inverness and District*, ed. L. Maclean (Inverness and Edinburgh, Inverness Field Club), pp. 109-22.

Review: W.L. Warren, *Henry II* (London, 1973), in *History*, lx, pp. 106-7.

1976 *Robert Bruce and the Community of the Realm of Scotland*: second edition (Edinburgh, Edinburgh University Press).

'Lothian in the first War of Independence, 1296-1328', *SHR*, lv, pp. 151-71.

'Introduction to part I', in *Medieval Settlement: Continuity and Change*, ed. P.H. Sawyer (London, Edward Arnold), pp. 11-14.

Reviews: G.W.O. Addleshaw, *The Pastoral Structure of the Celtic Church in Northern Britain* (Borthwick Papers, York, 1973), in *Northern History*, xi, pp. 237-8; A.A.M. Duncan, *Scotland: The Making of the Kingdom* (Edinburgh, 1975), and Ranald Nicholson, *Scotland: The Later Middle Ages* (Edinburgh, 1974), in *History*, lxi, pp. 252-3; A.P. Smyth, *Scandinavian York and Dublin*, vol. i (Dublin, 1975), in *Northern History*, xii, pp. 257-8; B. Webster, *Scotland from the Eleventh Century to 1603* (London, 1975), in *Journ. Eccles. Hist.*, xxvii, pp. 421-2.

1977 Encyclopaedia article: 'Abernethy (Vertrag von)', in *Lexikon des Mittelalters* (Munich, Artemis Verlag, 1977-), vol. i, col. 32.

Reviews: A.A.M. Duncan, *Scotland: The Making of the Kingdom* (Edinburgh, 1975), in *Historian*, xl, pp. 110-11; W. Ferguson, *Scotland's Relations with England* (Edinburgh, 1977), in *History*, lxii, pp. 486-7; *Calendar of Ancient Petitions Relating to Wales (13th to 16th centuries)*, ed. W. Rees (Cardiff, 1975), and *The Merioneth Lay Subsidy Roll, 1292-93*, ed. K. Williams-Jones (Cardiff, 1976), in *SHR*, lvi, pp. 196-7.

1978 'Some problems in twelfth- and thirteenth-century Scottish history - a genealogical approach', *The Scottish Genealogist*, xxv, pp. 97-112.

'The aftermath of war: Scotland and England in the late thirteenth and early fourteenth centuries', *Transactions of the Royal Historical Society*, 5th ser., xxviii, pp. 103-25 (The Prothero Lecture).

Encyclopaedia articles: 'Alexander I, Kg. v. Schottland (1107-24)', 'Alexander II, Kg. v. Schottland (1214-49)', and 'Alexander III, Kg. v. Schottland (1249-86)', in *Lexikon des Mittelalters*, i, cols. 367-8.

Reviews: *Scottish Society in the Fifteenth Century*, ed. J.M. Brown (London, 1977), in *History*, lxiii, pp. 445-6; *The Medieval Church of St Andrews*, ed. D. McRoberts (Glasgow, 1976), in *Journ. Eccles. Hist.*, xxix, p. 372; J. Durkan and J. Kirk, *The University of Glasgow, 1451-1577* (Glasgow, 1977), in *Journ. Eccles. Hist.*, xxix, pp. 490-1; D. Walker, *A New History of Wales: the Norman*

Conquerors (Swansea, 1977), in *SHR*, lvii, pp. 109-10.

1979 'The idea of freedom in late medieval Scotland', *Innes Review*, xxx, pp. 16-34.

'Robert the Bruce, 1329-1979', *History Today*, xxix, pp. 805-15.

Reviews: R.R. Davies, *Lordship and Society in the March of Wales, 1282-1400* (Oxford, 1978), in *Welsh History Review*, ix, pp. 497-9; *Boroughs of Medieval Wales*, ed. R.A. Griffiths (Cardiff, 1978), in *SHR*, lviii, p. 104; *Boswell, Laird of Auchinleck, 1778-1782*, ed. J.W. Reed and F.A. Pottle (New York, 1977), in *SHR*, lviii, pp. 202-3; R. Somerville, *Pope Alexander III and the Council of Tours (1163)* (Berkeley and London, 1977), in *SHR*, lviii, p. 196.

1980 *The Anglo-Norman Era in Scottish History* (Oxford, Oxford University Press).

Robert Bruce and Ayrshire (Ayr, Ayrshire Archaeological and Natural History Collections, 2nd ser., xiii, no. 2).

Encyclopaedia articles: 'Bamburgh, Dynastie von', and 'Bannockburn, Schlacht bei', in *Lexikon des Mittelalters*, i, cols. 1404, 1421.

Review: W.E. Kapelle, *The Norman Conquest of the North* (London, 1979), in *History*, lxv, pp. 462-3.

1981 *Kingship and Unity: Scotland, 1000-1306* (The New History of Scotland, vol. ii) (London, Edward Arnold; Toronto, University of Toronto Press).

'Popular courts in early medieval Scotland: some suggested place-name evidence', *Scottish Studies*, xxv, pp. 1-24.

'Wales and Scotland in the Middle Ages', *Welsh History Review*, x, pp. 302-19.

'The sources for the history of the Highlands in the Middle Ages', in *The Middle Ages in the Highlands*, ed. L. Maclean (Inverness, Inverness Field Club), pp. 11-22.

'Robert I', 'Scotland 1100-1488', and 'William Wallace', in *A Companion to Scottish Culture*, ed. D. Daiches (London, Edward Arnold), pp. 322-5, 311-2, 396-7.

1982 *The Extinction of Scotland*, Inaugural Lecture, University of Edinburgh, 1980 (Stirling).

'Das mittelalterliche englische und schottische Königtum: ein Vergleich', *Historisches Jahrbuch*, cii, pp. 362-89.

Reviews: J.M. Gilbert, *Hunting and Hunting Reserves in Medieval Scotland* (Edinburgh, 1979), in *Eng. Hist. Rev.* xcvii, pp. 621-3; N. MacDougall, *James III: A Political Study* (Edinburgh, 1982), in *Times Literary Supplement*, no. 4153, p. 1228.

1983 'The childhood of Scottish Christianity', *Scottish Studies*, xxvii, pp. 1-15.

'Sources for the history of the Border region in the middle ages', in *The Borders*, ed. P. Clack and J. Ivy (Durham, Ccouncil for British Archaeology Group 3), pp. 1-7.

'Popular courts in early medieval Scotland: some suggested place name evidence — additional note', *Scottish Studies*, xxvii, pp. 67-8.

'Celtic Church', and 'Picts, union with the Scots', in *The Companion to Gaelic Scotland*, ed. D.S. Thomson (Oxford, Basil Blackwell), pp. 37-9, 226.

Encyclopaedia articles: 'Carlisle, Stadt und Bistum', 'Chronik: [Section] H: Schottland', and 'Clann II: Schottland', in *Lexikon des Mittelalters*, ii, cols. 1508, 1988-90, 2120-1.

Reviews: R.A. Dodgshon, *Land and Society in Early Scotland* (Oxford, 1981), in *Cambridge Medieval Celtic Studies*, v, pp. 92-3; P. Herde, *Cölestin V. (1294) (Peter von Morrone), der Engelpapst* (Stuttgart, 1981), in *SHR*, lxii, p. 168; A.T. Simpson and S. Stevenson, *Historic Auchtermuchty*; *Historic Cupar*; *Historic Kinghorn* (Glasgow, 1981), in *SHR*, lxii, pp. 167-8; A.P. Smyth, *Scandinavian York and Dublin*, vol. ii (Dublin, 1979), in *Northern History*, xix, pp. 254-6.

1984   *Robert the Bruce and the Scottish Identity* (Saltire Society, Saltire Pamphlets, new ser. no. 4).

'Land routes: the medieval evidence', in *Loads and Roads in Scotland and Beyond: Land Transport over 6,000 Years*, ed. A. Fenton and G. Stell (Edinburgh, John Donald), pp. 49-66.

'Robert the Bruce, king of Scots, 1306-1329', *Swissair Gazette* (May 1984), pp. 16-17.

Encyclopaedia articles: 'Comyn', 'Cumberland', and 'David I', in *Lexikon des Mittelalters*, iii, cols. 109-10, 369-70, 599-600.

Reviews: M.T. Clanchy, *England and its Rulers, 1066-1272* (London, 1983), in *History Today*, xxxiv, pp. 52-3; *Welsh Society and Nationhood: Historical Essays Presented to Glanmor Williams*, ed. R.R. Davies, R.A. Griffiths, I.G. Jones and K.O. Morgan (Cardiff, 1984), in *Journ. Eccles. Hist.*, xxxv, pp. 620-1.

1985   *David I of Scotland (1124-1153): the Balance of New and Old* (The Stenton Lecture, 1984) (Reading, University of Reading).

G.W.S. Barrow and A. Royan, 'James fifth Stewart of Scotland, 1260-1309', in *Essays on the Nobility of Medieval Scotland*, ed. K.J. Stringer (Edinburgh, John Donald), pp. 166-94.

'The Scots charter', in *Studies in Medieval History Presented to R.H.C. Davis*, ed. H. Mayr-Harting and R.I. Moore (London, Hambledon Press), pp. 148-64.

'Midlothian - or the shire of Edinburgh?', *Book of the Old Edinburgh Club*, xxxv, pt. 2, pp. 141-8.

Encyclopaedia articles: 'Donald McAlpin', 'Donald, Kg. v. Schottland 889-900', 'Donald Bane', 'Dunbar' (with D.A. Bullough), 'Dumbarton', 'Duncan I', 'Duncan II', 'Dunkeld', 'Earl: II Schottland','Edgar, Kg. v. Schottland 1097-1107', 'Edinburgh', and 'Edward Bruce', in *Lexikon des Mittelalters*, iii, cols. 1230-1, 1448-9, 1452-4, 1459-60, 1504-5, 1572, 1575-6, 1593-4.

Reviews: J. Prebble, *John Prebble's Scotland* (London, 1984), in *SHR*, lxiv, pp. 201-2; *Documents on the Affairs of Ireland before the King's Council*, ed. G.O. Sayles (Dublin, 1979), *England and Ireland in the Later Middle Ages*, ed. J. Lydon (Dublin, 1981), and R. Frame, *English Lordship in Ireland, 1318-1361* (Oxford, 1982), in *History*, lxx, pp. 113-15.

1986   'The origins and early history of Edinburgh and Glasgow: two Scottish cities contrasted', in *Ekonomiczne i pozaekonomiczne czynniki' rozwoju miast* (Akademia Ekonomiczna w Krakowie, Kraków), pp. 5-22.

'Scotland and the Norman Conquest', in *Domesday: 900 years of England's Norman Heritage* (London, Millbank), pp. 125-6.

Encyclopaedia article: 'Erroll, Earls of', in *Lexikon des Mittelalters*, iii, col. 2185.

Reviews: A.D. Carr, *Medieval Anglesey* (Llangefni, 1982), and *The Scandinavians in Cumbria*, ed. J.R. Baldwin and I.D. Whyte (Edinburgh, 1985), in *SHR*, lxv, pp. 206-7; L.J. Macfarlane, *William Elphinstone and the Kingdom of Scotland, 1431-1514* (Aberdeen, 1985), in *Innes Review*, xxxvii, pp. 100-1; A. Macquarrie, *Scotland and the Crusades, 1095-1560* (Edinburgh, 1985), in *International History Review*, viii, pp. 636-8; A. Macquarrie, *Scotland and the Crusades* (as above), and K.J. Stringer, *Earl David of Huntingdon, 1152-1219* (Edinburgh, 1985), in *History Today*, xxxvi, pp. 58-9.

1987   Encyclopaedia article: 'Fife', in *Lexikon des Mittelalters*, iv, col. 438.

Reviews: A. Fisher, *William Wallace* (Edinburgh, 1986), in *Cambridge Medieval Celtic Studies*, xiii, p. 101; *Tradition and Change: Essays in Honour of Marjorie Chibnall*, ed. D. Greenway, C. Holdsworth and J. Sayers (Cambridge, 1985), in *Journ. Eccles. Hist.*, xxxviii, pp. 287-8; Walter Bower, *Scotichronicon*, vol. viii, ed. and trans. D.E.R. Watt (Aberdeen, 1987), in *Times Lit. Supp.*, no. 4411, p. 1148.

1988   *Robert Bruce and the Community of the realm of Scotland*: third edition (Edinburgh, Edinburgh University Press).

'Badenoch and Strathspey, 1130-1312, 1: Secular and Political', *Northern Scotland*, viii, pp. 1-15.

'Tannochbrae and all that: "Tamhnach" (Tannoch etc.) in Scottish
placenames as an indicator of early Gaelic-speaking settlement', in
*A Sense of Place: Studies in Scottish Local History (A Tribute to Eric
Forbes)*, ed. G. Cruickshank (Edinburgh, Scotland's Cultural
Heritage), pp. 1-4.

'The Norman factor', in *The Sunday Mail Story of Scotland*, i, pt. 4,
pp. 96-9.

'The Fight for Freedom', ibid., i, pt. 5, pp. 123-5.

Encyclopaedia articles: 'Fordun, John', and 'Glasgow', in *Lexikon
des Mittelalters*, iv, cols. 633-4, 1483-4.

Reviews: *Codex Diplomaticus Regni Siciliae, Series I, tomus II/1:
Rogerii II Regis Diplomata Latina*, ed. C. Brühl (Cologne and
Vienna, 1987), in *SHR*, lxvii, pp. 190-1; B.E. Crawford,
Scandinavian Scotland (Leicester, 1987), and L.J.F. Keppie,
Scotland's Roman Remains: An Introduction and Handbook
(Edinburgh, 1986), in *Albion*, xx, pp. 524-7; R.R. Davies,
Conquest, Co-existence and Change, 1063-1415 (Oxford and
Cardiff, 1987), in *Welsh History Review*, xiv, pp. 306-8; *Calendar
of Documents Relating to Scotland*, vol. v (Supplementary), *A.D.
1108-1516*, ed. G.G. Simpson and J.D. Galbraith (Edinburgh,
1986), in *Times Lit. Supp.*, no. 4452, p. 835; *Scottish Texts and
Calendars: An Analytical Guide to Serial Publications*, ed. D. and
W.B. Stevenson, (Edinburgh and London, 1987), in *Journal of the
Society of Archivists*, ix, pp. 160-1; A.J.L. Winchester, *Landscape
and Society in Medieval Cumbria* (Edinburgh, 1987), in *Journal of
Historical Geography*, xiv, pp. 204-5.

1989   *Kingship and Unity: Scotland, 1000-1306*: second edition (Edinburgh,
Edinburgh University Press).

'The lost Gàidhealtachd of medieval Scotland', in *Gaelic and
Scotland: Alba agus a' Ghàidlig*, ed. W. Gillies (Edinburgh,
Edinburgh University Press), pp. 67-88.

'Frontier and Settlement: Which influenced Which? England and
Scotland, 1100-1300,' in *Medieval Frontier Societies*, ed. R.
Bartlett and A. MacKay (Oxford, Clarendon Press), pp. 3-21.

'Badenoch and Strathspey, 1130-1312, 2: The Church', *Northern
Scotland*, ix, pp. 1-16.

'The tribes of North Britain revisited', *PSAS*, cxix, pp. 161-3.

Encyclopaedia articles: 'Halidon Hill', 'Hebriden', and 'Hexham', in
Lexikon des Mittelalters, iv, cols. 1876, 1983-4, 2204-5.

Reviews: S. Cruden, *Scottish Medieval Churches* (Edinburgh, 1986),
in *Innes Review*, xxxix, pp. 156-7; *The British Isles 1100-1500*, ed.
R.R. Davies (Edinburgh, 1988), in *Cambridge Medieval Celtic
Studies*, xvii, pp. 87-8; M. Prestwich, *Edward I* (London, 1988),
in *SHR*, lxviii, pp. 205-8.

1990 'A kingdom in crisis: Scotland and the Maid of Norway', *SHR*, lxix, pp. 120-41.

'The army of Alexander III's Scotland', in *Scotland in the Reign of Alexander III, 1249-1286*, ed. N.H. Reid (Edinburgh, John Donald), pp. 132-47.

'Earl David's burgh: Dundee in medieval Scotland', in *The Dundee Book*, ed. B. Kay (Edinburgh, Mainstream), pp. 19-26.

Encyclopaedia articles: 'Jakob I, Kg. v. Schottland 1406-37', 'Jakob II, Kg. v. Schottland 1437-60', 'Jakob III, Kg. v. Schottland 1460-88', 'Jakob IV, Kg. v. Schottland 1488-1513', 'Inverness', and 'Kelso', in *Lexikon des Mittelalters*, v, cols. 284-6, 475-6, 1099.

Reviews: Walter Bower, *Scotichronicon*, vol. ii, ed. and trans. J. and W. MacQueen (Aberdeen, 1989), in *Times Lit. Supp.*, no. 4542, p. 415; *The Scottish Medieval Town* (Edinburgh, 1988), ed. M. Lynch, M. Spearman and G. Stell, in *Journal of Historical Geography*, xvi, pp. 239-40; L.E. Schmidt, *Holy Fairs — Scottish Communions and American Revivals in the Early Modern Period* (Princeton, 1989), in *Times Lit. Supp.*, no. 4558, p. 852.

1991 Encyclopaedia articles: 'Lamberton', 'Lennox', 'Lindsay', and Lothian', in *Lexikon des Mittelalters*, v, cols. 1627, 1872-3, 2000-1, 2133-4.

Review: R. Frame, *The Political Development of the British Isles 1100-1400* (Oxford, 1990), in *Times Lit. Supp.*, no. 4579, p. 18.

1992 *Scotland and its Neighbours in the Middle Ages* (London and Rio Grande, Hambledon Press).

Robert Bruce and the Community of the Realm of Scotland: fourth edition (Edinburgh, Edinburgh University Press).

'The Charters of David I', in *Anglo-Norman Studies XIV: Proceedings of the Battle Conference, 1991* (Woodbridge, Boydell Press), pp. 25-37.

Index

This index is selective rather than exhaustive. Places in Great Britain are located by pre-197: county, as far as seems necessary. Kings (national and provincial), other than the kings of Scots are indexed territorially and arranged chronologically; similarly, named abbots, bishops, duke and earls appear chronologically within the entries for their respective abbeys, bishoprics dukedoms and earldoms.

Cadzow (Lanarkshire), 11, 19, 96n
cain, 38, 48, 74-6
Cairnbulg castle (Aberdeenshire), 188-9, 191, 198, 201-2
Caithness, 175; bishopric, bishops, 55n, 140; earl, 176; earl named, Harald Maddadson, 87, 95
Callendar (thanage, Stirlingshire), 43, 45, 50-2, 55n, 56-8, 61, 66, 69, 81; Malcolm of, 56-8, 81; Ness of, 56, 81;
calp, see tribute
calumpnie, 38, 278, 281, 290
Cambo, William of, 229, 237
Cambuskenneth (Stirlingshire), abbey, 53, 81; parliament, 218
Campania, Henry de, 113; Ralph de, 91n, 99-101, 103-5, 107, 113; Robert de, 113
Campbell, Clan, 30n; Archibald, 268
Cardeny, Marjorie, 67, 74
Carham (Northumberland), battle, 17, 47
Carlisle, 19, 88, 92, 105, 107; bishopric, bishops, 92, 105-6; castle, 90-1; Eudo of, 90-1, 105, 107
Carrick, earldom, province, 33, 35-6, 274-5, 277-84, 286, 288-9, 292-3, 296; bailie, 278, 284, 290-3, 295n; *judex*, 280; sergeants, 281, 295; steward, 282
Carrick, earls, lords, 224, 278, 281, 284, 294; earls named, Duncan, 87, 89, 98, 108-9, 278, 281, 284; Niall (or Neil), 33-4, 38, 278, 281, 284; Robert Bruce (I), 278-9, 281, 284; Robert Bruce (II), 34, 279-81, 284 (see also Robert I); Edward Bruce, 211-14, 279, 284, 286; David Bruce, 279, 284, 286 (see also David II); Alexander Bruce, 237, 279, 284, 286; John Stewart, 67, 74, 79, 284, 286 (see also Robert III); lord named, William Cunningham, 286; Eleanor, w. of Alexander Bruce, 286; Marjorie, countess, dau. of Earl Niall, 34, 278-9, 281, 284
Carrick, family, 284-6; Colin of, 282, 284; Gilbert of (1300s), 278, 284-5; Lachlan (or Roland) of, 33, 38, 278, 281, 284-5, 294; Malcolm son of Roland of, 284-5; others, 278, 284-5; see also Dunkeld
Castle Carrock, Robert of, 90, 91n, 105, 107
castles, 60-1, 65n, 67, 72, 75, 78, 80, 89-91, 97, 99, 159, 180, 183, 187-91, 193, 198, 201-2, 223, 227-9, 231-5, 253, 264, 266-7, 276n, 278, 282n, 284-6, 288, 291, 293
Cawdor (thanage, Nairnshire), 42, 55n, 57-8, 60n, 65n, 70-2; Donald of, 57, 72; William of, 58, 60n, 72
Celtic (Gaelic) society, lordship, concepts, 1ff, 20ff, 42, 45n, 47, 49, 57, 82, 84-5, 87, 96-9, 101, 174, 176, 179, 181-2, 185-6, 194, 242, 274-5, 278, 280, 288, 293-4, 296; succession customs, 18, 24, 31-2, 35,

36n, 96-7
Cenél nEógain, 28-9
chamberlains of Scotland, 100, 112, 227, 230-1, 234, 243-6, 248-54, 261-5; ayres, 245-6
chancellors of Scotland, 100, 134-6, 153, 244, 249-50, 268, 275, 285
chancery, royal, 139, 248-50
Château Gaillard (Normandy), 223, 232
Chester, 16, 88; constable, 104
Cheyne, family, 184, 186, 196, 199; Reginald, 55, 62, 74-5
Clackmannan, sheriffdom, sheriffs, 27, 195, 251
Clifton, Ralph of, 108-9, 112
cloth, clothing, manufacture, dyeing, trade, 156, 160, 166, 168-71, 186
Clova (thanage, Angus), 66, 76
Clydesdale, 19, 108; see also Strathclyde
Cockersand abbey (Lancashire), 113
Coldingham priory (Berwickshire), 117, 119; Coldinghamshire, 117
Coleraine (Co. Londonderry), 88-9, 93
community, of the realm, 131, 156, 173, 199, 203ff; of Scotland, 85, 98; of the burgh, 117, 156, 173; Celtic, 98, 101; of the earldom, 186, 199; of Galloway, 84, 99; local, 115, 120; of the sheriffdom, 57
Comyn earls of Buchan, see Buchan; lords of Badenoch, 194, 198; John, 198, 283; Walter, 101, 178, 181
Comyn, family, 98, 131, 134-6, 174-8, 181-3, 193, 198-9, 215, 268n, 283; Agnes, 182, 192; Alexander, 178, 198; Alice, 183; David, of Kilbride, 181; Idonea, 182, 193; Jordan, 182, 188, 192; Margaret, 182; Mary, 66; Richard, 176, 194; Roger, 196-7; Thomas, 238; Thomas Cumming, 191; William, of Kilbride, 283; others, 181-2
constables of Scotland, 82ff, 183, 188-9, 195, 235, 244, 248-9
Constantine I king of Scots, 6, 12, 13
Constantine II king of Scots, 6, 14-17
conveth, 48-9, 60, 76
Conveth (thanage, Banffshire), 48, 52, 55n, 58, 62, 73, 183-4, 193, 197, 201
Conveth (thanage, Kincardineshire), see Laurencekirk
Coull (barony, Aberdeenshire), 193; Upper, 182, 193
councils of Scottish Church, provincial, 140ff; conservators, 144-7, 150-1, 155; legislation, 142-3, 147-53
Coupar Angus (*manerium*, thanage, Perthshire), 43, 46, 52, 54n, 63, 78
Coupar Angus, abbey, abbots, 52, 78, 229, 234, 238
court, household, royal, 49, 56, 60, 62-4, 68, 83-4, 88-9, 94-8, 100-1, 112-13, 126, 160, 165, 180, 185-6, 246-9, 253, 263-5

DATE DUE

	GLX	APR 1 7 1995	
GL/Rec		APR 1 8 1995	
GLX		MAY 1 6 1995	
	GL/Rec	MAY 0 4 1995	
		MAY 3 1 1999	
FEB 1 5 2012			
		Printed in USA	